Fidele Ntie-Kang (Ed.)
Chemoinformatics of Natural Products

Also of interest

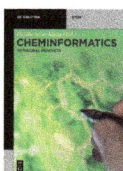

Chemoinformatics of Natural Products.
Volume 1: Fundamental Concepts
Ntie-Kang (Ed.), 2020
ISBN 978-3-11-057933-8, e-ISBN 978-3-11-057935-2

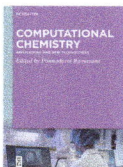

Computational Chemistry.
Applications and New Technologies
Ramasami (Ed.), 2021
ISBN 978-3-11-068200-7, e-ISBN 978-3-11-068204-5

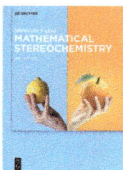

Mathematical Stereochemistry
2nd Edition
Fujita, 2021
ISBN 978-3-11-072818-7, e-ISBN 978-3-11-072823-1

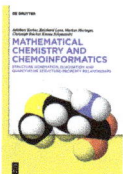

Mathematical Chemistry and Chemoinformatics.
Structure Generation, Elucidation and Quantitative Structure-Property
Relationships
Adalbert Kerber, Reinhard Laue, Markus Meringer, Christoph Rücker
and Emma Schymanski, 2014
ISBN 978-3-11-030007-9, e-ISBN 978-3-11-025407-5

Physical Sciences Reviews.
e-ISSN 2365-659X

Chemoinformatics of Natural Products

Volume 2: Advanced Concepts and Applications

Edited by
Fidele Ntie-Kang

DE GRUYTER

Editor
PD Dr. Fidele Ntie-Kang
Department of Chemistry, Faculty of Science,
University of Buea, P. O. Box 63,
Buea, Cameroon
fidele.ntie-kang@pharmazie.uni-halle.de

ISBN 978-3-11-066888-9
e-ISBN (PDF) 978-3-11-066889-6
e-ISBN (EPUB) 978-3-11-066906-0

Library of Congress Control Number: 2021944238

Bibliographic information published by the Deutsche Nationalbibliothek
The Deutsche Nationalbibliothek lists this publication in the Deutsche Nationalbibliografie;
detailed bibliographic data are available on the Internet at http://dnb.dnb.de.

© 2022 Walter de Gruyter GmbH, Berlin/Boston
Cover image: iPandastudio/iStock/Getty Images Plus
Typesetting: Integra Software Services Pvt. Ltd.
Printing and binding: CPI books GmbH, Leck

www.degruyter.com

To Prof. Simon M. N. Efange for a career dedicated to drug discovery in academia.

Contents

Lena Y. E. Ekaney, Donatus B. Eni and Fidele Ntie-Kang

Part II: Advanced Chemoinformatics Tools and Methods for Lead Compound Discovery and Development —— 75

Abdulkarim Najjar, Abdurrahman Olğaç, Fidele Ntie-Kang and Wolfgang Sippl

Part III: Chemoinformatics Tools for Natural Products Discovery in the Modern Age and Case Studies — 177

Ramsay Soup Teoua Kamdem, Omonike Ogbole, Pascal Wafo, Philip F. Uzor,
Zulfiqar Ali, Fidele Ntie-Kang, Ikhlas A. Khan and Peter Spiteller

Sergi Herve Akone, Cong-Dat Pham, Huiqin Chen, Antonius R. B. Ola,
Fidele Ntie-Kang and Peter Proksch

Part IV: Case Studies —— 219

Conrad V. Simoben, Fidele Ntie-Kang, Dina Robaa and Wolfgang Sippl

Lucas Paul, Celestin N. Mudogo, Kelvin M. Mtei, Revocatus L. Machunda
and Fidele Ntie-Kang

List of contributors

Sergi Herve Akone
Faculty of Science
Department of Chemistry
University of Douala
P.O. Box 24157
Douala, Cameroon
E-mail: sergiherve@yahoo.fr

Zulfiqar Ali
National Center for Natural Products
Research
School of Pharmacy
University of Mississippi
MS 38677, USA
E-mail: Zulfiqar@olemiss.edu

Fatima Baldo
Department of Chemistry
University of Cambridge
Cambridge CB2 1EW, UK
E-mail: fmhb2@cam.ac

Boris D. Bekono
Department of Physics
Ecole Normal Supérieure
University of Yaoundé I
P. O. Box 47
Yaoundé, Cameroon
E-mail: borisbekono@gmail.com

Huiqin Chen
Institute of Tropical Bioscience and
Biotechnology
Chinese Academy of Tropical Agricultural
Sciences
Haikou 571101, China

Lena Y. E. Ekaney
Faculty of Science
Department of Chemistry
University of Buea
P.O. Box 63
Buea, Cameroon
E-mail: lena2yoh@gmail.com

Donatus B. Eni
Faculty of Science
Department of Chemistry
University of Buea
P.O. Box 63, Buea, Cameroon;
Department of Inorganic Chemistry
Faculty of Science
University of Yaoundé I.
Yaoundé, Cameroon
E-mail: donatus.bekindaka@ubuea.cm

Stefan Günther
Institute of Pharmaceutical Sciences
Research Group Pharmaceutical
Bioinformatics
Albert-Ludwigs-University Freiburg
Hermann-Herder-Straße 9
Freiburg, 79104 Germany
E-mail: stefan.guenther@pharmazie.uni-frei
burg.de

Joelle Ngo Hanna
Department of Chemistry
University of Buea
P. O. Box 63
Buea, Cameroon;
Department of Chemistry
University of Douala
P. O. Box 24157
Douala, Cameroon
E-mail: ngo_hanna@yahoo.fr

Ricardo Bruno Hernández-Alvarado
Chemical Physics
Instituto de Química
Universidad Nacional Autónoma de México
Avenida Universidad 3000
Mexico City, Mexico
Email: hernandez-alvarado@ ciencias.unam.mx

Linda Jahn
Institute of Botany
Technische Universität Dresden
Zellescher Weg 20b
01062 Dresden, Germany
E-mail: Linda.Jahn@tu-dresden.de

https://doi.org/10.1515/9783110668896-203

Ramsay Soup Teoua Kamdem
Institut für Organische und Analytische
Chemie
Universität Bremen
Leobener Strasse 7 (NW2C)
Bremen 28359, Germany;
Department of Organic Chemistry
Higher Teachers' Training College
University of Yaounde I.
P. O. Box 47
Yaoundé, Cameroon
E-mail: ramsaykamdem@gmail.com

Ikhlas A. Khan
National Center for Natural Products
Research
School of Pharmacy
University of Mississippi
MS 38677, USA
E-mail: khan@olemiss.edu

Jutta Ludwig-Müller
Institute of Botany
Technische Universität Dresden
Zellescher Weg 20b
01062 Dresden, Germany
E-Mail: jutta.ludwig-mueller@tu-dresden.de

Revocatus L. Machunda
The Department of Water and Environmental
Science and Engineering
The Nelson Mandela African Institution of
Science and Technology
P.O. Box 447, Arusha, Tanzania
E-mail: revocatus.machunda@nm-aist.ac.tz

Abraham Madariaga-Mazón
Chemical Physics
Instituto de Química
Universidad Nacional Autónoma de México
Avenida Universidad 3000
Mexico City, Mexico
Email: madariaga@ciencias. unam.mx

Karina Martinez-Mayorga
Chemical Physics
Instituto de Química
Universidad Nacional Autónoma de México
Avenida Universidad 3000
Mexico City, Mexico
E-mail: kmtzm@unam.mx

James A. Mbah
Department of Chemistry
University of Buea
P. O. Box 63
Buea, Cameroon

José L. Medina-Franco
DIFACQUIM research group
Department of Pharmacy, School of Pharmacy
Universidad Nacional Autónoma de México
Mexico City 04510, México
E-mail: jose.medina-franco@gmail.com

Kelvin M. Mtei
The Department of Water and Environmental
Science and Engineering
The Nelson Mandela African Institution of
Science and Technology
P.O. Box 447
Arusha, Tanzania
E-mail: kelvin.mtei@nm-aist.ac.tz, revocatus.
machunda@nm-aist.ac.tz

Celestin N. Mudogo
Biochemistry and Molecularbiology
University of Hamburg Institute of
Biochemistry and Molecularbiology
Hamburg, Germany;
Department of Basic Sciences
School of Medicine
University of Kinshasa
Kinshasa, Congo
E-mail: cmudogo@gmail.com

Abdulkarim Najjar
Institute of Pharmacy
Martin-Luther University of Halle-Wittenberg
Department of Medicinal Chemistry
06120 Halle (Saale), Germany
E-mail: a.k.najjar@hotmail.com

Karla Olivia Noriega-Colima
Chemical Physics
Instituto de Química
Universidad Nacional Autónoma de México
Avenida Universidad 3000
Mexico City, Mexico

Fidele Ntie-Kang
Faculty of Science
Department of Chemistry
University of Buea
P.O. Box 63
Buea, Cameroon;
Department of Pharmaceutical Chemistry
Martin-Luther University Halle-Wittenberg
Kurt-Mothes-Str. 3
Halle (Saale), 06120 Germany;
Institute of Botany
Technische Universität Dresden
Zellescher Weg 20b
Dresden, 01217 Germany
E-mail: ntiekfidele@gmail.com

Omonike Ogbole
Department of Pharmacognosy
University of Ibadan
Ibadan, Nigeria
E-mail: nikeoa@yahoo.com

Antonius R. B. Ola
Chemistry Department
Nusa Cendana University
Kupang, Indonesia

Abdurrahman Olgaç
Laboratory of Molecular Modeling
Evias Pharmaceutical R&D Ltd.
Ankara 06830, Turkey;
Department of Pharmaceutical Chemistry
Faculty of Pharmacy
Gazi University
Ankara 06330, Turkey

Adriana Osnaya-Hernández
Chemical Physics
Instituto de Química
Universidad Nacional Autónoma de México
Avenida Universidad 3000
Mexico City, Mexico

Luc C. O. Owono
Department of Physics
Ecole Normal Supérieure
University of Yaoundé I
P. O. Box 47
Yaoundé, Cameroon
E-mail: lcowono@yahoo.fr

Lucas Paul
The Department of Materials and Energy
Science & Engineering
The Nelson Mandela African Institution of
Science and Technology
P.O. Box 447 Arusha, Tanzania;
Department of Chemistry
Dar es Salaam University College of
Education
P.O. Box 2329, 255 Dar es Salaam, Tanzania
E-mail: lucaspaul33@gmail.com,
lucasp@nm-aist.ac.tz

Cong-Dat Pham
Department of Chemistry
Kyoto Prefectural University of Medicine
1-5 Shimogamohangi-cho
Sakyo-ku, Kyoto 606-0823, Japan

Peter Proksch
Institute for Pharmaceutical Biology and
Biotechnology
Heinrich-Heine-University Düsseldorf
Universitätsstrasse 1, Geb. 26.23,
40225 Duesseldorf, Germany
E-mail: Peter.Proksch@uni-duesseldorf.de

Dina Robaa
Department of Medicinal Chemistry (AG
Sippl)
Institute of Pharmacy
Martin-Luther-Universität Halle-Wittenberg,
Kurt-Mothes-Str. 3
06120 Halle (Saale), Germany
E-mail: dina.robaa@pharmazie.uni-halle.de

Daniel M. Shadrack
Department of Chemistry
St. John's University of Tanzania
P. O. Box 47
Dodoma, Tanzania
E-mail: dmssjut@gmail.com

Conrad V. Simoben
Department of Medicinal Chemistry (AG
Sippl)
Institute of Pharmacy
Martin-Luther-Universität Halle-Wittenberg,
Kurt-Mothes-Str. 3
06120 Halle (Saale), Germany
E-mail: veranso.conrad@gmail.com

Wolfgang Sippl
Institute of Pharmacy
Martin-Luther University of Halle-Wittenberg
Department of Medicinal Chemistry
06120 Halle (Saale), Germany
Email: wolfgang.sippl@pharmazie.uni-halle.
de

Peter Spiteller
Institut für Organische und Analytische
Chemie
Universität Bremen
Leobener Strasse 7 (NW2C)
Bremen 28359, Germany
E-mail: psp@uni-bremen.de

Daniel Svozil
Department of Informatics and Chemistry
University of Chemistry and Technology
Prague
Technická 5 166 28 Prague 6, Dejvice Praha
Czech Republic
E-mail: daniel.svozil@vscht.cz

Kiran K. Telukunta
International Solar Energy Society eV, IT
Wiesentalstr 50
79115 Freiburg, Germany
E-mail: kiran.telukunta@indiayouth.info

Flavien A. A. Toze
Department of Chemistry
University of Douala
P. O. Box 24157
Douala, Cameroon

Pascal Wafo
Department of Organic Chemistry
Higher Teachers' Training College
University of Yaounde I.
P. O. Box 47
Yaoundé, Cameroon
E-mail: wafopascal@yahoo.fr

Philip F. Uzor
Department of Pharmaceutical and Medicinal
Chemistry
Faculty of Pharmaceutical Sciences
University of Nigeria
Nsukka, 410001 Nigeria

About the Editor

PD Dr. Fidele Ntie-Kang studied Chemistry at the University of Douala in Cameroon between 1999 and 2004, leading to BSc and MSc degrees. His Ph.D. work in Physical Sciences at the Centre for Atomic Molecular Physics and Quantum Optics (CEPAMOQ) obtained in 2014 was based on the computer-aided design of anti-tubercular agents. He later obtained a "Habilitation" in pharmaceutical chemistry from Martin-Luther University Halle-Wittenberg, Germany, after having carried out a period of extensive research on chemoinformatics applications for natural products drug discovery from African medicinal plants under Prof. Wolfgang Sippl. He is experienced in molecular modeling and has been involved in the design and management of databases of natural products from African flora for virtual screening. Fidele has formerly worked as a Scientific Manager/Senior Instructor at the Chemical and Bioactivity Information Centre (CBIC), hosted at the Chemistry Department of the University of Buea, Cameroon. PD Dr. Ntie-Kang currently teaches physical chemistry, computational chemistry and medicinal chemistry at the undergraduate and postgraduate levels in the University of Buea, Cameroon and in several summer schools, since 2008. Fidele has been awarded several fellowships and distinctions, including a doctoral fellowship from the German Academic Exchange Services (DAAD), Our Common Future Fellowship by the VolksWagen Foundation (Germany), UNIDO Fellowship in ICS-UNIDO Trieste (Italy), Commonwealth Professional Fellowship in Leeds (UK), Georg Forster Postdoctoral Fellowship by the Alexander von Humboldt Foundation (Germany), and ChemJets fellowship from the Czech Ministry of Youth, Sports and Education (Czech Republic). He is and has been a member of several scientific organizations, including the African-German Network for Excellence in Science (AGNES), the African Scientific Institute (ASI), the German Chemical Society, the American Chemical Society (ACS), the African Network for Drug Diagnostics and Innovation (ANDI) and the Cameroon Academy of Young Scientists (CAYS). Fidele has published about 90 journal articles and book chapters, with about 2000 citations. He is a reviewer for several funding organizations, several high impact factor journals, editorial board member of several journals, and member of the scientific committee of international conferences, as well as a guest co-editor of special issues focused on natural products in the journals *Molecules* (MDPI) and *Frontiers in Pharmacology*.

https://doi.org/10.1515/9783110668896-204

Part I: **Advanced Chemoinformatics Concepts
for Natural Product-based Drug Discovery**

Fidele Ntie-Kang, Daniel Svozil

1 An enumeration of natural products from microbial, marine and terrestrial sources

Abstract: The discovery of a new drug is a multidisciplinary and very costly task. One of the major steps is the identification of a lead compound, i.e. a compound with a certain degree of potency and that can be chemically modified to improve its activity, metabolic properties, and pharmacokinetics profiles. Terrestrial sources (plants and fungi), microbes and marine organisms are abundant resources for the discovery of new structurally diverse and biologically active compounds. In this chapter, an attempt has been made to quantify the numbers of known published chemical structures (available in chemical databases) from natural sources. Emphasis has been laid on the number of unique compounds, the most abundant compound classes and the distribution of compounds in terrestrial and marine habitats. It was observed, from the recent investigations, that ~500,000 known natural products (NPs) exist in the literature. About 70 % of all NPs come from plants, terpenoids being the most represented compound class (except in bacteria, where amino acids, peptides, and polyketides are the most abundant compound classes). About 2,000 NPs have been co-crystallized in PDB structures.

Keywords: drug discovery, functional groups, natural products, numbers, scaffolds

1.1 Introduction

In the quest to discover new drugs, researchers have often resorted to natural sources, e. g. plants, marine organisms, bacteria and fungi [1, 2]. This is because these organisms are known to host sophisticated metabolic pathways that have led to complex and intriguing chemical structures. The existence of such structures could never have been figured out by any chemist, had nature not synthesized them. Besides, some naturally occurring compounds have required several decades to be synthesized, even after the entire chemical structure had been elucidated [3–5]. Such compounds are often the products of secondary metabolism in higher organisms, fungi, and microbes. Thus, they are often referred to as secondary metabolites (or natural products).

An attempt has been made to provide a classification of secondary metabolites (SMs) according to their structural diversity, bioactivity and ecological functions

This article has previously been published in the journal *Physical Sciences Reviews*. Please cite as: Ntie-Kang, F., Svozil, D. An enumeration of natural products from microbial, marine and terrestrial sources. *Physical Sciences Reviews* [Online] 2019 DOI: 10.1515/psr-2018-0121

https://doi.org/10.1515/9783110668896-001

[6]. By so doing, an examination of the main natural product (NP) classes was carried out according to their metabolic building blocks, e. g. alkaloids, fatty acids, polyketides, phenylpropanoids, and aromatic polyketides, and terpenoids. This included a discussion of the structural diversity of natural product classes using the scaffold approach while focusing on the characteristic carbon frameworks.

Several key questions still, however, remain to be answered:
- How many naturally occurring compounds are currently known?
- How diverse are they?
- What are the most frequently occurring chemical scaffolds and functional groups (FGs) among secondary metabolites?
- Which of the main pool of NPs (marine, terrestrial or microbial) is the most promising source of new and biologically active compounds?
- What proportion of the biosphere is still unexplored in terms of organisms and in terms of their metabolite contents?

In summary, what future lies ahead of us in terms of the coverage of the chemical space of secondary metabolites?

It must be mentioned that, although the investigation of several NP datasets *versus* synthetically obtained chemical libraries (SOCLs) have shown that NPs often occupy a much wider chemical space [7–13], the number of known unique compounds obtained from synthetic chemistry laboratories far outnumbers those of natural sources [14].

An illustration would be by comparing the number of synthetically derived compounds in the ZINC [15] and PubChem [16] databases, to those obtained biosynthetically, i. e. NPs (Table 2.1).

Table 2.1: The number of NPs *versus* synthetically obtained compounds currently in ZINC and PubChem databases.

Database	# SCs[a]*	# NPs[b]*	Ratio	Reference
ZINC	94,774,466	856,096	110:1	[15]
PubChem	97,997,240	2,760	35,500:1	[16]

[a]Approximate number of synthetic compounds; [b]Approximate number of natural products; *These are not the number of unique compound entries, since the number of unique compounds numbers might be much lower that these.

The ratios are 110:1 for ZINC and 35,500:1 for PubChem, showing that chemists have gone a long way in mastering the art of making new desired compounds much more than mastering how to identify and obtain compounds made by nature's machinery. Besides, although the use of traditional medicinal preparations from source organisms containing NPs is so widespread, the art of engineering organisms to make the required compounds in the quantities relevant for drug discovery with the goal of treating several millions of patients is relatively new [17, 18].

With the renewed interest in NPs as sources of new compounds for the discovery of drugs, agrochemicals, cosmetics, etc., several databases and datasets have been collected and made available to the scientific community, either freely or commercially. Table 2.2 provides a summary of selected NP databases with the most abundant compound annotations (~3,000 or more).

Although Table 2.2 provides a rough idea of the number of known NPs available in the literature, it should be mentioned that this is only based on data that has been curated and included in chemical databases. The non-curated chemical compound data (or data is the process of being curated), as well as compounds in uninvestigated species, have not been included in such estimations. A clearer picture would only be possible if parallel information of the number of known terrestrial, marine, fungal and microbial species which have been completely investigated is made available. This is rather far beyond the scope of this review.

There have been several attempts to quantify the number of NPs available in nature, by first attempting to provide an approximation of the number of NPs available in the literature:

– In his investigation of bioactive microbial metabolites János Bérdy first intuitively estimated the number of NPs to be about 1 million [44].
– Although not supported by strong experimental evidence and without providing details of the approximation method, this same author later estimated the number of published NPs to be between 300,000 and a double the number, i. e. 600,000 [45].
– Blunt et al. also conducted and published a similar estimate the same year, based on information available in NP databases [46].
– Based on information currently available in the most advanced NP databases like the CAS registry [23], DNP [26], Reaxys [36], SuperNatural II [38] and the UNPD [43] (see Table 2.2), such a number could be expected to be >300,000 compounds, knowing that the compound annotations from NPs in ZINC (UEFS) rather represent data donated collectively from several vendors (without removing duplicates and including compounds derived from semi-synthesis).
– A recent attempt by Chen et al. [35] to collect all known NPs in commercial and freely available virtual databases, including vendor libraries and removing duplicates, led to about 250,000 NPs, not including compound annotations from SuperNatural II and Reaxys. If one would also add compounds from fossils and define them as NPs [47], we could easily place the number at Bérdy's original estimate of between 300,000 and 600,000 NPs (*still an estimate!*).
– The most recent analysis of NPs which have been included currently available (free and commercial) databases by Zeng et al. placed the number at ~470,000 NPs [32], which is still within Bérdy's original estimate.

Table 2.2: Selected NP databases including ≥3,000 molecular annotations.

Database	# mol.[a]	Weblink	Origin	Reference
Ambinter and Greenpharma*	~6 k	www.ac-discovery.com/	Diverse	[19]
Analyticon Discovery MEGx*	~3 k	www.ambinter.com/	Diverse	[20]
AntiBase**	~40 k	http://www.user.gwdg.de/~hlaatsc/antibase	Fungi and microbes	[21]
AntiMarin** (AntiBase + MarinLit)	~60 k	https://omictools.com/marinlit-tool/	Marine, fungi and microbes	[22]
Chemical Abstracts Services (CAS)	~283 k	–	Diverse	[23]
ConMedNP	~3 k	http://www.african-compounds.org/conmednp/	Mostly plants	[24]
DMNP**	~55 k	dmnp.chemnetbase.com/	Marine	[25]
DNP**	~300 k	http://dnp.chemnetbase.com/	Diverse	[26]
HIT	~5 k	http://lifecenter.sgst.cn/hit/	Plants	[27]
iSMART	~20 k	http://ismart.cmu.edu.tw/	Mostly plants	[28]
Marinlit	~29 k	http://pubs.rsc.org/marinlit/	Marine	[29]
NANPDB	~5 k	www.african-compounds.org/nanpdb/	Mostly plants	[30]
Natural Products Atlas	~21 k	https://www.npatlas.org/joomla/index.php/	Fungi and microbes	[31]
NPASS	~35 k	http://bidd2.nus.edu.sg/NPASS/	Mostly plants	[32]
NPs in PubChem Substance	~3 k	–	Diverse	[16, 33]
NPs in ZINC (UEFS)	~900 k	http://zinc.docking.org/catalogs/uefsnp/	Diverse	[15]
NPs with known producing organism	~186 k	–	Diverse	[34]
OpenNP	~67 k	–	Diverse	[35]
Readily obtainable NPs	~26 k	–	Diverse	[35]
Reaxys**	~220 k	https://www.reaxys.com/	Diverse	[36]
StreptomeDB	~4 k	www.pharmaceutical-bioinformatics.de/streptomedb/	Bacteria	[37]
SuperNatural II***	~326 k	http://bioinf-applied.charite.de/supernatural_new/	Diverse	[38]
TCM Database@Taiwan	~53 k	http://tcm.cmu.edu.tw/	Mostly plants	[39]
TCMID	~13 k	www.megabionet.org/tcmid/	Plants	[40]
TIPdb	~9 k	http://cwtung.kmu.edu.tw/tipdb/	Plants	[41, 42]
UNPD	~229 k	http://pkuxxj.pku.edu.cn/UNPD/	Mostly plants	[43]

*Compound samples available; **Commercial database; ***Data could be available for collaborative projects; [a]Approximate number of molecular or sample annotations.

In this chapter, an attempt to review the literature with the view of providing recent numbers of known or published SMs from the various aforementioned major sources of NPs has been carried out. The discussion is subdivided into several parts, the first summarizing the major efforts from industry and academia towards a universal collection (hence enumeration) of NPs. We shall then attempt an enumeration by countries/regions of the world, then by biological activities, compound types and according to the various major sources of NPs (plants, marine, fungal and microbial), together with the geographical distribution of organisms producing SMs in the marine environment. The last sections will focus on the enumeration of NPs present in the major data resources for food chemicals and metabolites in the human body. The discussion is solely based on information available in databases or data collected, analysed and published by various research groups.

1.2 Attempts to obtain numbers by the generation of universal natural compound libraries

1.2.1 Enumeration of compound annotations in commercial libraries

1.2.1.1 The Dictionary of Natural Products (DNP)
Until now, the DNP (although only commercially available) is being regarded as one of the largest and most comprehensive compilations of compounds from natural sources [35, 48]. The latest version of the DNP (v. 27.1) provides data for ~300,000 NPs, including their physicochemical and biological properties, their systematic and common names, literature references, molecular structures, and origins (e. g. family, genus, and species names) [48]. An analysis of the current content of the DNP showed that:
- Of the ~195,000 SMs with available information on the compound origin, most SMs are derived from plants (almost 70 %), animals and bacteria (Figure 2.1a).
- Terpenoids and alkaloids are the two most abundant chemical classes of NPs in plants, animals, fungi, and bacteria put together, representing more than half of all compounds isolated from the plants (Figure 2.1b).
- Compositae and Leguminosae are the plant families with the highest number of SMs identified (Figure 2.2).
- Among all kingdoms, NPs isolated from *Streptomyces* spp. were largely represented
- The most abundant bioactive NPs are of bacterial, botanical and fungal sources (Figure 2.3).

However, it could be noted that certain classes of NPs were conspicuously absent from certain kingdoms, e. g. steroids were seldom noted in bacteria, while flavonoids and lignans were almost exclusively seen in plant species and polyketides were almost exclusively found in bacteria, fungi, and protists (Figure 2.1b) [48].

a

Bacteria (17,531)

Protista (2,402)

Archaebacteria (184)

Fungi (19,869)

Animalia (25,064)

Plantae (133,881 NP)

b

Archaebacteria
Protista
Bacteria
Fungi
Animalia
Plantae

0% 10% 20% 30% 40% 50% 60% 70% 80% 90% 100%

■ Terpenoids
■ Amino Acid and Peptide Natural Products
■ Simple Aromatic Natural Products
■ Steroid Natural Products
■ Polyketides
■ Carbohydrate Natural Products

■ Alkaloids
■ Flavonoids
■ Aliphatic Natural Products
■ Lignans
■ Polycyclic Arimatic Natural Products
■ Others

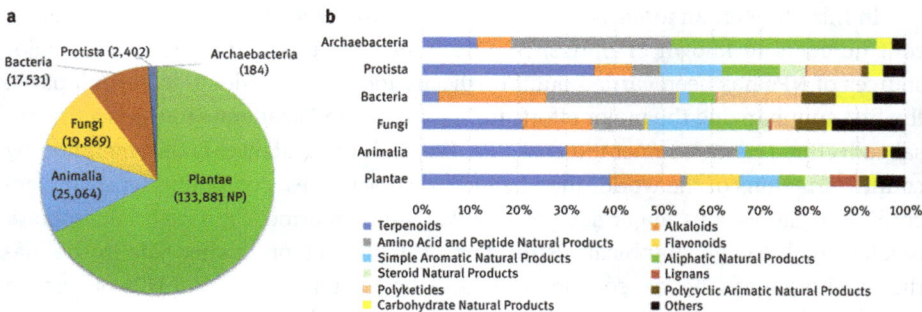

Figure 2.1: Summary of the contents of the DNP; (a) distribution of natural products per kingdom of life; (b) distribution of the main chemical classes of natural products in each kingdom of life [48]. Figure reproduced by permission.

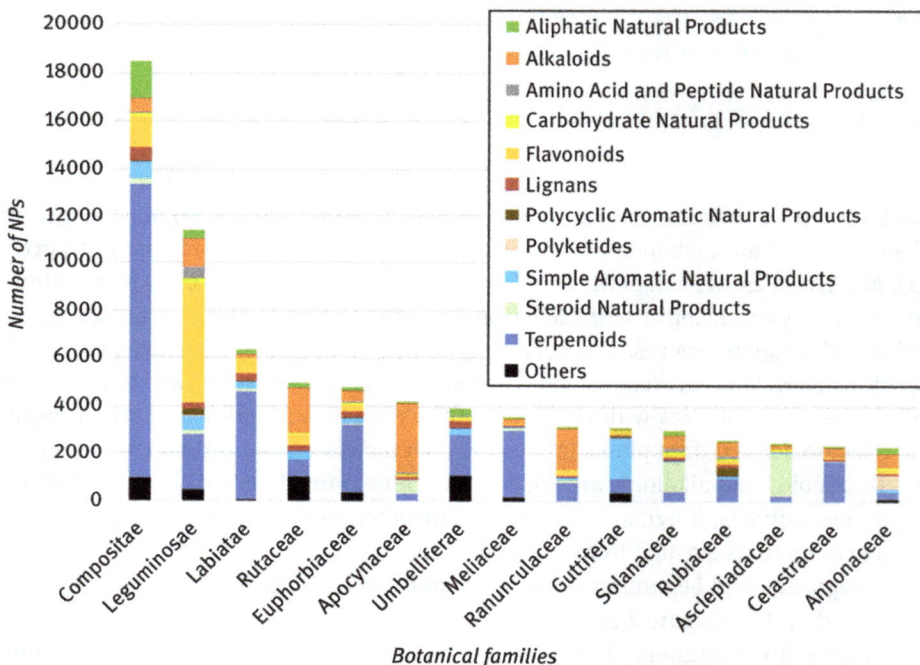

■ Aliphatic Natural Products
■ Alkaloids
■ Amino Acid and Peptide Natural Products
■ Carbohydrate Natural Products
■ Flavonoids
■ Lignans
■ Polycyclic Aromatic Natural Products
■ Polyketides
■ Simple Aromatic Natural Products
■ Steroid Natural Products
■ Terpenoids
■ Others

Number of NPs

Botanical families

Figure 2.2: The top 15 plant families containing NPs and the distribution of the different chemical classes [48]. Figure reproduced by permission.

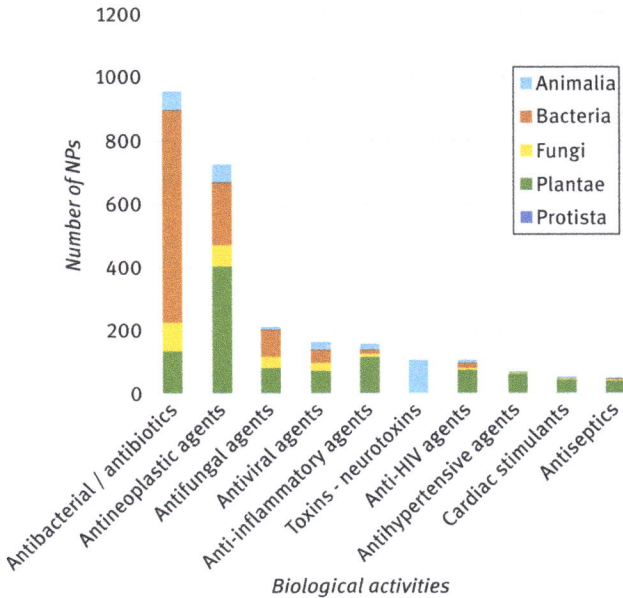

Figure 2.3: The top 10 biological activities exhibited by NPs and their distribution by kingdom of life [48]. Figure reproduced by permission.

Being the second kingdom to produce abundant NPs (about 25 % of DNP compounds with source species information), most compounds produced by animals were found in snake venoms. The reader is referred to the recently published Snake Venom Database (SVDB) [49] for further reading.

1.2.1.2 Natural products in the Chemical Abstracts Services (CAS) directory

With about 285,000 NPs, the CAS Registry (the search engine of this database is known as SciFinder) has been the gold standard for chemical substance information. This database currently includes >149 million unique and diverse organic and inorganic molecules and substances, e. g. alloys, coordination compounds, minerals, mixtures, polymers and salts, and >67 million sequences, more than any other similar database [23]. The CAS Registry Number® is used to identify your substance of interest. This is universally used to provide a unique, unmistakable identifier for chemical substances. The CAS registry information includes daily updated information on literature references to the substance experimental and predicted property data (e. g. boiling and melting points, etc.), CA Index Names and synonyms, commercial availability of compounds/substances, preparative methods, spectra, regulatory information from international sources, etc.

1.2.1.3 Reaxys

This is a vast commercial database that includes about 220,000 NP annotations [36]. Apart from supporting chemistry research, including pharmaceutical development, the chemicals industry, and academic research, Reaxys provides integrated access to the eMolecules database containing more than 8 million unique molecules, including screening compounds and building blocks, from over 150 commercial suppliers, which is updated weekly.

1.2.2 Academic and other open-access efforts

The quantification of available NPs has also been assisted by efforts from academic groups, often providing the information in open access. A collection of freely accessible small molecule databases is available [50]. The most important efforts focused on NPs are described herein.

1.2.2.1 The Berlin collection

This is the SuperNatural II database, published and regularly updated by Robert Preissner's group in Charité Berlin [38]. Published in Open Access, the authors also included information of modes of action, pathways and clusters for the entire dataset of currently 325,508 natural products extracted from various resources. It also provides the toxicity prediction for the database compounds. For instance, a substructure search can be performed to identify compounds containing this substructure. Additionally, possible target proteins and pathways are predicted for the natural compounds, based on 2D structural similarity search to known drug molecules.

1.2.2.2 The Hamburg collection

The efforts of researchers from Johannes Kirchmair's group in Hamburg led to a collection of close to 250,000 unique NPs from a wide range of commercial and freely available virtual databases, excluding data from SuperNatural II and Reaxys [35]. It was reported that only about 10% of the unique NPs had readily available samples either from compound vendors of *via* collaborations.

1.2.2.3 The UNPD collection

This database, currently including about 229,000 NP entries, was originally derived as a collection, aimed at getting a "universal" database of NPs [43]; with ~118,000 unique compounds from Reaxys [36], ~42,000 from Chinese Natural Product Database (CNPD) [51], ~7,500 from Traditional Chinese Medicines Database (TCMD) [52] and ~30,000 from the Chinese Herbal Drugs Database (CHDD) [53], after removing duplicates. Although the goal was to get a "universal" collection of NPs, this dataset ended up mostly including compounds from plants used in Traditional Chinese Medicine (TCM) and does not include NPs from bacterial and marine sources.

1.2.2.4 The Natural Product Activity and Species Source (NPASS) collection

NPASS is a freely accessible database focused on the biological activities and source species of NPs, currently including >35,000 NPs [32]. The strength of NPASS is the availability of information on NPs derived from 25,041 species with activities against 5,863 targets (i. e. 2,946 proteins, along with 1,352 microbial species and 1,227 cell lines). The database also includes 446,552 quantitative activity records (e. g. IC_{50}, K_i, EC_{50}, GI_{50} or MIC values) in relation to 222,092 NP-target pairs and 288,002 NP-species pairs

1.2.2.5 The PubChem and ZINC collection

PubChem [16] and ZINC [15] are the most frequently used and cited small molecule databases with the computer-aided drug design and virtual screening community, because of the availability of compound vendor or supplier information, in addition to the fact that the included data are freely accessible. In numbering the NPs in ZINC and PubChem, care must be taken because their open access policy makes duplicates and errors in compound structures inevitable.

1.2.3 Enumerating natural product sample collections

Several NP sample collections are available within academic groups and from vendors [35], e. g. Ambinter and Greenpharma [19], Analyticon Discovery MEGx [20], etc. However, the number and individual quantities of available samples are often very limited [17]. Although Chen et al. stated that about 10 % of NPs could be directly available through vendors and academic collaborations [35], a major challenge towards getting the exact number of NP samples available is that such information is either seldom published, not properly organized, inaccessible in existing NP databases or are not constantly being updated as the samples are being used up. A reference case for the efficient development and management of NPs samples are Nature Bank and Queensland Compound Library (QCL), hosted at the Eskitis Institute for Drug Discovery, Australia [54], which currently holds about 3,250 compounds [55], started off as a collection of about 800 samples [56]. Nature Bank, for example, is a comprehensive collection of plants and marine invertebrates, mainly terrestrial plants (from Queensland, China, and Papua New Guinea) and marine invertebrates (from the Great Barrier Reef and Tasmania). Meanwhile, QCL is the Australian national resource for compound management and logistics. The two institutions are working closely together. The Bioinformatics Institute (Singapore) currently hosts 2,500 compounds and 340,000 crude extracts from 37,000 plants, while the ChromaDex library includes about 3,000 compounds from 1,640 plants [55]. The largest currently available screening library with compounds of natural origin is the InterBioScreen (Moscow) [57], which includes over 68,000 NPs. Originally the main contributors of natural compounds and derivatives were research institutes of the former Soviet Union but now includes contributions from Japan, Europe, and the USA. About 13,000 natural and synthetic in-stock-available building blocks are also available [55, 57].

1.3 Number of unique molecules in the country and regional datasets

The number of NPs included in the country or regional databases have been summarized in Table 2.3. These rather represent much smaller numbers and most of the data have already been included in the majority of the aforementioned databases [35, 38].

1.4 Number of unique molecules in disease or therapeutic use datasets

These have been summarized in Table 2.4 and also represent much smaller numbers, almost never exceeding 1,000 compound annotations per dataset, apart from the datasets related to Chinese traditional medicine, which could also be classified under the country/regional databases. This is because the compounds are not related to a specific therapeutic class but are derived from plants with many diverse uses.

1.5 Enumeration by compound types

Apart from the previously described classification of NPs in the DNP into the various compound classes [48], very little effort has been put into the development of databases focused on compound classes, apart from the Carotenoid database [86] and the on-going project for the development of a database of flavonoids [87]. At the moment, it would be quite tedious to get exact numbers of NPs per compound class, as such information is only included in a few datasets [26, 30, 72–74].

Table 2.3: Number of NPs included in the country/regional datasets.

Database	Country/region	# mol.[a]	Weblink	Origin	Reference
AfroDb	Africa	~1 k	http://african-compounds.org/about/afrodb/	Plants	[58]
BIOFACQUIM	Mexico	~400	https://biofacquim.herokuapp.com	Diverse	[59]
ChemDP	Pakistan	~1 k	www.chemdp.com		[60]
CamMedNP	Cameroon	~1,5 k	http://african-compounds.org/about/cammednp/	Mostly plants	[61]
ConMedNP	Central Africa	~3 k	http://african-compounds.org/about/conmednp/	Mostly plants	[24]
EARNPDB	Eastern Africa		http://african-compounds.org/about/earnpdb/	Plants	unpublished
Indonesian NPs	Indonesia	~6,800	http://herbaldb.farmasi.ui.ac.id/v3/	Plants	[62]
MAPS Database	India	>1,200	http://www.mapsdatabase.com/	Plants	[63]
Mitishamba	Kenya		http://mitishamba.uonbi.ac.ke/	Plants	[64]
NANPDB	Northern Africa	~5 k	http://african-compounds.org/nanpdb/	Mostly plants	[30]
NUBBE$_{DB}$	Brazil	~3 k	https://nubbe.iq.unesp.br/portal/nubbe-search.html	Mostly plants	[65, 66]
Panama NPs	Panama	~400	–	Diverse	[67]
SANCDB	South Africa	~700	https://sancdb.rubi.ru.ac.za/	Diverse organisms	[68]
Phytochemica	Himalaya	~1 k	home.iitj.ac.in/bagler/webservers/Phytochemica	Mostly plants	[69]
TCM Database@Taiwan	Taiwan	~53 k	http://tcm.cmu.edu.tw/	Mostly plants	[39]
TCMID	China	~10 k	www.megabionet.org/tcmid/	Plants	[40]
TIPdb	Taiwan	~9 k	http://cwtung.kmu.edu.tw/tipdb/	Plants	[41, 42]
TM-MC	Northeast Asia	~26 k	http://informatics.kiom.re.kr/compound/	Medicinal materials	[70]
VIETHERB	Vietnam	~11 k	http://vietherb.com.vn/	Plants	[71]
WADB	West Africa	~1 k	http://african-compounds.org/about/wadb/	Plants	unpublished

Table 2.4: Number of NPs included in therapeutic use datasets.

Database	Disease/use/characteristic	# mol.[a]	Weblink	Origin	Reference
AfroCancer	Cancer	~400	http://african-compounds.org/about/afrocaner/	Plants	[72]
AfroMalariaDB	Malaria	~500	http://african-compounds.org/about/afromalariadb/	Plants	[73]
Afrotryp	Human African trypanosiomiasis	~300	http://african-compounds.org/about/afrotryp/	Plants	[74]
Antimalarial NPs	Malaria	~1 k	–	Diverse	[75–77]
Ayurveda	Ayurvedic medicine	~1 k	https://omictools.com/ayurveda-tool/		[78]
BioPhytMol	Mycobacterial infections	~600	http://ab-openlab.csir.res.in/biophytmol/		[79]
CHMIS-C	Bind to anticancer targets	~9 k	http://sw16.im.med.umich.edu/chmis-c/		[80]
CVDHD	Cardiovascular diseases	~35 k	http://pkuxxj.pku.edu.cn/CVDHD/		[81]
NPACT	Cancer	~1.5 k	http://crdd.osdd.net/raghava/npact/	Mostly plants	[82]
NPCARE	Cancer	~6.5 k	http://silver.sejong.ac.kr/npcare/		[83]
SVDB	Snake venom	~700	https://www.snakevenomdb.org/	Snake venom	[49]
TCM-ID	Traditional Chinese medicine	~13 k	http://www.megabionet.org/tcmid/		[40]
TCMSP	Traditional Chinese medicine	~29 k	http://sm.nwsuaf.edu.cn/lsp/tcmsp.php/		[84, 85]
TIPdb	Antiplatelet, anticancer and antitubercular	~9 k	http://cwtung.kmu.edu.tw/tipdb/	Plants	[41, 42]

[a]Approximate number of molecular annotations.

1.6 Global enumeration by major natural product pools

1.6.1 The challenge of classifying natural products into terrestrial, marine and microbial origins

It is quite challenging to get a clear cut demarcation of natural product sources into terrestrial, marine and microbial for several reasons:
- Plants animals and microbes are both terrestrial and aquatic.
- Apart from the challenging effort to develop "universal" databases [23, 26, 32, 36, 38, 43] and marine-based metabolites [25, 29], NP databases are often developed based on types of species, e. g. plant-based [24, 28], animal-based [50], microbial [31, 37], etc., by geographical regions (Table 2.3) by disease/therapeutic uses (Table 2.4) or by compound classes (Table 2.5).

Table 2.5: Number of NPs included in therapeutic use datasets.

Database	Compound class	# mol.[a]	Weblink	Origin	Reference
Carotenoids Database	Carotenoids	~1,100	http://carotenoiddb.jp/	Mostly plants	[86]
Flavonoid Database	Flavonoids	Not determined	Ón going	Plants	[87]
ProCarDB	Bacterial carotenoids	304	http://bioinfo.imtech.res.in/servers/procardb/	Bacteria	[88, 89]

[a]Approximate number of molecular annotations.

1.6.1.1 Databases of natural products from terrestrial plants
Apart from the enumeration of plant-based SMs from diverse countries (Table 2.3), as well as including the molecular activities of useful plants, the Collective Molecular Activities of Useful Plants (CMAUP), was recently published to cover plants growing on the terrestrial habitat [90]. However, no database with a universal collection of plant-based SMs exists [91]. The CMAUP database currently includes 5,645 plants (i. e. 2,567 medicinal plants, 170 food plants, 1,567 edible plants, 3 agricultural plants, and 119 garden plants). These were collected from 153 countries and regions and includes 47,645 plant ingredients active against 646 targets in 234 KEGG pathways associated with 2473 gene ontologies and 656 diseases.

1.6.1.2 Databases of natural products from fungal species
Most databases which could enable the enumeration of NPs from fungal species are combined with those from microbes (Table 2.6), e. g. AntiBase [21], AntiMarin [22], and Natural Product Atlas [31]. A specialized dataset of compounds exclusive to

Table 2.6: Number of NPs from fungal sources.

Database	# mol.[a]	Weblink	Reference
Antibase	~40 k	http://wwwuser.gwdg.de/~hlaatsc/antibase	[21]
AntiMarin	~60 k	https://omictools.com/marinlit-tool/	[22]
Natural Products from Mushrooms	~1,100	–	[92]
Natural Product Atlas	~21 k	https://www.npatlas.org/joomla/index.php/	[31]

[a]Approximate number of molecular annotations.

saprophytic fungi like mushrooms only currently includes about 1,100 NPs. However, this does not include compounds from plant endophytes and parasitic fungi.

1.6.2 Natural products of microbial origin

Compounds from bacteria and protists do not currently represent a significant proportion of NPs included in the DNP [48]. However, Natural Products Atlas was designed to cover all microbially derived natural products published in the peer-reviewed primary scientific literature [31]. This includes about 21,000 bacterial, fungal and cyanobacterial compounds, as well as NPs from lichens and mushrooms and other higher fungi. This excludes compounds from plants, invertebrates or other higher organisms, except if these have also been explicitly identified from a microbial source. Compounds from marine macroalgae and diatoms are also excluded. Pye et al. [91] collected and analysed such data by combining a dataset of a dataset comprising all published microbial and marine-derived natural products from the period 1941–2015, which were obtained from the commercial database AntiMarin [22]. They then combined this with data for the period 2012–2015 were through manual curation of all published articles from a large panel of journals in the chemistry and chemical biology arena. This resulted in a collection of 40,229 NPs. It was not, however, mentioned how many of these compounds were of microbial origin. The StreptomeDB, for example, currently contains 4,040 NPs that have been biosynthesized by 2,584 bacterial strains from the genus *Streptomyces* [37]. A recent collection of compounds from cyanobacteria alone led to the identification of 578 NPs distributed between the three major environmental sources, i. e. marine, terrestrial and freshwater [93]. Crüsemann et al. were recently able to develop a rapid method, based on molecular networks, comprising of 603 samples from 146 marine *Salinispora* and *Streptomyces* strains [94]. The method was capable of generating ~1.8 million mass spectra, although it wasn't specified how many SMs this might correspond to [94].

1.6.3 Natural products of marine origin

Marine organisms are quite varied and include; phytoplankton, green, brown, and red algae, sponges, cnidarians, bryozoans, molluscs, tunicates, echinoderms, mangroves, and other intertidal plants and microorganisms. Again, it is quite challenging to enumerate marine separately from microbial NPs, since many NP-producing microbes are marine-based, so the most advanced databases focusing on such metabolites, e. g. AntiBase [21], AntiMarin [22], DMNP [25], and MarinLit [29] are often combined. SMs from marine organisms are known to have several implications in medicinal chemistry [95–100], although only 8 marine NPs have to date been approved as drugs and while 12 marine-derived metabolites are currently in different phases of clinical trials [1, 101–103]. Since, for example, most currently used antibiotics have been isolated from terrestrial microbes, accounting for more than 75 % of all antibiotics discovered [44, 104], the marine environment remains an untapped source of new bioactive molecules. For this reason, several thousands of SMs have been collected from marine sources and could enable us to enumerate the available NPs from marine sources.

1.6.3.1 Marine databases of natural products
In addition to terrestrial organisms that still remain a promising source of new bioactive metabolites, the marine environment (which covers approximately 70 % of the earth's surface) represents largely unexplored biodiversity [105]. Apart from the aforementioned commercial databases specialized in marine metabolites, Lei and Zhou had also published a dataset of 6,000 chemical compounds derived from over 10,000 marine-derived material, including information on the source organisms (mainly coelenterates, sponges and blue algae) and biological activities of each compound [98]. Another collection focused on metabolites from red algae of the genus *Laurencia* showed that for data published until 2015, a total of 1,047 secondary metabolites with carbocyclic skeletons (sesquiterpenes, diterpenes, triterpenes, acetogenins, indoles, aromatic compounds, steroids), and miscellaneous compounds were already published for this genus [105]. Detailed analysis and enumeration of NPs from cyanobacteria by biological activities have also been provided by Burja et al. [106]. Davis et al. also published a publicly accessible Seaweed Metabolite Database (SWMD) currently containing >1,100 compounds, mostly from the red algae of the genus *Laurencia* (Ceramiales, Rhodomelaceae) [107]. The authors made an extra effort to include the geographical origin of the seaweed, the extraction method and detailed chemical information on each metabolite.

1.6.3.2 An analysis of the geographical distribution marine source organism
A detailed comparison between the physicochemical properties of NPs from marine and terrestrial sources has been recently published [108] and a summary will be provided in the next chapter. In addition, Principe and Fisher recently reviewed a

collection of information data associated with 298 pharmacological products originating from marine biota during the past 47 years [97]. The products were developed from 232 different marine species belonging to 15 phyla, i. e. 1,296 collections
of specimens from 69 countries and from all 7 continents (Table 2.7). An investigation of the spatial distribution of the geographical locations of sample collections
(Figure 2.4) provides a sort of map of where and when the specimens were collected
that yielded MNPs with pharmacological potential (for which the clinical is reported)
were collected. The goal of the study was not to have a representative sample of
chemical structures or geographic locations, but rather to identify locations that yield
the MNPs with demonstrated value or potential value. This also led to the identification of species from which those MNPs had been isolated and the locations where the
specimens yielding those MNPs had been collected. The data collected covered 298
MNPs (i. e. 16 the FDA-approved drugs, 55 compounds in clinical, 51 compounds
in preclinical testing and 176 lead compounds or probes) from 1,296 specimen collections. The spatial distributions of such data around the Bramble reef (North
East Australia) have been shown in Figure 2.4 [97].

Table 2.7: Collections by Phylum[a] [97].

Phyla	Number in collections	Percentage
Actinobacteria[b]	40	3.1 %
Bryozoa[c]	18	1.4 %
Cnidaria[d]	31	2.4 %
Cyanobacteria[e]	77	5.9 %
Dinoflagellata[f]	13	1.0 %
Echinodermata[g]	12	0.9 %
Fish	29	2.2 %
Fungi	3	0.2 %
Green Algae	8	0.6 %
Hemichordata[h]	2	0.2 %
Mollusca	195	15.1 %
Nemertea[i]	3	0.2 %
Porifera[j]	716	55.2 %
Rhodophyta[k]	4	0.3 %
Tunicata[l]	145	11.2 %

[a]Fucoxanthin is found in most or all species of the classes Phaeophyceae
(brown algae), Chrysophyceae (golden algae) and Bacillariophyceae
(diatoms) in the phylum Ochrophyta plus some species in the phyla
Dinoflagellata and Haptophyta, possibly as many as 16,000 species in
total, not included in this analyses. [b]Gram-positive bacteria. [c]Mostly
colonial filter feeders. [d]Includes octocorals. [e]Ex-blue-green algae. [f]Mostly
marine plankton. [g]Includes starfish, sea urchins, and sea cucumbers.
[h]Acorn worms. [i]Ribbon worms. [j]Sponges. [k]Red algae. [l]Tunicates.

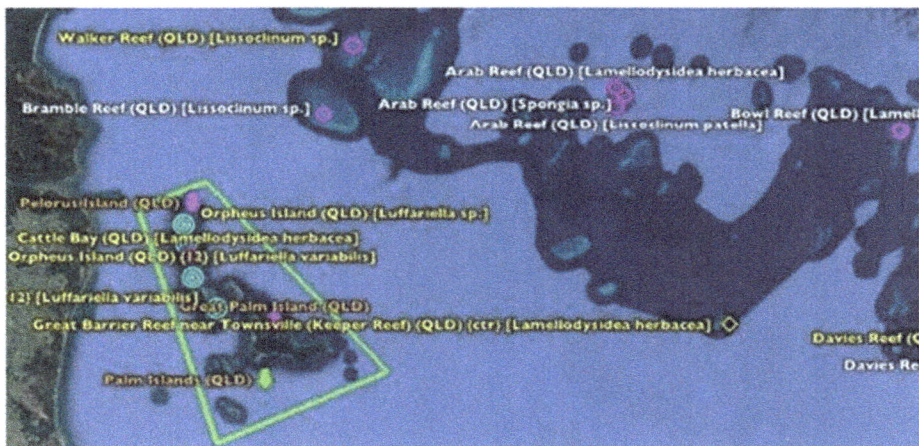

Figure 2.4: An example of the Google Earth Database showing collections made from the Great Barrier Reef near Townsville, Queensland, Australia [97]. Figure reproduced by permission.

1.6.4 Natural products from food plants

Food-based chemicals include primary metabolites and secondary metabolites. Table 2.8 shows some databases of food-based compounds and their approximate numbers of components.

The commercially available Dictionary of Food Compounds (DFC) is the most advanced collection of compounds contained in foods, currently holding more than 30,000 compounds [109] and the FooDB with about the same number of compounds [111]. Although the chemoinformatics analysis of components of foods, particularly phytochemicals and natural additives present or added in foods [123–132], the most advanced open access resources for natural compounds in foods is the NutriChem 1.0 server [114], a database generated by text mining of 21 million MEDLINE abstracts, with information that links plant-based foods with their small molecule components and human disease phenotypes. This server contains text-mined data corresponding to 18,478 pairs of 1,772 plant-based foods and 7,898 phytochemicals, along with 6,242 pairs of 1,066 plant-based foods and 751 diseases. Predicted associations for 548 phytochemicals and 252 diseases are also included. This tool provides the latest foundation for a mechanistic understanding of the consequences of eating behaviours on health [115, 116].

Table 2.8: Number of NPs from some food sources.

Database	Origin	# mol.[a]	Weblink	Reference
DFC	Natural food components and additives	~30k	http://dfc.chemnetbase.com	[109]
EuroFIR-BASIS	Phytochemicals in foods	~260	http://ebasis.eurofir.org/Default.asp	[110]
FooDB	Diverse food sources	~28 k	http://foodb.ca/compounds	[111]
GRAS	Flavour chemicals	~2,300	https://www.fda.gov/food/food-ingredients-packaging/generally-recognized-safe-gras	[112, 113]
NutriChem	Plant-based foods and phytochemicals	~8 k	http://www.cbs.dtu.dk/services/NutriChem-1.0/	[114–116]
Phenol-Explorer	Polyphenols in foods	~500	http://phenol-explorer.eu/	[117, 118]
SuperScent	Components of flavours and aromas	~2,300	http://bioinf-applied.charite.de/superscent/	[119]
SuperSweet	Sweet compounds	~8 k	http://bioinf-applied.charite.de/sweet/	[120]
TMDB	Tea (*Camellia* spp.)	~1,400	http://pcsb.ahau.edu.cn:8080/TCDB/index.jsp	[121]
USDA	Food Components	–	https://www.nal.usda.gov/fnic/food-composition	[122]

[a]Approximate number of molecular annotations.

1.6.5 Human metabolites

Metabolites in humans have been collected into the Human Metabolome Database (HMDB) [133–136], a freely available resource containing detailed information on small molecule metabolites found in the human body, mainly:
- chemical data,
- clinical data and
- molecular biology/biochemistry data.

It currently contains 114,083 water-soluble and lipid-soluble metabolite entries, as well as metabolites that would be regarded as either abundant (>1 μM) or relatively rare (<1 nM). Moreover, the metabolite entries have been linked to 5,702 protein sequences, chemical/clinical data, and enzymatic or biochemical data. Many data fields are hyperlinked to other databases (e. g. KEGG [137], PubChem [16], MetaCyc [138], ChEBI [139], PDB [140], UniProt [141] and GenBank [142]) and a variety of structure and pathways. Another utility of the HMDB is its links with additional databases like DrugBank [143], T3DB [144, 145], SMPDB [146, 147] and FooDB [111], which contain information on ~2,280 drug and drug metabolites, ~3,670 common toxins and environmental pollutants, pathway diagrams for ~25,000 human metabolic and disease pathways, food components, and food additives, respectively [136].

1.7 Conclusions

Due to the important roles that NPs play in the pharmaceutical, cosmetics and food industries, their potential for exploration in these areas has always come to question. Several attempts have been made to quantify the number of molecules made by nature. This chapter has been focused on estimating this approximate number. The discussion has been based on recently published data on compounds produced in plants, bacteria, marine organisms, humans or contained food substances. It has been estimated that a total number of more than 450,000 NPs exist in the literature, the majority being of plant origin. However, most bioactive NPs are of marine origin, although the marine environment still remains largely unexplored. A major limitation to the exploitation of NPs in large scale drug discovery remains the availability of samples since NPs are quite hard to synthesize and only a proportion of about 10 % of known NPs have available samples ready for screening. It is hoped that the automatic structure elucidation of metabolites [148] and the exploration of genomic data from NP-producing organisms would revolutionalize the world of NP drug discovery.

Acknowledgements: F.N.K. would also like to acknowledge the European Structural and Investment Funds, OP RDE-funded project "ChemJets" (No. CZ.02.2.69/0.0/0.0/ 16_027/0008351). D.S. was supported from the Ministry of Education, Youth and Sports of the Czech Republic (project number LM2018130) and by RVO 68378050-KAV-NPUI.

References

[1]   Harvey AL. Natural products in drug discovery. Drug Discov Today. 2008;13:894–901.
[2]   Rodrigues T, Reker D, Schneider P, Schneider G. Counting on natural products for drug design. Nat Chem. 2016;8:531.
[3]   Carreira EM. Natural products synthesis: a personal retrospective and outlook. Israel J Chem. 2018;58:114–21.
[4]   Lear MJ, Hirai K, Ogawa K, Yamashita S, Hirama M. A convergent total synthesis of the kedarcidin chromophore: 20-years in the making. J Antibiot (Tokyo). 2019. DOI: 10.1038/ s41429-019-0175-y.
[5]   Ortholand J-Y, Ganesan A. Natural products and combinatorial chemistry: back to the future. Curr Opin Chem Biol. 2004;8:271–80.
[6]   Abegaz BM, Kinfe HH. Secondary metabolites, their structural diversity, bioactivity, and ecological functions: an overview. Phys Sci Rev. 2018. DOI: 10.1515/psr-2018-0100.
[7]   Feher M, Schmidt JM. Property distributions: differences between drugs, natural products, and molecules from combinatorial chemistry. J Chem Inf Comput Sci. 2003;43:218–27.
[8]   Larsson J, Gottfries J, Muresan S, Backlund A. ChemGPS-NP: tuned for navigation in biologically relevant chemical space. J Nat Prod. 2007;70:789–94.
[9]   Singh N, Guha R, Giulianotti MA, Pinilla C, Houghten RA, Medina-Franco JL. Chemoinformatic analysis of combinatorial libraries, drugs, natural products, and molecular libraries small molecule repository. J Chem Inf Model. 2009;49:1010–24.
[10] Rosén J, Gottfries J, Muresan S, Backlund A, Oprea TI. Novel chemical space exploration via natural products. J Med Chem. 2009;52:1953–62.
[11] Gu J, Chen L, Yuan G, Xu X. A drug-target network-based approach to evaluate the efficacy of medicinal plants for type II diabetes mellitus. Evid Based Complement Alternat Med. 2013;2013:203614.
[12] Lachance H, Wetzel S, Kumar K, Waldmann H. Charting, navigating, and populating natural product chemical space for drug discovery. J Med Chem. 2012;55:5989–6001.
[13] López-Vallejo F, Giulianotti MA, Houghten RA, Medina-Franco JL. Expanding the medicinally relevant chemical space with compound libraries. Drug Discov Today. 2012;17:718–26.
[14] Harvey AL, Edrada-Ebel R, Quinn RJ. The re-emergence of natural products for drug discovery in the genomics era. Nat Rev Drug Discov. 2015;14:111–29.
[15] Sterling T, Irwin JJ. ZINC 15–ligand discovery for everyone. J Chem Inf Model. 2015;55:2324 –37. Available at: http://zinc.docking.org/catalogs/uefsnp/. Accessed: 16 May 2019.
[16] Kim S, Thiessen PA, Bolton EE, Chen J, Fu G, Gindulyte A, et al. PubChem substance and compound databases. Nucleic Acids Res. 2016;44:D1202–13.
[17] Pan L, Chai HB, Kinghorn AD. Discovery of new anticancer agents from higher plants. Front Biosci (Schol Ed). 2013;4:142–56.
[18] Akone SH, Pham C-D, Chen H, Ola AR, Ntie-Kang F, Proksch P. Epigenetic modification, co-culture and genomic methods for natural product discovery. Phys Sci Rev. 2018. DOI: 10.1515/psr-2018-0118.
[19] AnalytiCon Discovery. Available at: www.ac-discovery.com. Accessed: 16 May 2019.

[20] Ambinter. Available at: www.ambinter.com. Accessed: 16 May 2019.

[21] Laatsch H. Antibase, version 4.0 – the natural compound identifier (for microbial secondary metabolites as well as higher fungi). Weinheim: Wiley-VCH Verlag GmbH & Co. KGaA, 2012.

[22] Blunt JW, Munro MH, Laatsch H. AntiMarin database. Christchurch, New Zealand: University of Canterbury, 2006.

[23] CAS REGISTRY - The gold standard for chemical substance information. Available at: https://www.cas.org/support/documentation/chemical-substances. Accessed: 9 May 2019.

[24] Ntie-Kang F, PA O, Scharfe M, Owono Owono LC, Megnassan E, LM M, et al. ConMedNP: a natural product library from Central African medicinal plants for drug discovery. RSC Adv. 2014;4:409–19. Available at: http://www.african-compounds.org/conmednp. Accessed: 16 May 2019.

[25] Blunt JW, Munro MH, editors. Dictionary of marine natural products with CD-ROM. Boca Raton, FL: Chapman and Hall/CRC, 2007.

[26] Dictionary of Natural Products (DNP). Available at: http://dnp.chemnetbase.com. Accessed: 16 May 2019.

[27] Ye H, Ye L, Kang H, Tao L, Tang K, Liu X, et al. HIT: linking herbal active ingredients to targets. Nucleic Acids Res. 2011;39:D1055–9.

[28] Chang K-W, Tsai T-Y, Chen K-C, Yang S-C, Huang H-J, Chang T-T, et al. iSMART: an integrated cloud computing web server for traditional Chinese medicine for online virtual screening, de novo evolution and drug design. J Biomol Struct Dyn. 2011;29:243–50.

[29] Blunt JW, Munro MH MarinLit: A database of the marine natural products literature. Available at: http://pubs.rsc.org/marinlit/. Accessed: 20 Nov 2017.

[30] Ntie-Kang F, Telukunta KK, Döring K, Simoben CV, Moumbock AF, Malange YI, et al. NANPDB: a resource for natural products from Northern African sources. J Nat Prod. 2017;80:2067–76. Available at: www.african-compounds.org/nanpdb. Accessed: 16 May 2019.

[31] Natural Product Atlas. Available at: https://www.npatlas.org/joomla/index.php.

[32] Zeng X, Zhang P, He W, Qin C, Chen S, Tao L, et al. NPASS: natural product activity and species source database for natural product research, discovery and tool development. Nucleic Acids Res. 2018;46:D1217–22.

[33] Hao M, Cheng T, Wang Y, Bryant HS. Web search and data mining of natural products and their bioactivities in PubChem. Sci China Chem. 2013;56:1424–35.

[34] Ertl P, Schuhmann T. A systematic cheminformatics analysis of functional groups occurring in natural products. J Nat Prod. 2019;82:1258–63.

[35] Chen Y, de Bruyn Kops C, Kirchmair J. Data resources for the computer-guided discovery of bioactive natural products. J Chem Inf Model. 2017;57:2099–111.

[36] Reaxys; Elsevier: New York. Available at: https://www.reaxys.com (accessed Jul 17, 2017). Accessed: 9 May 2019.

[37] Klementz D, Döring K, Lucas X, Telukunta KK, Erxleben A, Deubel D, et al. StreptomeDB 2.0 - an extended resource of natural products produced by streptomycetes. Nucleic Acids Res. 2016;44:D509–14. Available at: www.pharmaceutical-bioinformatics.de/streptomedb. Accessed: 16 May 2019.

[38] Banerjee P, Erehman J, Gohlke B-O, Wilhelm T, Preissner R, Dunkel M. Super Natural II – a database of natural products. Nucleic Acids Res. 2015;43:D935–9. Available at: http://bioinf-applied.charite.de/supernatural_new. Accessed: 9 May 2019.

[39] Chen CY. TCM database@Taiwan: the world's largest traditional chinese medicine database for drug screening in silico. PLoS One. 2011;6:e15939. Available at: http://tcm.cmu.edu.tw.

[40] Xue R, Fang Z, Zhang M, Yi Z, Wen C, Shi T. TCMID: traditional Chinese medicine integrative database for herb molecular mechanism analysis. Nucleic Acids Res. 2013;41:D1089–95. Available at: www.megabionet.org/tcmid. Accessed: 16 May 2019.

[41] Lin Y-C, Wang C-C, Chen I-S, Jheng J-L, Li J-H, Tung C-W. TIPdb: a database of anticancer, antiplatelet, and antituberculosis phytochemicals from indigenous plants in Taiwan. Sci World J. 2013;2013:736386.

[42] Tung C-W, Lin Y-C, Chang H-S, Wang C-C, Chen I-S, Jheng J-L, et al. TIPdb-3D: the three-dimensional structure database of phytochemicals from Taiwan indigenous plants. Database. 2014;2014:bau055. Available at: http://cwtung.kmu.edu.tw/tipdb. Accessed: 16 May 2019.

[43] Gu J, Gui Y, Chen L, Yuan G, Lu H-Z, Xu X. Use of natural products as chemical library for drug discovery and network pharmacology. PLoS One. 2013;8:e62839. Available at: http://pkuxxj.pku.edu.cn/UNPD. Accessed: 16 May 2019.

[44] Bérdy J. Bioactive microbial metabolites. J Antibiot. 2005;58:1–26.

[45] Bérdy J. Thoughts and facts about antibiotics: where we are now and where we are heading. J Antibiot. 2012;65:385–95.

[46] Blunt J, Munro M, Upjohn M. The role of databases in marine natural products research. In: Fattorusso E, Gerwick WH, Taglialatela-Scafati O, editors. Handbook of marine natural products. Dordrecht: Springer, 2012:389–421.

[47] Falk H, Wolkenstein K. Natural product molecular fossils. In: Kinghorn D, Falk H, Gibbons S, Kobayashi J, editors. Progress in the chemistry of organic natural products, vol. 104. Cham: Springer, 2017:1–126.

[48] Chassagne F, Cabanac G, Hubert G, David B, Marti G. The landscape of natural product diversity and their pharmacological relevance from a focus on the dictionary of natural products®. Phytochem Rev. 2019. DOI: 10.1007/s11101-019-09606-2.

[49] Hossain M, Haque A, Mazid ZS, Khan A, Ullah TR, Rumee TA, Jesmin. SVDB: a comprehensive domain specific database of snake venom toxins generated through NCBI. Preprints. 2019. DOI: 10.20944/preprints201809.0454.v1.

[50] Sixty-Four Free Chemistry Databases. Available at:https://depth-first.com/articles/2011/10/12/sixty-four-free-chemistry-databases/. Accessed: 09 May 2019.

[51] Shen JH, Xu XY, Cheng F, Liu H, Luo XM, Shen J, et al. Virtual screening on natural products for discovering active compounds and target information. Curr Med Chem. 2003;10:2327–42.

[52] He M, Yan XJ, Zhou JJ, Xie GR. Traditional Chinese medicine database and application on the Web. J Chem Inf Comput Sci. 2001;41:273–7.

[53] Qiao XB, Hou TJ, Zhang W, Guo SL, XJ A X. 3D structure database of components from Chinese traditional medicinal herbs. J Chem Inf Comput Sci. 2002;42:481–9.

[54] Camp D, Newman S, Pham NB, Quinn RJ. Nature Bank and the Queensland Compound Library: unique international resources at the Eskitis Institute for drug discovery. Comb Chem & High Throughput Screen. 2014;17:201–9.

[55] Ng SB, Kanagasundaram Y, Fan H, Arumugam P, Eisenhaber B, Eisenhaber F. The 160K Natural Organism Library, a unique resource for natural products research. Nat Biotechnol. 2018;36:570–3. Available at: http://www.bii.a-star.edu.sg.

[56] Quinn RJ, Carroll AR, Pham NB, Baron P, Palframan ME, Suraweera L, et al. Developing a drug-like natural product library. J Nat Prod. 2008;71:464–8.

[57] InterBioScreen (Moscow). Available at: https://www.ibscreen.com/. Accessed: 16 May 2019.

[58] Ntie-Kang F, Zofou D, Babiaka SB, Meudom R, Scharfe M, Lifongo LL, et al. AfroDb: a select highly potent and diverse natural product library from African medicinal plants. PLoS One. 2013;8:e78085. Available at: http://zinc.docking.org/catalogs/afronp. Accessed: 9 May 2019.

[59] Pilón-Jiménez BA, Saldívar-González FI, Díaz-Eufracio BI, Medina-Franco JL. BIOFACQUIM: A Mexican compound database of natural products. Biomolecules. 2019;9:E31.

[60] Mirza SB, Bokhari H, Fatmi MQ. Exploring natural products from the biodiversity of Pakistan for computational drug discovery studies: collection, optimization, design and development of a chemical database (ChemDP). Curr Comput Aided Drug Des. 2015;11:102–9.

[61] Ntie-Kang F, Mbah JA, Mbaze LM, Lifongo LL, Scharfe M, Hanna JN, et al. CamMedNP: building the Cameroonian 3D structural natural products database for virtual screening. BMC Complement Altern Med. 2013;13:88.

[62] Yanuar A, Mun'im A, Lagho AB, Syahdi RR, Rahmat M, Suhartanto H. Medicinal plants database and three dimensional structure of the chemical compounds from medicinal plants in Indonesia. Int J Comput Sci. 2011;8:180–3.

[63] Ashfaq UA, Mumtaz A, Qamar TU, Fatima T. MAPS database: medicinal plant activities, phytochemical and structural database. Bioinformation. 2013;9:993–5.

[64] Mitishamba Database: A database of natural products from Kenya for drug discovery. Available at: http://mitishamba.uonbi.ac.ke/. Accessed: 08 May 2019.

[65] Valli M, Dos Santos RN, Figueira LD, Nakajima CH, Castro-Gamboa I, Andricopulo AD, et al. Development of a natural products database from the biodiversity of Brazil. J Nat Prod. 2013;76:439–44.

[66] Pilon AC, Valli M, Dametto AC, Pinto ME, Freire RT, Castro-Gamboa I, et al. NuBBE$_{DB}$: an updated database to uncover chemical and biological information from Brazilian biodiversity. Sci Rep. 2017;7:7215.

[67] Olmedo DA, González-Medina M, Gupta MP, Medina-Franco JL. Cheminformatic characterization of natural products from Panama. Mol Divers. 2017;21:779–89.

[68] Hatherley R, Brown DK, Musyoka TM, Penkler DL, Faya N, Lobb KA, et al. SANCDB: a South African natural compound database. J Cheminform. 2015;7:29.

[69] Pathania S, Ramakrishnan SM, Bagler G. Phytochemica: a platform to explore phytochemicals of medicinal plants. Database (Oxford). 2015;2015:bav075.

[70] Kim SK, Nam S, Jang H, Kim A, Lee JJ. TM-MC: a database of medicinal materials and chemical compounds in Northeast Asian traditional medicine. BMC Complement Altern Med. 2015;15:218.

[71] Nguyen-Vo TH, Le T, Pham D, Nguyen T, Le P, Nguyen A, et al. VIETHERB: a database for vietnamese herbal species. J Chem Inf Model. 2019;59:1–9.

[72] Ntie-Kang F, Nwodo JN, Ibezim A, Simoben CV, Karaman B, Ngwa VF, et al. Molecular modeling of potential anticancer agents from African medicinal plants. J Chem Inf Model. 2014;54:2433–50.

[73] Onguéné PA, Ntie-Kang F, Mbah JA, Lifongo LL, Ndom JC, Sippl W, et al. The potential of anti-malarial compounds derived from African medicinal plants, part III: an in silico evaluation of drug metabolism and pharmacokinetics profiling. Org Med Chem Lett. 2014;4:6.

[74] Ibezim A, Debnath B, Ntie-Kang F, Mbah CJ, Nwodo NJ. Binding of anti-*Trypanosoma* natural products from African flora against selected drug targets: a docking study. Med Chem Res. 2017;26:562–79.

[75] Egieyeh S, Syce J, Christoffels A, Malan SF. Exploration of scaffolds from natural products with antiplasmodial activities, currently registered antimalarial drugs and public malarial screen data. Molecules. 2016;21:104.

[76] Egieyeh SA, Syce J, Malan SF, Christoffels A. Prioritization of anti-malarial hits from nature: chemo-informatic profiling of natural products with in vitro antiplasmodial activities and currently registered anti-malarial drugs. Malar J. 2016;15:50.

[77] Egieyeh SA, Syce J, Malan S, Christoffels A. Antimalarial drug development from phytomedicine: cheminformatic and pharmacological studies. Int J Infect Dis. 2014;21S:1–460.

[78] Lagunin AA, Druzhilovsky DS, Rudik AV, Filimonov DA, Gawande D, Suresh K, et al. Computer evaluation of hidden potential of phytochemicals of medicinal plants of the traditional Indian ayurvedic medicine. Biomed Khim. 2015;61:286–97.

[79] Sharma A, Dutta P, Sharma M, NK R, Dodiya B, Georrge JJ, et al. BioPhytMol: a drug discovery community resource on anti-mycobacterial phytomolecules and plant extracts. J Cheminform. 2014;6:46.

[80] Fang X, Shao L, Zhang H, Wang S. CHMIS-C: a comprehensive herbal medicine information system for cancer. J Med Chem. 2005;48:1481–8.

[81] Gu J, Gui Y, Chen L, Yuan G, Xu X. CVDHD: a cardiovascular disease herbal database for drug discovery and network pharmacology. J Cheminform. 2013;5:51.

[82] Mangal M, Sagar P, Singh H, Raghava GPS, Agarwal SM. NPACT: naturally occurring plant-based anti-cancer compound-activity-target database. Nucleic Acids Res. 2013;41:D1124–9.

[83] Choi H, Cho SY, Pak HJ, Kim Y, Choi JY, Lee YJ, et al. NPCARE: database of natural products and fractional extracts for cancer regulation. J Cheminform. 2017;9:2.

[84] Chen X, Zhou H, Liu YB, Wang JF, Li H, Ung CY, et al. Database of traditional Chinese medicine and its application to studies of mechanism and to prescription validation. Br J Pharmacol. 2006;149:1092–103.

[85] Ru J, Li P, Wang J, Zhou W, Li B, Huang C, et al. TCMSP: a database of systems pharmacology for drug discovery from herbal medicines. J Cheminform. 2014;6:13.

[86] Kinoshita T, Lepp Z, Kawai Y, Terao J, Chuman H. An integrated database of flavonoids. Biofactors. 2006;26:179–88.

[87] Yabuzaki J. Carotenoids Database: structures, chemical fingerprints and distribution among organisms. Database (Oxford). 2017;2017:bax004.

[88] Nupur LN, Vats A, Dhanda SK, Raghava GP, Pinnaka AK, Kumar A. ProCarDB: a database of bacterial carotenoids. BMC Microbiol. 2016;16:96.

[89] Johnson EA, Schroeder WA. Microbial carotenoids. Adv Biochem Eng Biotechnol. 1996;53:119–78.

[90] Zeng X, Zhang P, Wang Y, Qin C, Chen S, He W, et al. CMAUP: a database of collective molecular activities of useful plants. Nucleic Acids Res. 2019;47:D1118–27.

[91] Pye CR, Bertin MJ, Lokey RS, Gerwick WH, Linington RG. Retrospective analysis of natural products provides insights for future discovery trends. Proc Natl Acad Sci U S A. 2017;114:5601–6.

[92] Maruca A, Moraca F, Rocca R, Molisani F, Alcaro F, Gidaro MC, et al. Chemoinformatic database building and in silico hit-identification of potential multi-targeting bioactive compounds extracted from mushroom species. Molecules. 2017;22:1571.

[93] González-Medina M, Medina-Franco JL. Chemical diversity of cyanobacterial compounds: a chemoinformatics analysis. ACS Omega. 2019;4:6229–37.

[94] Crüsemann M, O'Neill EC, Larson CB, Melnik AV, Floros DJ, Da Silva RR, et al. Prioritizing natural product diversity in a collection of 146 bacterial strains based on growth and extraction protocols. J Nat Prod. 2017;80:588–97.

[95] Jiménez C. Marine natural products in medicinal chemistry. ACS Med Chem Lett. 2018;9:959–61.

[96] Miller JH, Field JJ, Kanakkanthara A, Owen JG, Singh AJ, Northcote PT. Marine invertebrate natural products that target microtubules. J Nat Prod. 2018;81:691–702.

[97] Principe PP, Fisher WS. Spatial distribution of collections yielding marine natural products. J Nat Prod. 2018;81:2307–20.

[98] Lei J, Zhou J. A marine natural product database. J Chem Inf Comput Sci. 2002;42:742–8.

[99] Timmermans ML, Paudel YP, Ross AC. Investigating the biosynthesis of natural products from marine proteobacteria: a survey of molecules and strategies. Mar Drugs. 2017;15:E235.

[100] Pereira F, Aires-de-Sousa J. Computational methodologies in the exploration of marine natural product leads. Mar Drugs. 2018;16:E236.

[101] Mayer AM, Rodriguez AD, Taglialatela-Scafati O, Fusetani N. Marine pharmacology in 2009–2011: Marine compounds with antibacterial, antidiabetic, antifungal, anti-inflammatory, antiprotozoal, antituberculosis, and antiviral activities; affecting the immune and nervous systems, and other miscellaneous mechanisms of action. Mar Drugs. 2013;11:2510–73.

[102] Choudhary A, Naughton LM, Montanchez I, Dobson AD, Rai DK. Current status and future prospects of marine natural products (MNPs) as antimicrobials. Mar Drugs. 2017;15:272.

[103] Blunt JW, Copp BR, Keyzers RA, Munro MH, Prinsep MR. Marine natural products. Nat Prod Rep. 2016;33:382–431.

[104] Wohlleben W, Mast Y, Stegmann E, Ziemert N. Antibiotic drug discovery. Microb Biotechnol. 2016;9:541–8.

[105] Harizani M, Ioannou E, Roussis V. The *Laurencia* paradox: an endless source of chemodiversity. Prog Chem Org Nat Prod. 2016;102:91–252.

[106] Burja AM, Banaigs B, Abou-Mansour E, Burgess JG, Wright PC. Marine cyanobacteria - a prolific source of natural products. Tetrahedron. 2001;57:9347–77.

[107] Davis GD, Vasanthi AH. Seaweed metabolite database (SWMD): A database of natural compounds from marine algae. Bioinformation. 2011;5:361–4. Available at: http://www.swmd.co.in. Accessed: 16 May 2019.

[108] Shang J, Hu B, Wang J, Zhu F, Kang Y, Li D, et al. Cheminformatic insight into the differences between terrestrial and marine originated natural products. J Chem Inf Model. 2018;58:1182–93.

[109] The Dictionary of Food Compounds. Available at:http://dfc.chemnetbase.com. Accessed: 16 May 2019.

[110] Gry J, Black L, Eriksen FD, Pilegaard K, Plumb J, Rhodes M, et al. EuroFIR-BASIS a combined composition and biological activity database for bioactive compounds in plant-based foods. Trends Food Sci Technol. 2007;18:434–44.

[111] FooDB (The Food Database) (version 1.0). Available at: http://foodb.ca/compounds. Accessed: 16 May 2019.

[112] Burdock GA, Carabin IG. Generally Recognized as Safe (GRAS): history and description. Toxicol Lett. 2004;150:3–18.

[113] Martinez-Mayorga K, Peppard TL, López-Vallejo F, Yongye AB, Medina-Franco JL. Systematic mining of generally recognized as safe (GRAS) flavor chemicals for bioactive compounds. J Agric Food Chem. 2013;61:7507–14.

[114] Jensen K, Panagiotou G, Kouskoumvekaki I. NutriChem: a systems chemical biology resource to explore the medicinal value of plant-based foods. Nucleic Acids Res. 2015;43:D940–5.

[115] Jensen K, Panagiotou G, Kouskoumvekaki I. Integrated text mining and chemoinformatics analysis associates diet to health benefit at molecular level. PLoS Comput Biol. 2014;10:10.1371.

[116] The NutriChem 1.0 server. Available at: http://www.cbs.dtu.dk/services/NutriChem-1.0/. Accessed: 16 May 2019.

[117] Neveu V, Perez-Jimenez J, Vos F, Crespy V, Du Chaffaut L, Mennen L, et al. Phenol-Explorer: an online comprehensive database on polyphenol contents in foods. Database. 2010;2010:bap024.

[118] Perez-Jimenez J, Neveu V, Vos F, Scalbert A. Systematic analysis of the content of 502 polyphenols in 452 foods and beverages: an application of the Phenol-Explorer database. J Agric Food Chem. 2010;58:4959–69.

[119] Dunkel M, Schmidt U, Struck S, Berger L, Gruening B, Hossbach J, et al. SuperScent–a database of flavors and scents. Nucleic Acids Res. 2009;37:D291–4.

[120] Ahmed J, Preissner S, Dunkel M, Worth CL, Eckert A, Preissner R. SuperSweet–a resource on natural and artificial sweetening agents. Nucleic Acids Res. 2011;39:D377–82.

[121] Yue Y, Chu GX, Liu XS, Tang X, Wang W, Liu GJ, et al. TMDB: a literature-curated database for small molecular compounds found from tea. BMC Plant Biol. 2014;14:243.

[122] United States Department of Agriculture National Agricultural Library. Available at: https://www.nal.usda.gov/fnic/food-composition. Accessed: 16 May 2019.

[123] Peña-Castillo A, Méndez-Lucio O, Owen JR, Martínez-Mayorga K, Medina-Franco JL. Chemoinformatics in food science. In: Engel T, Gasteiger J, editors. Applied chemoinformatics: achievements and future opportunities. ISBN:9783527342013 |Online ISBN:9783527806539 |DOI:10.1002/9783527806539. Wiley-VCH Verlag GmbH & Co. KGaA, 2018:501–25.

[124] Naveja JJ, Rico-Hidalgo MP, Medina-Franco JL. Analysis of a large food chemical database: chemical space, diversity, and complexity. F1000 Res. 2018;7:993.

[125] Minkiewicz P, Darewicz M, Iwaniak A, Bucholska J, Starowicz P, Czyrko E. Internet databases of the properties, enzymatic reactions, and metabolism of small molecules - search options and applications in food science. Int J Mol Sci. 2016;17:2039.

[126] Holton TA, Vijayakumar V, Khaldi N. Bioinformatics: current perspectives and future directions for food and nutritional research facilitated by a food-wiki database. Trends Food Sci Technol. 2013;34:5–17.

[127] Minkiewicz P, Iwaniak A, Darewicz M. Using internet databases for food science organic chemistry students to discover chemical compound information. J Chem Educ. 2015;92:874–6.

[128] Iwaniak A, Minkiewicz P, Darewicz M, Protasiewicz M, Mogut D. Chemometrics and cheminformatics in the analysis of biologically active peptides from food sources. J Functional Foods. 2015;16:334–51.

[129] Martínez-Mayorga K, Peppard TL, Medina-Franco JL. Software and online resources: perspectives and potential applications. In: Martínez-Mayorga K, Medina-Franco JL, editors. *Foodinformatics*. Applications of chemical information to food chemistry. Cham, Switzerland: Springer International Publishing AG, 2014:233–48.

[130] Scalbert A, Andres-Lacueva C, Arita M, Kroon P, Manach C, Urpi-Sarda M, et al. Databases on food phytochemicals and their health-promoting effects. J Agric Food Chem. 2011;59:4331–48.

[131] Malkaram SA, Hassan YI, Zempleni J. Online tools for bioinformatics analyses in nutrition sciences. Adv Nutr. 2012;3:654–65.

[132] Minkiewicz P, Miciński J, Darewicz M, Bucholska J. Biological and chemical databases for research into the composition of animal source foods. Food Rev Int. 2013;29:321–51.

[133] Wishart DS, Tzur D, Knox C, Eisner R, Guo AC, Young N, et al. HMDB: the human metabolome database. Nucleic Acids Res. 2007;35:D521–6.

[134] Wishart DS, Knox C, Guo AC, Eisner R, Young N, Gautam B, et al. HMDB: a knowledgebase for the human metabolome. Nucleic Acids Res. 2009;37:D603–10.

[135] Wishart DS, Jewison T, Guo AC, Wilson M, Knox C, Liu Y, et al. HMDB 3.0 — the human metabolome database in 2013. Nucleic Acids Res. 2013;41:D801–7.

[136] Wishart DS, Feunang YD, Marcu A, Guo AC, Liang K, Vázquez-Fresno R, et al. HMDB 4.0 — The human metabolome database for 2018. Nucleic Acids Res. 2018;46:D608–17. Available at: http://www.hmdb.ca/. Accessed: 16 May 2019.

[137] Kanehisa M, Furumichi M, Tanabe M, Sato Y, Morishima K. KEGG: new perspectives on genomes, pathways, diseases and drugs. Nucleic Acids Res. 2017;45:D353–61.

[138] Caspi R, Billington R, Ferrer L, Foerster H, Fulcher CA, Keseler IM, et al. The MetaCyc database of metabolic pathways and enzymes and the BioCyc collection of pathway/genome databases. Nucleic Acids Res. 2016;44:D471–80.

[139] Hastings J, Owen G, Dekker A, Ennis M, Kale N, Muthukrishnan V, et al. ChEBI in 2016: Improved services and an expanding collection of metabolites. Nucleic Acids Res. 2016;44: D1214–9.

[140] Burley SK, Berman HM, Bhikadiya C, Bi C, Chen L, Di Costanzo L, et al. RCSB protein data bank: biological macromolecular structures enabling research and education in fundamental biology, biomedicine, biotechnology and energy. Nucleic Acids Res. 2019;47:D464–74.

[141] UniProt Consortium T. The UniProt Consortium UniProt: the universal protein knowledgebase. Nucleic Acids Res. 2017;45:D158–69.

[142] Benson DA, Cavanaugh M, Clark K, Karsch-Mizrachi I, Lipman DJ, Ostell J, et al. GenBank. Nucleic Acids Res. 2017;45:D37–42.

[143] Law V, Knox C, Djoumbou Y, Jewison T, Guo AC, Liu Y, et al. DrugBank 4.0: shedding new light on drug metabolism. Nucleic Acids Res. 2014;42:D1091–7.

[144] Lim E, Pon A, Djoumbou Y, Knox C, Shrivastava S, Guo AC, et al. T3DB: a comprehensively annotated database of common toxins and their targets. Nucleic Acids Res. 2010;38:D781–6.

[145] Wishart D, Arndt D, Pon A, Sajed T, Guo AC, Djoumbou Y, et al. T3DB: the toxic exposome database. Nucleic Acids Res. 2015;43:D928–34.

[146] Frolkis A, Knox C, Lim E, Jewison T, Law V, Hau DD, et al. SMPDB: the small molecule pathway database. Nucleic Acids Res. 2010;38:D480–7.

[147] Jewison T, Su Y, Disfany FM, Liang Y, Knox C, Maciejewski A, et al. SMPDB 2.0: big improvements to the small molecule pathway database. Nucleic Acids Res. 2014;42:D478–84.

[148] Jayaseelan KV, Steinbeck C. Building blocks for automated elucidation of metabolites: natural product-likeness for candidate ranking. BMC Bioinform. 2014;15:234.

Joelle Ngo Hanna, Boris D. Bekono, Luc C. O. Owono,
Flavien A. A. Toze, James A. Mbah, Stefan Günther
and Fidele Ntie-Kang

2 A chemoinformatic analysis of atoms, scaffolds and functional groups in natural products

Abstract: In the quest to know why natural products (NPs) have often been considered as privileged scaffolds for drug discovery purposes, many investigations into the differences between NPs and synthetic compounds have been carried out. Several attempts to answer this question have led to the investigation of the atomic composition, scaffolds and functional groups (FGs) of NPs, in comparison with synthetic drugs analysis. This chapter briefly describes an atomic enumeration method for chemical libraries that has been applied for the analysis of NP libraries, followed by a description of the main differences between NPs of marine and terrestrial origin in terms of their general physicochemical properties, most common scaffolds and "drug-likeness" properties. The last parts of the work describe an analysis of scaffolds and FGs common in NP libraries, focusing on huge NP databases, e.g. those in the Dictionary of Natural Products (DNP), NPs from cyanobacteria and the largest chemical class of NP – terpenoids.

Keywords: functional groups, drug discovery, scaffolds, natural products, numbers

2.1 Introduction

Natural products (NPs) are known to represent unique classes of compounds which result from secondary metabolism in bacteria, fungi and higher organisms [1, 2]. What makes NPs of major importance is their diverse applications in the pharmaceutical, food and cosmeceutical industries [3–5]. An attempt to provide the exact number of NPs based on the information currently available in the literature shows that there are slightly below 500,000 known naturally occurring compounds, the majority being from terrestrial sources, although the marine environment is increasingly becoming a source of novel bioactive secondary metabolites (SMs) [2, 6]. The increasing important roles of NPs in drug discovery [4, 7] has led to some authors to coin the term "golden age" for drug discovery based on NPs to our era [8].

This article has previously been published in the journal *Physical Sciences Reviews*. Please cite
as: Ngo, J. H., Bekono, B. D., Owono, L. C. O., Toze, F. A. A., Mbah, J. A., Günther, S., Ntie-Kang,
F. A chemoinformatic analysis of atoms, scaffoldsand functional groups in natural products .
Physical Sciences Reviews [Online] 2021 DOI: 10.1515/psr-2019-0096

https://doi.org/10.1515/9783110668896-002

Newman and Craig showed that a majority of known drugs, e.g. 50% of known anticancer agents and more than 70% of known antibacterials are either NPs or their derivatives [6]. Meanwhile, Koehn and Carter showed that, of the 877 small molecules published as new chemical entities by pharmaceutical companies between 1981 and 2002, about 49% were either pure NPs or semi-synthetic NP analogues and semi-synthetic compounds based on NP pharmacophores [7].

Several attempts have been made to distinguish NPs from synthetic drugs (SDs), particularly in terms of their respective coverage of chemical space, their atomic compositions and most frequent fragments [9–11]. Bade et al. [10], for example, showed that polycyclic molecules were more common among NPs than SDs, while values of the statistical means and standard deviations for physicochemical properties of NPs were higher than SDs for in terms of their coverage of chemical space. Meanwhile, Feher and Schmidt showed that NPs exhibit major structural differences when compared with SDs (e.g. NPs have more complex ring systems and chiral centers, aromatic rings, as well as higher degrees of the saturation and ratios of different heteroatoms). It was shown, for example, that the number of N-atoms, halogens and S-atoms were generally fewer in NPs while having generally more O-atoms than classical drugs, particularly those of synthetic origin. The authors explained that such properties are only introduced in SDs in order to enhance their synthesis feasibility [11].

Grabowski and Schneider [12] conducted a similar study by investigating the diversity of chemotypes as well as the molecular properties of drug molecules, alongside NPs, and NP-derived compounds. The authors compared pure NPs with drug molecules,e.g. from the COBRA collection (Collection of Bioactive Reference Analogues) [13]. The pure NPs were collected from the MEGAbolite database provided by AnalytiCon Discovery containing isolates from plants and terrestrial microorganisms [14], along with compounds from the Interbioscreen database [15]. The study showed that the pure NPs were quite distinct, containing on average more chiral centers, more O-atoms, and less aromatic atoms [12]. The authors also showed that more than one thousand scaffolds found in the pure NP library were not present in the other analyzed compound libraries (including semi-synthetic NPs and NP-derived combinatorial compounds). From these results, the authors were able to suggest design approaches for new NP-based compound libraries. This proves that NPs contain unique molecular frameworks in "drug space". This implies that NPs could be ideal starting points for molecular design considerations [12].

It, therefore, becomes important to quantify and enumerate the differences between NPs and SDs at the structural level, i.e. the numbers or proportions of atoms, chemical scaffolds and major functional groups (FGs) currently available in NP libraries that distinguish NPs from SDs. This shall be the major focus of this chapter.

2.2 An atom-based enumeration of natural product-like virtual libraries

Evolutionary algorithms [16, 17] and atom-based enumeration [18] approaches have been used to analyse building blocks of compounds in chemical libraries. These approaches depend on enumeration conditions and have demonstrated that NP-like structures can contain a broad variety of heterocyclic and alicyclic rings. For example, a molecular enumerator allowing diverse NP-like and drug-like structures (from a chemical and architectural viewpoint) to be generated, beginning from a single core structure (e.g. a carbon atom or a polycyclic ring) has been developed [19].

In this approach, connection points for each newly added fragment (node), shown by an asterisk in Figure 2.1, are terminated when predefined by criteria chosen by the user have been fulfilled, e.g. maximum molecular weight (MW) or when all nodes have been filled.

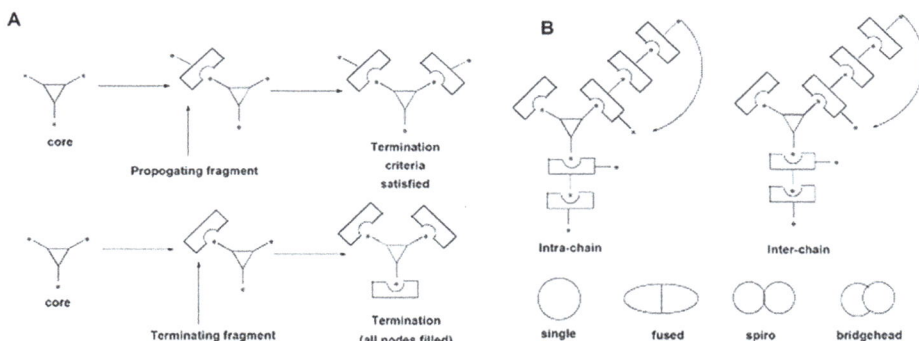

Figure 2.1: Summary of an enumeration method; (A) Connection points to add fragments (or nodes) have been shown (highlighted by an asterisk). When certain predefined criteria by the user are satisfied (e.g., the require molecular weight has been achieved) or when no unoccupied nodes exist, termination occurs (B) Showing the possible ring forming modes [19]. Figure adapted by permission.

The above enumeration method has been used to analyze the components of three chemical libraries, including the Dictionary of Natural Products (DNP), Table 2.1.

It was observed that if one begins from a relatively small library of building blocks, such (an atom-based) enumeration method could potentially generate diverse virtual compounds (from a chemical and architectural viewpoint). Since NPs are a unique source of building blocks for the identification and design of new potential drug molecules [4, 9, 11], with the ability to mimick the topological and structural diversity in nature, NP-like structures can be generated by such a method, depending on enumeration conditions. Additionally, applying this on a virtual compound library may enhance the generation of novel ideas for lead structures located in non-explored regions of chemical space.

Table 2.1: Datasets used as the training set [19].

Dataset	Original #[a]	Final #[b]
ACD[c]	661,820	545,377
MDDR	180,784	157,023
DNP	230,107	141,613

[a] The original number of entries,
[b] The final number of structures,
[c] The final number of structures representing the complete library prior to deleting any NP or a drug structures.

2.3 Chemoinformatic comparison between natural products of terrestrial and marine origins

At the atomic level, Hedner had already shown the relatively high abundance of halogens in sea water (particularly Br, Cl, and I) to be a major profound difference between the marine and terrestrial environments [20]. This could be based on the fact that marine organisms, generally, contain unique NPs (with rare or diverse chemical skeletons and structures from those contained in organisms living on land) [21, 22]. Such compounds usually include halogen-containing covalent organic structures [23]. Contrary to marine organisms, terrestrial plants are known for the presence of basic, nitrogen-containing NPs (e.g. alkaloids), which are comparably rare among marine organisms [24]. A summary of the comparison of representative chemical structures from marine (**7–9**) and terrestrial (**1–3**) organisms when compared with SDs (**4–6**) has been shown in Figure 2.2.

However, the first attempt to compare compounds derived from marine, terrestrial and synthetic origins by chemoinformatics, was by the use of the ChemGPS-NP tool that often used for the "navigation" of chemical space [24]. This investigation showed that the terrestrial environment has been most extensively investigated as a source of new NP scaffolds and compounds, meanwhile marine organisms are much relatively unexplored. The authors compared the chemical space of datasets of compounds from marine, terrestrial and synthetic origins, with respect to their physical and chemical properties, as well as their diversity in terms of biologically-relevant chemical space. In this study 3,802 marine NPs, 29,620 terrestrial NPs, and 58,856 compounds of synthetic origin were included [24].

The results showed that marine NPs:
- are more flexible, in contrast to terrestrial NPs,
- have higher MWs,
- include more halogen atoms, N atoms, and amide functional groups (FGs).

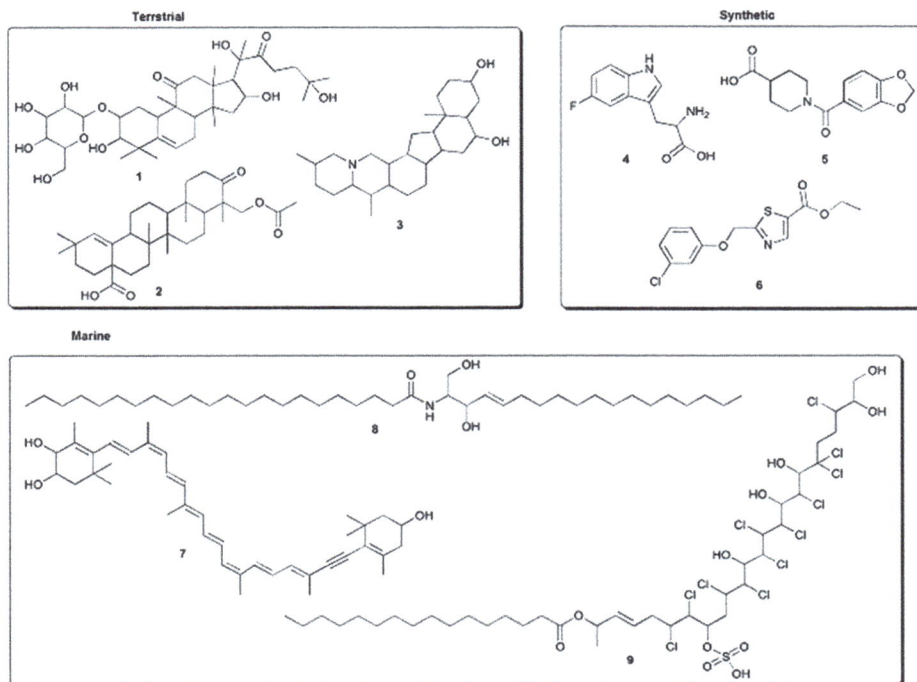

Figure 2.2: Summary of typical chemical structures of terrestrial (e.g. compounds **1–3**), of synthetic (e.g. compounds **4–6**), and of marine-derived (e.g. compounds **7–9**) origin [24]. Figure adapted by permission.

On the other hand, synthetic compounds are often rather small, some being highly flexible. Generally speaking, the study revealed that a typical and unique synthetic compound (Figures 3.2 and 3.3) would be relatively small, when compared with NPs, and are often characterised by the presence of aromatic rings as well as non-aromatic rings. This may partly explain why compounds of synthetic origin are often less polar when compared with their NP counterparts [24].

Shang et al., most recently, carried out a detailed analysis of NPs included in the DNP in comparison with marine NPs included in the Dictionary of Marine Natural Products (DMNP), with the view of providing further insight into the reasons behind the observed differences in physicochemical properties between NPs from marine and terrestrial environments [25]. Marine natural products (MNPs) were regarded as compounds from the DMNP (version 2015), while the terrestrial natural products (TNPs) were simply obtained by removing compounds found in the DMNP from those in the DNP. For the purpose of standardization, only the largest fragments of the structures in MNPs and TNPs were, respectively, kept. Inorganic and tiny molecules (with MW < 80 Da) were removed, H atoms were added, while duplicate molecules were deleted.

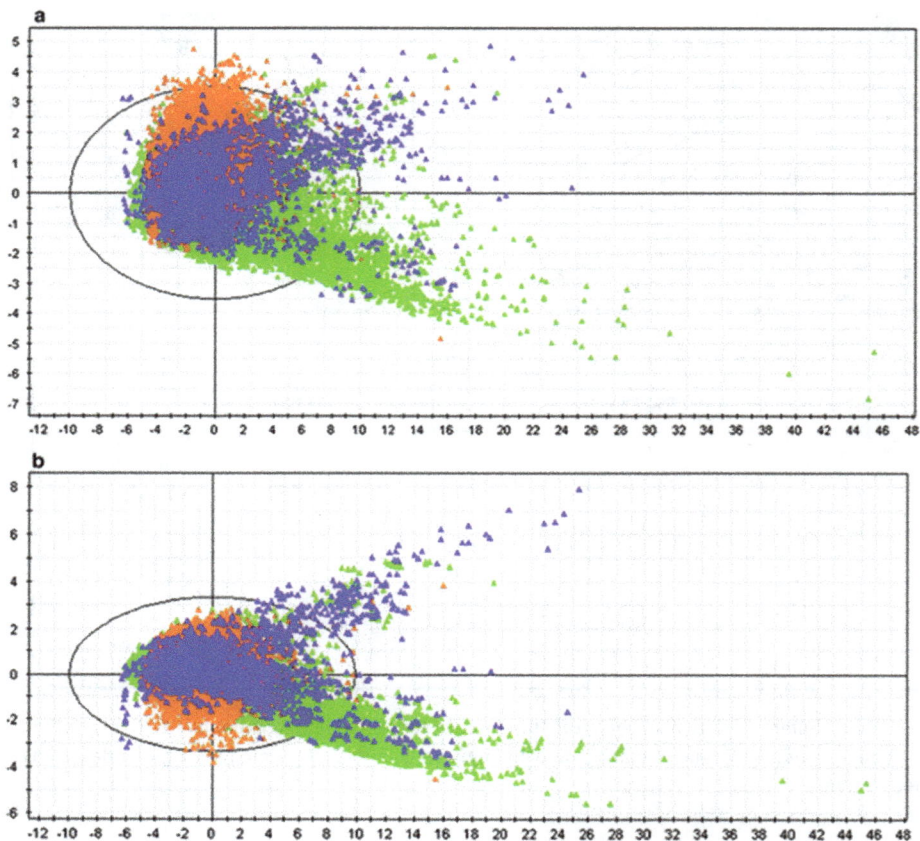

Figure 2.3: Predicted score plots for (a) MW against flexibility, and (b) MW against a score that represents mainly the number of halogen atoms, nitrogen atoms, amide FGs. In both curves, marine NPs are shown in blue, terrestrial NPs in green, while synthetic compounds are shown in orange [24]. Figure reproduced by permission.

The two retained datasets (TNPs and MNPs), respectively, contained 151,609 and 35,883 molecules. Afterwards, 60 computed molecular descriptors were used in the characterization of the physical and chemical properties of the NPs. These properties included hydrophobicity, H-bonding capacity, bulkiness, bonds, rings, solubility, elemental compositions, etc. The comparison of the two datasets (TNPs and MNPs) was focused on their physicochemical properties, structural features (e.g. scaffolds and fragments), and drug-likeness.

The most significant differences could be summarised below:

- MNPs are less soluble and are often larger than TNPs.
- Apart from the O-atoms, MNPs had higher average atom numbers of the other elements than the TNPs.

- Apart from F-atoms, major differences exist between the MNPs and the TNPs in terms of the number of atoms of the other elements.
- more non-F halogens occur in MNPs than in TNPs (the oceans represent the largest natural source of known biogenic organic halogens).
- MNPs generally include unique fragments and scaffolds, e.g. more long chains and large rings in particular (including 8- membered to 10-membered rings).
- MNPs include more N-atoms and halogens (particularly Br), but fewer O-atoms (implying that NPs of marine origin may have different biosynthetic pathways from those of terrestrial origin).

In terms of chemical scaffolds, a comparison of the most frequent Murcko frameworks in NPs from marine and terrestrial sources showed major differences (see Figure 2.4). In summary:
- MNPs have longer scaffolds than TNPs.
- MNPs often include 10-membered rings having ester bonds.

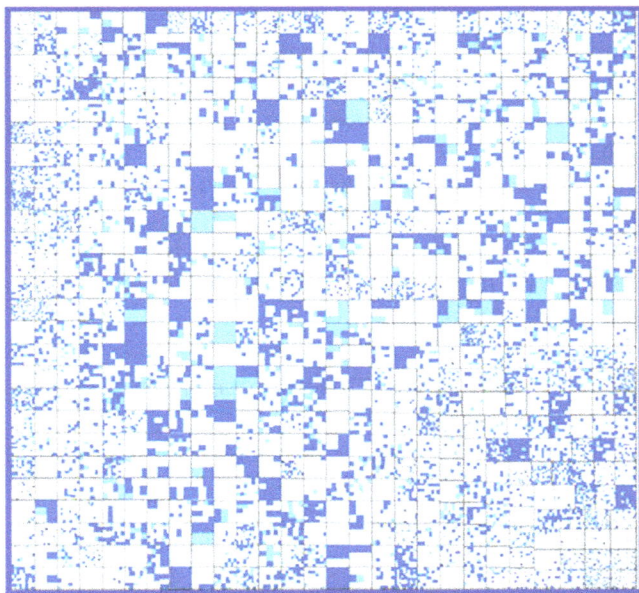

Figure 2.4: Comparison of the differences between MNP scaffolds (in blue) and TNP scaffolds (in white). Light blue indicates scaffolds found in both TNPs and MNPs. Clusters of scaffolds are represented by squares surrounded by a grey perimeter [25]. Figure reproduced by permission.

In terms of molecular sizes, it was shown that:
- MNPs have much higher volumes and surface areas than in TNPs (see Table 2.2). This means that molecules from marine sources are generally larger than those from terrestrial sources.

- MNPs also had relatively larger MW values, thus supporting the previous observation.
- It could be predicted that the larger (and heavier) MNPs may be more problematic in terms of their ability to penetrate cell membranes, hence be absorbed by the intestines.

The above reasons are also valid from the number of rings (*NR*) count, Table 2.2.

Regarding the fragments, it was observed that the structural features of the NPs in the MNPs dataset and those of the TNPs dataset represented four types of fragments (i.e. ring assemblies, chain assemblies, Retrosynthetic Combinatorial Analysis Procedure (RECAP) fragments [26] and Murcko frameworks [27], see Table 2.3). Kong et al. had previously analyzed the Murcko frameworks as well as ClogP values of NP of marine and terrestrial origins, showing that MNPs have the majority of (over 70%) unique scaffolds [28]. However, it was observed by the authors that the drug development potential of compounds from marine sources might be problemmatic because of higher hydrophobicities, in comparison with those from terrestrial sources [28, 29].

Let us consider a few definitions:
- A ring assembly is a continuous set of rings in a molecule or fragment which does not include linker (e.g. fused rings, bridged ring systems, etc.).
- A chain assembly is defined as a set of contiguous chain atoms in a molecule or fragment (e.g. a ring atom at the end of a chain).
- A Murcko framework is defined as the union of the linkers and ring systems in a molecule [26].
- RECAP fragments are defined as those fragments that are cleaved from a molecule at the bonds based on some 11 pre-defined bond cleavage rules (derived from known chemical reactions) [27].

A comparison of these fragments from MNPs and TNPs from the Shang et al. study [25] has been shown in Table 2.3. It was observed that:
- Apart from RECAP fragments, MNPs have higher percentages of unduplicated fragments than TNPs.
- The percentages of unique fragments in MNPs vary from ~ 70% (in ring assemblies) to ~ 93% (in RECAP fragments).
- MNPs have more new fragments and scaffolds in (particularly RECAP fragments), showing the diversity of marine biosynthetic pathways.
- The most common chain assemblies in marine-derived compounds include haloalkanes, multiple hydroxyls, alkenyls, esters, ethers, and sulfates.
- MNPs more O and Br (O and halogens are very abundant oceans and seas).
- The most common chain assemblies in terrestrial NPs are esters, carboxyl groups, hydroxyl groups, ketones, amines, alkynyls, and alkenyl FGs.
- Ester groups are more abundant in TNPs than in MNPs.
- Phosphate groups are more common in marine NPs.

Table 2.2 Selected molecular descriptors used for the analysis of the MNPs and TNPs described in the study by Shang et al. [25]. Table adapted.

Descriptors[a]	MNPs						TNPs						
	Mean	Std dev[m]	Min[n]	Max[o]	Median	Mode	Mean	Std dev[m]	Min[n]	Max[o]	Median	Mode	u[p]
Mol. vol.[b]	294.11	155.99	25.38	3535.64	263.76	238.04	273.89	133.80	25.03	2261.05	245.93	180.76	21.63
Surf. area[c]	419.90	222.05	43.13	5519.49	375.31	253.17	391.93	191.89	49.49	3445.90	350.06	340.06	20.99
Pol. surf. area[d]	92.87	81.87	0.00	2250.36	72.83	20.23	96.39	80.06	0.00	1743.71	79.15	46.53	−7.02
fPSA[e]	0.22	0.13	0.00	1.00	0.20	0.00	0.25	0.12	0.00	1.00	0.23	0.00	−28.46
AlogP[f]	4.07	3.39	−8.81	53.47	3.69	4.37	3.14	2.84	−19.90	41.42	2.86	2.45	45.82
logD[g]	3.68	3.51	−36.32	53.47	3.39	3.81	2.76	2.96	−22.28	41.42	2.58	2.67	44.14
logS[h]	−5.89	3.90	−60.29	5.45	−5.32	−4.83	−4.68	3.15	−45.81	15.74	−4.09	−3.88	−52.62
Mol. weight[i]	420.70	221.30	32.04	5,033.75	379.41	304.47	396.06	199.32	32.04	3748.60	357.36	264.32	18.43
Numb. rot. bonds[j]	14.82	10.89	0.00	171.00	15.00	0.00	17.29	10.98	0.00	204.00	17.00	6.00	−36.75
Numb. rings[k]	2.64	1.94	0.00	32.00	2.00	2.00	3.28	2.09	0.00	32.00	3.00	3.00	−52.23
Numb. arom. rings[l]	0.77	1.16	0.00	20.00	0.00	0.00	1.01	1.28	0.00	20.00	1.00	0.00	−32.79

[a] Molecular descriptors are shown in bold, which include the molecular solvent accessible surface area (SASA), the number of hydrogen donors, the number of hydrogen acceptors, number of N and O atoms, donors, etc. were among the descriptors used in the study [25];
[b] the molecular volume descriptor;
[c] the surface area descriptor;
[d] the polar surface area descriptor;
[e] the sum of surface area on the positive parts of molecule/total molecular surface area descriptor;
[f] the n-octanol-water partition coeffient descriptor computed from atomic contributions;
[g] the octanol/water partition coeffient descriptor which considers the dissociation of polar species;
[h] the predicted solubility descriptor;

[i] the molecular weight descriptor;

[j] the number of rotatable bonds;

[k] the number of rings;

[l] the number of aromatic rings;

[m] the standard deviation;

[n] the minimum value of a given descriptor within the given dataset;

[o] the maximum value of a given descriptor within the given dataset;

[p] the u-test, used to evaluate the statistical significance of the difference between the means of two populations for a given molecular property (taking the significance level of $\alpha = 0.05$ and the critical value $Z_{0.025} = 1.96$). This was defined as:

$$u = \frac{\bar{x}_1 - \bar{x}_2}{\sqrt{\frac{s_1^2}{n_1} + \frac{s_2^2}{n_2}}}$$

in the case where $\bar{x}_1$ and $\bar{x}_2$, repectively, represent the means of a given molecular property for MNPs and TNPs; s_1 and s_2, respectively, represent the standard deviations of a given property for MNPs and TNPs; n_1 and n_2, respectively, represent the total numbers of molecules in the MNPs and TNPs datasets. A positive u value implies that the property values for the MNPs dataset are higher than those for the TNPs dataset. On the contrary, a negative u-value would imply that the properties for the MNPs dataset are lower than those for the TNPs dataset.

Table 2.3: Four types of fragments found in NPs of marine and terrestrial origins [25].

		Statistics of fragments[a]					
Fragments	**Sources**	**Total**	**Non-duplicated**	**Non-duplicated _P**[a]	**Unique**[b]	**Common**	**Unique_P**[c]
chain assemblies	MNPs	367,156	10,188	2.77%	7,919	2,269	77.73%
	TNPs	1,621,518	20,844	1.29%	18,575		89.11%
ring assemblies	MNPs	42,279	4,868	11.51%	3,456	1,412	70.99%
	TNPs	190,454	17,109	8.98%	15,697		91.75%
RECAP fragments	MNPs	1,777,882	142,123	7.99%	131,872	10,251	92.79%
	TNPs	6,915,803	718,662	10.39%	708,411		98.57%
Murcko frameworks	MNPs	28,833	8,066	27.97%	6,197	1,869	76.83%
	TNPs	121,975	29,985	24.58%	28,116		93.77%

[a] **Non-duplicated_P** = number of unduplicated fragments or total number of the same kind of fragments;
[b] **Unique** = number of unique fragments (e.g. the number of the unduplicated fragments or the number of the same kind of common fragments);
[c] **Unique_P** = number of unique fragments or unduplicated of the same kind.

- All MNP and TNP fragments contain 5- or 6-membered rings (e.g. benzene and cyclic ethers), which are building blocks for hormones and vital substances for living organisms (in both terrestrial and marine environments).
- Fragments produced by marine life may be relatively more reactive efficiently use oxygen in an anaerobic environment.
- A large proportion of MNP-based unique ring assemblies include 10-membered rings.
- TNP-based unique ring assemblies contain 5- or 6-membered rings (e.g. benzene and condensed rings), contributing to their structural stability, when compared with MNP-based unique ring assemblies.
- Most of the 10 MNP-based unique scaffolds contain long linkers or 10-membered rings, suggesting that these MNP-based unique frameworks are more flexible. These help to facilitate adaptation to marine life.
- Contrary to MNPs, TNP-based unique scaffolds contain more stable structures having lower complexity.
- In terms of elemental compositions, there are substantial differences between NPs of marine and terrestrial origins (e.g. their chemical groups, their chains, and the rings attached to their scaffolds).

The drug-likeness of NPs from marine and terrestrial sources were further evaluated by the use of a drug-likeness classifier implemented by the use of a naïve Bayesian classification technique [30, 31].

It was observed that:

- TNPs often bear Bayesian drug-likeness classification model with more stable bond types and ring systems.
- Most MNPs and TNPs are drug-like. However, NPs from seas and oceans are narrowly more drug-like than those from land.

2.4 Scaffold analysis in natural products

2.4.1 The case of the Dictionary of Natural Products (DNP)

A scaffold could be defined as "the basic or core structure of a molecule to which functional groups are attached" [32]. Many of NP scaffolds represent privileged or preferable architectures, which could facilitate the synthesis of new compounds that could serve as ligands by binding to the biological targets. Thus, rather than to perform molecular properties calculations, scaffold analysis of NPs might relevant for the navigation of the chemical space. In this context, Witzel and colleagues used the DNP as the basis to carry out a chemoinformatic approach to structurally classify NPs according to the scaffolds they contain [32]. A parent-child relationships between the scaffolds was established by grouping the scaffold hierarchically based on a particular set of rules. The tree diagrams below (Figure 2.5) show the relationships, some of which are carefully explained in the original reference [32].

This displays a general overview of NPs structure space available in the DNP. The tree diagram describes the more frequent scaffolds, which represented a minimum of 0.2% of the dataset. Their relationships in a genealogy-like way, following some rules, such as,

- the parent core structure must be a substructure of the child scaffold,
- the parent core structure must have fewer rings than the child scaffold,
- the breaking of ring bonds must not be allowed,
- only the parent with the largest number of heteroatoms must be chosen,
- only the largest parent scaffold must be selected,
- only the most frequent parent scaffold (e.g. the one representing more NPs) must be selected.

It was observed that, in terms of scaffold abundance, the carbocyclic section was most populated. This was followed by the O-heterocycles and the N-heterocycles. Meanwhile, mainly 3- and 4- ringed scaffolds could be seen in the carbocyclic section [32]. It was also observed that the O-heterocyclic section contained mainly 1- and 2-ringed scaffolds. Among the N-heterocycles only some branches extended

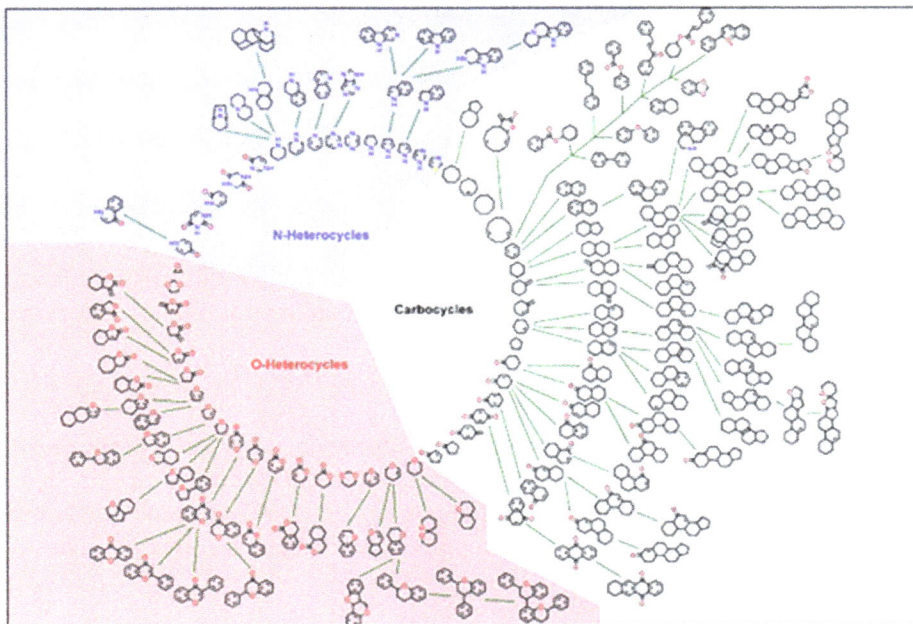

N-Heterocycles

Carbocycles

O-Heterocycles

Figure 2.5: Scaffold tree generated from the DNP [32]. Figure reproduced by permission.

beyond the first ring. Additionally, the scaffolds with three rings were most often found in NPs followed by scaffolds that contain 2 or 4 rings. These three scaffold types represented more than 50% of all scaffolds found within the DNP [32].

Yongye et al. also investigated 5 NP collections from the public domain (e.g. Specs, Analyticon, Timtec, Traditional Chinese medicine, and ZINC) [33]. The authors focused their study on analysing the molecular scaffold content and their diversity. It was noted that the NP databases showed different scaffold diversity. Favones, coumarins, and flavanones were noted to be the most frequent molecular scaffolds among the NPs investigated, apart from benzene and acyclic compounds.

2.4.2 The case of natural products from cyanobacteria

An investigation of chemical scaffolds from specific classes of NP-producing organisms has been of interest lately. For example, cyanobacterial metabolites are NPs having attractive biotechnological applications and distinctive scaffolds. Recently González et al. performed a chemoinformatic analysis of cyanobacterial SMs [34]. The goal was to investigate the diversity of chemical structures (and scafolds) from marine, terrestrial and freshwater cyanobacteria. This included 279 freshwater and 281 marine NPs from cyanobacterial sources. The authors used a Shannon entropy

(*SE*) approach to carry out this task. This consists in analysing the frequency distributions of the various scaffolds, which is a measure of the scaffold diversity. The *SE* could be defined for a population of *P* compounds, which are distributed in *n* systems as follows [35]:

$$SE = - \sum_{i=1}^{n} p_i \log_2 p_i \tag{2.1}$$

with $p_i = \dfrac{c_i}{P}$

- in this case, p_i represents the relative frequency of any cyclic system *i* found within a population of *P* compounds which contain a total of *n* distinct cyclic systems;
- c_i represents the absolute number of molecules which contain a particular cyclic system *i*,
- the *SE* values range from 0 and $\log_2 n$ and so depend on *n*, but not explicitly on *P*,
- when *SE* is equal to 0, all *P* compounds contain only a single cyclic system, whereas when *SE* is equal to $\log_2 n$, the *P* compounds are uniformly distributed among the *n* cyclic systems which represent maximum cyclic system diversity on the data set.

The scaled *SE* (*SSE*) is often defined to normalize the SE values for different values of *n*, as in the following equation:

$$SSE = \frac{SE}{\log_2 n} \tag{2.2}$$

The values of *SSE* range from 0 to 1 (i.e. if *SSE* = 0, all *P* compounds are contained in one cyclic system, while the closer the value of *SSE* is 1 is an indicator of large scaffold diversity within the *n* most populated cyclic systems).

In the study by Gonzalez et al. [34], it was shown for each library, that the *SSE* for the first 60 most frequent chemotypes in each library were close to 1 (see Table 2.4).

Table 2.4: Computed SSE values for the 10–60 most populated scaffolds in the study by González et al. [34].

Data set	SSE10[*]	SSE20[*]	SSE30[*]	SSE40[*]	SSE50[*]	SSE60[*]
Marine	0.937	0.929	0.919	0.914	0.916	0.908
Freshwater	0.917	0.882	0.869	0.870	0.863	0.856

[*]Computed SSE values for the 10, 20, 30, 40, 50 and 60 most frequent chemotypes, respectively.

For marine cyanobacterial compounds (with *SSE* values ranging from 0.937 to 0.908), compared with freshwater cyanobacterial compounds (with *SSE values* ranging from 0.917 to 0.856). This indicates that the compounds of marine origin were more diverse than those from freshwater. The most frequent scaffolds found in two datasets are shown in Figure 2.6.

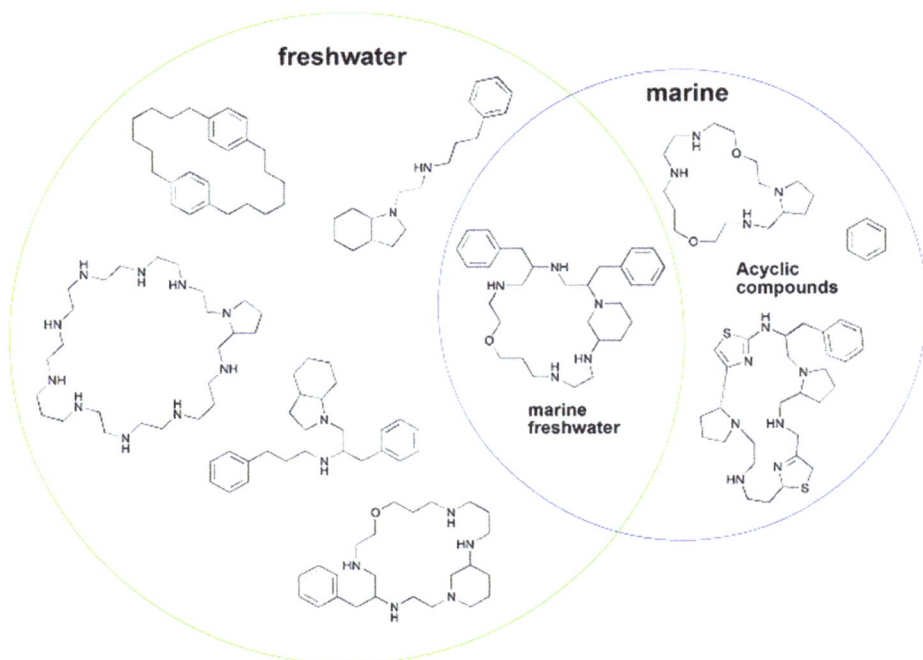

Figure 2.6: Most frequent scaffolds contained in NPs from freshwater and marine cyanobacteria [34]. Figure adapted by permission.

The scaffold analysis revealed only 5% of scaffold overlap between the marine and freshwater metabolites dataset investigated, implying that in general, different scaffolds occur within marine and freshwater cyanobacterial NPs. In addition, it was shown that the compounds of marine origin had the tendency to be structurally more diverse than those from freshwater (showing higher *SSE* values, see Table 2.4). Besides, it was shown that the scaffolds from cyanobacterial compounds possess pharmaceutical interesting properties, thus making them an ideal source of molecules which are structurally unique and bioactive [34].

2.4.3 The case of the terpenoid chemical class

Terpenoids are known to be the largest and most distributed class of NPs [24, 36]. Zeng recently analysed 77,317 terpenoid NPs deposited in the DNP, among which 76,155 NPs had source-availability information [37]. Structurally, terpenoids are derived from isoprene subunits (a 5 carbon system) and are often classified based on the number of C-atoms in their skeletons, e.g. triterpenoids (30 carbons), diterpenoids (20 carbons), sesquiterpenoids (15 carbons), etc. This study showed that the majority of terpenoids in the DNP are sesquiterpenoids (constituting 84% of all terpenoids in the DNP), Figure 2.7(a). Besides, it was observed that the majority of terpenoid NPs were biosynthesized by plants (~ 85.5%) and the minority by bacteria (~ 1.2%), Figure 2.7(b). Moreover, it was shown that:

- the majority of the terpenoids were shown to be derived from a single source organism,
- < 900 of them were obtainable from different species,
- only 19 terpenoids had been derived from 3 kinds of species,
- none of the terpenoids had ever been found in all four kingdoms (i.e. plants, animals, fungi, and bacteria).

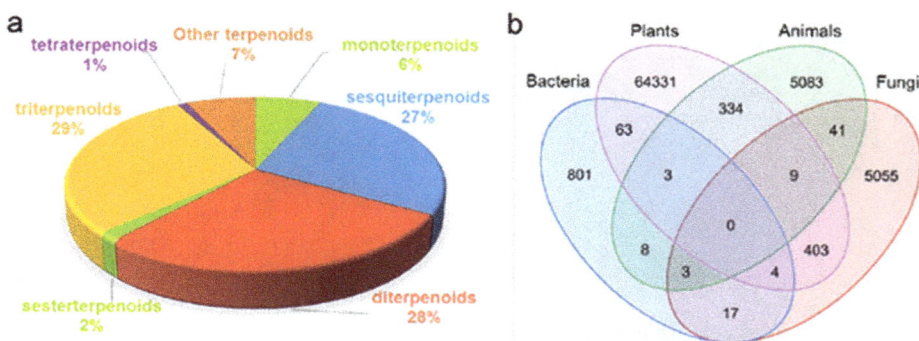

Figure 2.7: Distribution of terpenoids by (a) compound sub-classes and (b) species kingdom of origin [37]. Figure reproduced by permission.

An analysis of the scaffold and rings showed;

- terpenoids have more ring scaffolds (~ 98%) when compared with non-terpenoids (~ 90%, Table 2.5),
- that ringed compounds were observed to be even more prevailing in terpenoids, when compared with non-terpenoids,
- terpenoids have more aliphatic rings than non-terpenoids
- terpenoids have less aromatic rings than non-terpenoids.

A study of the ring systems in terpenoids showed that;

- they contain 3- and 4-membered rings more frequently,
- cyclic terpenoids have more bridge- or spiro- rings,
- this is an indication that cyclic terpenoids have more complex scaffolds.

Table 2.5: The scaffolds and ring-systems of terpenoid and non-terpenoid datasets [37].

Data sets	[a]F_{ring}	[b]F_{ring_3}	[c]F_{ring_4}	[d]F_{bridge}	[e]F_{spiro}	[f]N_s
Non-terpenoid NPs	89.987%	4.222%	0.779%	9.720%	8.828%	32,850
Terpenoid NPs	98.281%	15.015%	2.033%	19.826%	22.033%	20,953

[a] F_{ring} = fraction of compounds with rings;
[b] F_{ring_3} = fraction of compounds with 3-membered rings;
[c] F_{ring_4} = fraction of compounds with 4-membered rings;
[d] F_{bridge} = fraction of cyclic compounds with bridgehead atoms;
[e] F_{spiro} = fraction of cyclic compounds with spiro atoms;
[f] N_s = number of scaffolds.

Besides, $F_{sp}{}^3$ (the fraction of sp^3 C-atoms) and the number of chiral centers could be used as measures to quantify the complexity of molecules, an important property in drug design [38]. In this case, a high molecular complexity would imply that such a molecule would have a higher selectivity towards a given biological target [39]. Selective compounds towards a given drug target are expected to show minor side effects in the clinic. It was shown also that terpenoid scaffolds generally have higher molecular complexity when compared with non-terpenoids, apart from their greater stereochemical complexity.

2.5 Functional group analysis in natural products

Ertl and Schuhmann recently carried out an analysis of the common FGs in NPs [40]. The approach was used to analyse several datasets, including the DNP [41]. However, only included molecules whose source organisms are annotated in the Taxonomy Database [42] were included. All in all 186,000 NPs could be classified into the four source organism classes (i.e. animals, plants, fungi, or bacteria). Also included in the analysis was a collection of 67,000 NPs known as the OpenNP (Open Natural Products collection), which resulted from cleaning and normalizing compounds Natural Products Atlas (containing fungal and microbial metabolites) [43], added to compounds from the TCM Database@Taiwan (mostly plant metabolites). This dataset of NPs was then compared with about 13 million commonly synthetic compounds from the ZINC database [44].

The computed similarity (eq. (3.3)) was as the cosine coefficient between the vectors that represent the %FG occurrences within the different classes, defined as follows:

$$similarity = \cos(\theta) = \frac{\mathbf{A} - \mathbf{B}}{\| \mathbf{A} \| \| \mathbf{B} \|} = \frac{\sum_{i=1}^{n} A_i B_i}{\sqrt{\sum_{i=1}^{n} A_i^2} \sqrt{\sum_{i=1}^{n} B_i^2}} \quad (2.3)$$

and was used as a measure of similarity between two data sets that may be characterised by vectors of numbers. In this case, the vectors A_i and B_i were the vectors containing the percentage of occurrences of various FGs in the two different sets. The results can be summarized as follows:

- The > 186,000 compound in the DNP contain 2,785 unique FGs.
- The distribution of the FGs showed that there are a few groups common among NPs, but a large number of them are non-frequent.
- A total of 25 FGs are present in > 1% of the NPs and 93 FGs are present in > 0.1% of the NPs.
- A total of 1,214 (i.e. ~ 43.6%) of the FGs were singletons (i.e. were present in only one molecule).
- The most common NP-based FGs in order of frequency are the -OH group of alcohol (61.1%), the alkene FG (39.9%), the ether FG (35.2%), the ester FG (28.7%), and the -OH group of phenols (28.4%).

The commonest FGs are represented in Figure 2.8.

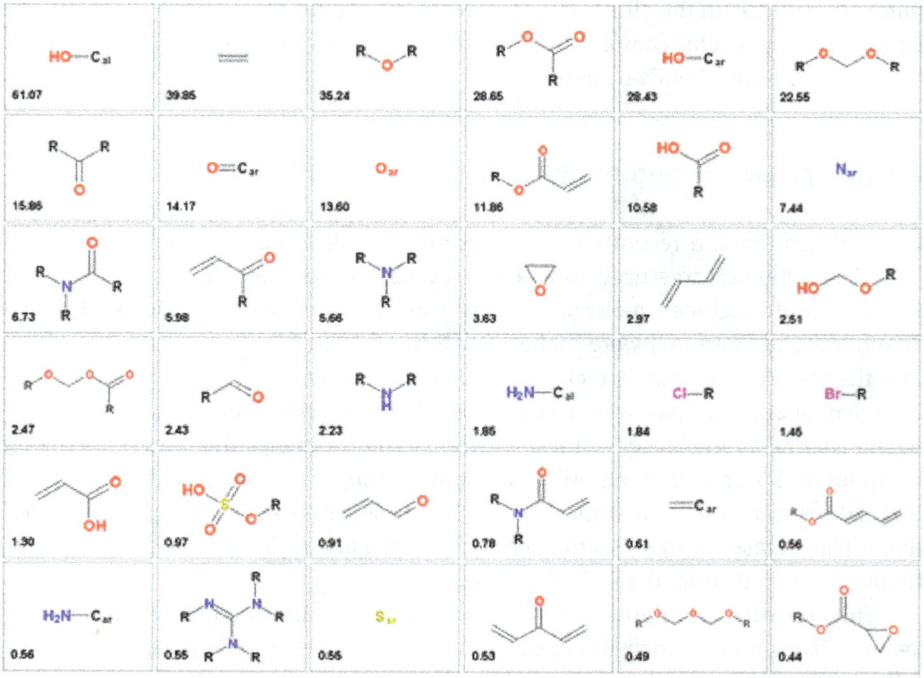

Figure 2.8: Commonest FGs in NPs. The number in each box represents the % of NPs containing the given group [40]. Figure reproduced by permission.

Several quite exotic FGs (Figure 2.9) were shown in the frequency distribution, showing a wide structural diversity that biosynthetic machinery involved in source organisms containing the NPs.

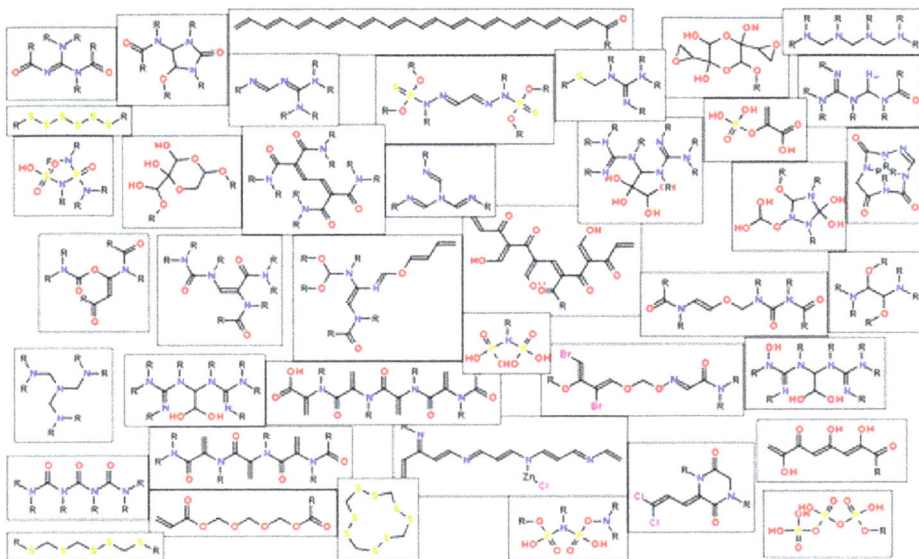

Figure 2.9: Some examples of randomly ordered exotic FGs found in NPs [40]. Figure reproduced by permission.

Futhermore, the frequencies of occurrence of specific FGs in the DNP, in the OpenNP, and in the library of synthetic molecule for this study were analyzed [44]. The 50 most common FGs within each dataset have been shown in Figure 2.10. It was observed that:

– FGs in NPs contain mostly O-atoms (e.g. hydroxy, ester, peroxide, polyglycol, epoxide rings). The only exceptions are ethylene-derived groups and various α, β-unsaturated systems.

– N-containing (e.g. amide, urea) and chemically more easily accessible FGs (e.g. sulfone, sulfonamide, imide functionalities, fluoro or nitro groups) are more present in synthetic molecules.

The authors also sought to compare the distribution of FGs in NPs in relation to their source organisms (i.e. animals, plants, fungi, or bacteria) and further carried out a comparison with FGs found in synthetic molecules (Figure 2.11). The heights of bars show the relative frequency of the class each given FG. In some FGs (e.g. keto groups, esters, alcohols, aldehydes) a fairly similar distribution was seen for

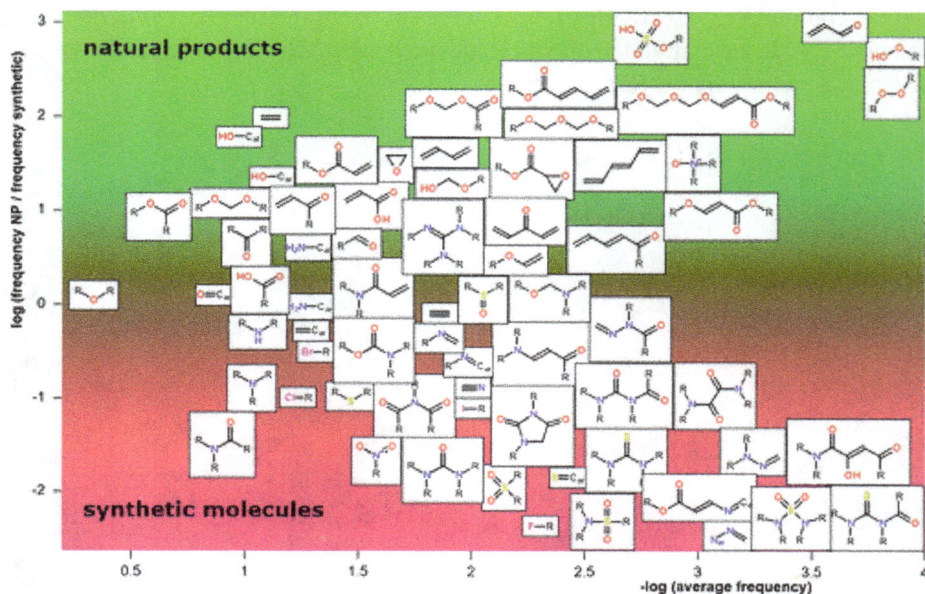

Figure 2.10: Plot of the common FGs in NPs (represented in the green area) and in synthetic molecules (represented in the red area) as the logarithm of the ratio between the respective frequencies in NPs and molecules of synthetic origin. The horizontal axis is proportional to the frequency of the FGs expressed as the negative logarithm of the average frequencies in the respective datasets. The most common FGs are shown to the left, while the less common ones are shown to the right [40]. Figure reproduced by permission.

all types of source organisms, while major differences were observed in most other cases (e.g. for amide, cyano, nitro, fluorine, sulfonamide, and sulfone groups).

The similarities between the distribution of FGs in NPs from the DNP by classes of source of organisms has been summary in Table 2.6. It was also observed that the majority of FGs was found in bacteria (a total of 148), followed by animals (a total of 121), fungi (a total of 106), and plants (a total of 85). Similar results were obtained with the analysis of the 67,000 NPs from the OpenNP, apart from the fact that the observed differences in the distribution of a few FGs (e.g. alkynes and secondary amines) [40].

Another study has been recently carried out on NP scaffolds on the same dataset [45]. This study focused on comparing the properties of scaffolds of NPs produced by organisms from different kingdoms (plants, animals, fungi and bacteria). It was shown that NP scaffolds from plant sources are the most complex, while NP scaffolds from bacterial sources are quite distinct structurally from those produced by fungi, plants and animals [45].

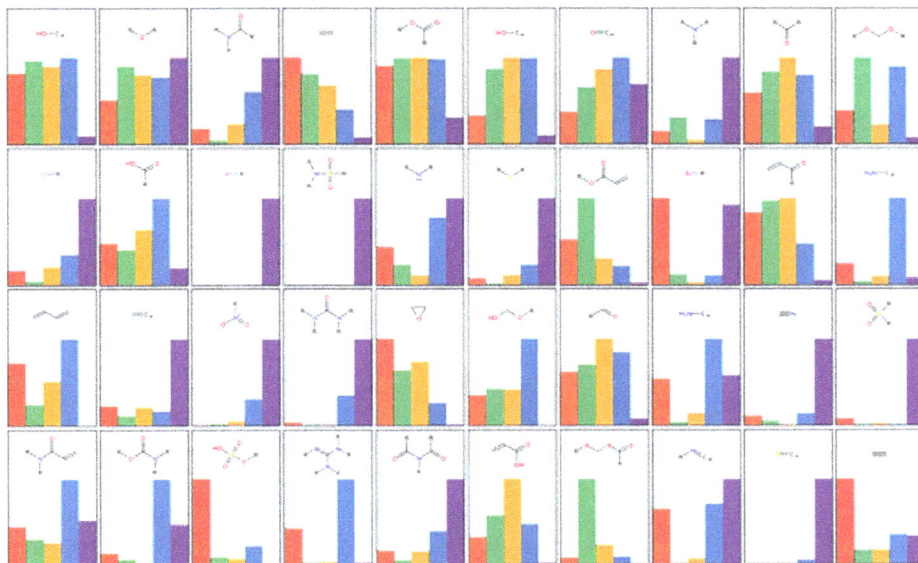

Figure 2.11: Relative frequencies of common FGs contained in NPs from animals (shown in red), plants (shown in green), fungi (shown in orange), and bacteria (shown in blue). FGs present in synthetic molecules have been shown in magenta [40]. Figure reproduced by permission.

Table 2.6: Similarity between different classes of NPs and synthetic molecules based on the frequency of FG occurrences[a] [40].

Source	Plants	Fungi	Bacteria	Synthetic
Animals (21,007)	56.9	57.4	48.6	20.7
Plants (133,480)		68.0	55.3	20.1
Fungi (18,412)			58.9	21.9
Bacteria (13,575)				28.9

[a]The figures in brackets represent the number of NPs analysed.

2.6 Conclusions

NPs have always represented a major source for the design of novel molecules with interesting biological activities in rational drug discovery. Several previous chemo-informatic analyses have been carried out aimed at determining the fundamental differences that exist between NPs and molecules of synthetic origin. A few of such studies were based on the atomic compositions of the compounds [11], while others focused on scaffold analysis [32] and most recently on the frequency distribution of the various FGs [40]. In the beginning, the goal was to slightly modify the original NPs so as to enhance the biological activities or their physical and chemical properties which could have effects on their metabolism and pharmacokinetics. NPs from plants [46, 47] and marine sources have been important sources of bioactive metabolites [48–50]. More recently, the analysis of large NP libraries by chemoinformatics approaches have provided deeper insight into the structural features that could enhance their target-specific interactions, so as to derive knowledge that could be of importance in molecular design [51–58]. This includes the analysis of the most common atom types, the most frequent scaffolds and FGs found in NPs, etc, and could be used as one way of assessing NPs by machine learning or evolutionary algorithms. These results summarized in this chapter will be applicable in both *in silico* and wet lab screening of NP libraries with the aim of identifying new lead compounds for drug discovery. As an example, NP scaffold trees could be use for charting the chemical space covered by naturally occurring compounds for designing NP-inspired compound libraries. FGs in NPs have found interest due to their importance in synthetic organic chemistry, drug discover, toxicology, and spectroscopy. This is because the presence of FGs often explains the properties of the parent molecule, including their chemical reactivities and their ability to elucidate their biological activities by establishing specific interactions within their targets. Besides, FGs are also responsible for vital aspects of drug properties like the absorption, bioavailability and metabolic stability of the parent molecule. The knowledge of the frequencies of FGs in NPs, for example, could provide insights for tailoring reaction conditions for synthesising the desired NP-derivatives or reconstituting the original NPs in the synthesis lab. Besides, a knowledge of the enzymatic processes *via* which the biosynthesis of the respective FGs and scaffolds included in the NPs are built is a core theme in NP research today. This includes searching for the genes encoding specific FGs in large bacterial, fungal, and plant genomes. This new field promises a bright future for NP-based drug discovery, since SM genes are often clustered into biosynthetic gene clusters.

Acknowledgements: FNK acknowledges a return fellowship from the Alexander von Humboldt accompanied by BDB and JNH. FNK would also like to acknowledge the European Structural and Investment Funds, OP RDE-funded project 'ChemJets' (No. CZ.02.2.69/0.0/0.0/16_027/0008351).

References

[1]   Abegaz BM, Kinfe HH. Secondary metabolites, their structural diversity, bioactivity, and ecological functions: an overview. Phys Sci Rev. 2018. DOI:10.1515/psr-2018-0100.

[2]   Ntie-Kang F, Svozil D. An enumeration of natural products from microbial, marine and terrestrial sources. Phys Sci Rev. 2019. DOI:10.1515/psr-2018-0121.

[2]   Ribes S, Fuentes A, Talens P, Barat JM. Prevention of fungal spoilage in food products using natural compounds: a review. Crit Rev Food Sci Nutr. 2018;58:2002–16.

[4]   Harvey AL. Natural products in drug discovery. Drug Discov Today. 2008;13:894–901.

[5]   Martins A, Vieira H, Gaspar H, Santos S. Marketed marine natural products in the pharmaceutical and cosmeceutical industries: tips for success. Mar Drugs. 2014;12: 1066–101.

[6]   Altmann KH. Drugs from the oceans: marine natural products as leads for drug discovery. Chimia (Aarau). 2017;71:646–52.

[7]   Koehn FE, Carter GT. The evolving role of natural products in drug discovery. Nat Rev Drug Discov. 2005;4:206–20.

[8]   Shen B. A new golden age of natural products drug discovery. Cell. 2015;163:1297–300.

[9]   Newman DJ, Cragg GM. Natural products as sources of new drugs from 1981 to 2014. J Nat Prod. 2016;79:629–61.

[10] Bade R, Chan HF, Reynisson J. Characteristics of known drug space. natural products, their derivatives and synthetic drugs. Eur J Med Chem. 2010;45:5646–52.

[11] Feher M, Schmidt JM. Property distributions: differences between drugs, natural products, and molecules from combinatorial chemistry. J Chem Inf Comput Sci. 2003;43:218–27.

[12] Grabowski K, Schneider G. Properties and architecture of drugs and natural products revisited. Curr Chem Biol. 2007;1:115–27.

[13] Schneider P, Schneider G. Collection of bioactive reference compounds for focused library design. QSAR Comb Sci. 2003;22:713–8.

[14] AnalytiCon Discovery GmbH. Hermannswerder Haus 17, D-14473 Potsdam, Germany. http://www.ac-discovery.com.

[15] InterBioScreen Ltd. 121019 Moscow, P.O. Box 218, Russia. http://www.ibscreen.com/.

[16] Schneider G, Lee M-L, Stahl M, Schneider P. *De novo* design of molecular architectures by evolutionary assembly of drugderived building blocks. J Comput-Aided Mol Des. 2000;14:487–94.

[17] Pegg SC-, Haresco JJ, Kuntz ID. A genetic algorithm for structure-based *de novo* design. J Comput-Aided Mol Des. 2001;15:911–33.

[18] Lameijer E-W, Kok JN, Bäck T, IJzerman AP. The molecule evoluator. an interactive evolutionary algorithm for the design of drug-like molecules. J Chem Inf Model. 2006;46:545–52.

[19] Yu MJ. Natural product-like virtual libraries: recursive atom-based enumeration. J Chem Inf Model. 2011;51:541–57.

[20] Hedner E. Bioactive compounds in the chemical defence of marine sponges: structure-activity relationships and pharmacological targets. Uppsala: Uppsala University, Interfaculty Units, Acta Universitatis Upsaliensis, 2007:54.

[21] Faulkner DJ. Marine natural products. Nat Prod Rep. 2002;9:1–48.

[22] Rinehart KL. Secondary metabolites from marine organisms. Ciba Found Symp. 1992;171:236–49.

[23] Fenical W. Natural products chemistry in the marine environment. Science. 1982;215:923–8.

[24] Muigg P, Rosén J, Bohlin L, Backlund A. In silico comparison of marine, terrestrial and synthetic compounds using ChemGPS-NP for navigating chemical space. Phytochem Rev. 2013;12:449–57.

[25] Shang J, Hu B, Wang J, Zhu F, Kang Y, Li D, et al. Cheminformatic insight into the differences between terrestrial and marine originated natural products. J Chem Inf Model. 2018;58: 1182–93.

[26] Lewell XQ, Judd DB, Watson SP, Hann MM. RECAP - Retrosynthetic combinatorial analysis procedure: a powerful new technique for identifying privileged molecular fragments with useful applications in combinatorial chemistry. J Chem Inf Comput Sci. 1998;38:511–22.

[27] Bemis GW, Murcko MA. The properties of known drugs 0.1. molecular frameworks. J Med Chem. 1996;39:2887–93.

[28] Kong D-X, Jiang -Y-Y, Zhang H-Y. Marine natural products as sources of novel scaffolds: achievement and concern. Drug Discov Today. 2010;15:884–6.

[29] Kong D-X, Guo M-Y, Xiao Z-H, Chen -L-L, Zhang H-Y. Historical variation of structural novelty in a natural product library. Chem Biodiv. 2011;8:1968–77.

[20] Tian S, Wang J, Li Y, Xu X, Hou T. Drug-likeness analysis of traditional Chinese medicines: prediction of drug-likeness using machine learning approaches. Mol Pharm. 2012;9: 2875–86.

[21] Tian S, Wang J, Li Y, Li D, Xu L, Hou T. The application of in silico drug-likeness predictions in pharmaceutical research. Adv Drug Delivery Rev. 2015;86:2–10.

[22] Wetzel S, Schuffenhauer A, Roggo S, Ertl P, Waldmann H. Cheminformatic analysis of natural products and their chemical space. Chimia. 2007;61:355–60.

[23] Yongye AB1, Waddell J, Medina-Franco JL. Molecular scaffold analysis of natural products databases in the public domain. Chem Biol Drug Des. 2012;80:717–24.

[24] González-Medina M, Medina-Franco JL. Chemical diversity of cyanobacterial compounds: a chemoinformatics analysis. ACS Omega. 2019;4:6229–37.

[25] Medina-Franco JL, Martinez-Mayorga K, Bender A, Scio T. Scaffold diversity analysis of compound data sets using an entropy-based measure. QSAR Comb Sci. 2009;28:1551–60.

[26] Brahmkshatriya PP, Brahmkshatriya PS. Terpenes: chemistry, biological role, and therapeutic applications. In: Ramawat K, Mérillon JM, editors. Natural products. Berlin, Heidelberg: Springer, 2013.

[27] Zeng T, Liu Z, Liu H, He W, Tang X, Xie L, et al. Exploring chemical and biological space of terpenoids. J Chem Inf Model. 2019;59:3667–78.

[28] Lovering F, Bikker J, Humblet C. Escape from flatland: increasing saturation as an approach to improving clinical success. J Med Chem. 2009;52:6752–6.

[29] Clemons PA, Bodycombe NE, Carrinski HA, Wilson JA, Shamji AF, Wagner BK, et al. Small molecules of different origins have distinct distributions of structural complexity that correlate with protein-binding profiles. Proc Natl Acad Sci USA. 2010;107:18787–92.

[40] Ertl P, Schuhmann T. A systematic cheminformatics analysis of functional groups occurring in natural products. J Nat Prod. 2019;82:1258–63.

[41] Dictionary of Natural Products 27.1. Boca Raton, FL, USA: CRC Press, Taylor & Francis Group, 2018.

[42] Federhen S. The NCBI taxonomy database. Nucleic Acids Res. 2012; 40: D136–43. https://www.ncbi.nlm.nih.gov/taxonomy.

[43] Natural Product Atlas. https://www.npatlas.org/joomla/index.php. Accessed: 19 Aug 2019.

[44] Sterling T, Irwin JJ. ZINC 15–ligand discovery for everyone. J Chem Inf Model. 2015;55: 2324–37. http://zinc.docking.org/catalogs/uefsnp/. Accessed: 16 May 2019.

[45] Ertl P, Schuhmann T. Cheminformatics analysis of natural product scaffolds: comparison of scaffolds produced by animals, plants, fungi and bacteria. Mol Inf. 2020. DOI:10.1002/minf.202000017.

[46] Seca AM, Pinto DC. Plant secondary metabolites as anticancer agents: successes in clinical trials and therapeutic application. Int J Mol Sci. 2018;19:E263.

[47] Fang J, Wu Z, Cai C, Wang Q, Tang Y, Cheng F. Quantitative and systems pharmacology 1. In silico prediction of drug-target interaction of natural products enables new targeted cancer therapy. J Chem Inf Model. 2017;57:2657–71.

[48] Gerwick WH, Moore BS. Lessons from the past and charting the future of marine natural products drug discovery and chemical biology. Chem Biol. 2012;19:85–98.

[49] Ding W, Gu J, Cao L, Li N, Ding G, Wang Z, et al. Traditional Chinese herbs as chemical resource library for drug discovery of anti-infective and anti-inflammatory. J Ethnopharmacol. 2014;155:589–98.

[50] Buenz EJ, Verpoorte R, Bauer BA. The Ethnopharmacologic contribution to bioprospecting natural products. Annu Rev Pharmacol Toxicol. 2018;58:509–30.

[51] Grisoni F, Merk D, Consonni V, Hiss JA, Tagliabue SG, Todeschini R, et al. Scaffold hopping from natural products to synthetic mimetics by holistic molecular similarity. Comm Chem. 2018;1:44.

[52] Muegge I, Mukherjee P. An overview of molecular fingerprint similarity search in virtual screening. Expert Opin Drug Discov. 2016;11:137–48.

[53] Saldívar-González FI, Valli M, Andricopulo AD, Bolzani VS, Medina-Franco JL. Chemical space and diversity of the NuBBE database: a chemoinformatic characterization. J Chem Inf Model. 2019;59:74–85.

[54] Krier M, Bret G, Rognan D. Assessing the scaffold diversity of screening libraries. J Chem Inf Model. 2006;46:512–24.

[55] Shang J, Sun H, Liu H, Chen F, Tian S, Pan P, et al. Comparative analyses of structural features and scaffold diversity for purchasable compound libraries. J Cheminform. 2017;9:25.

[56] Olğaç A, Orhan IE, Banoglu E. The potential role of in silico approaches to identify novel bioactive molecules from natural resources. Future Med Chem. 2017. DOI:10.4155/fmc-2017-0124.

[57] Skinnider MA, Magarvey NA. Statistical reanalysis of natural products reveals increasing chemical diversity. Proc Natl Acad Sci USA. 2017;114:E6271–2.

[58] Ho TT, Tran QT, Chai CL. The polypharmacology of natural products. Future Med Chem. 2018;10:1361–8.

Lena Y. E. Ekaney, Donatus B. Eni and Fidele Ntie-Kang

3 Chemical similarity methods for analyzing secondary metabolite structures

Abstract: The relation that exists between the structure of a compound and its function is an integral part of chemoinformatics. The similarity principle states that "structurally similar molecules tend to have similar properties and similar molecules exert similar biological activities". The similarity of the molecules can either be studied at the structure level or at the descriptor level (properties level). Generally, the objective of chemical similarity measures is to enhance prediction of the biological activities of molecules. In this article, an overview of various methods used to compare the similarity between metabolite structures has been provided, including two-dimensional (2D) and three-dimensional (3D) approaches. The focus has been on methods description; e.g. fingerprint-based similarity in which the molecules under study are first fragmented and their fingerprints are computed, 2D structural similarity by comparing the Tanimoto coefficients and Euclidean distances, as well as the use of physiochemical properties descriptor-based similarity methods. The similarity between molecules could also be measured by using data mining (clustering) techniques, e.g. by using virtual screening (VS)-based similarity methods. In this approach, the molecules with the desired descriptors or /and structures are screened from large databases. Lastly, SMILES-based chemical similarity search is an important method for studying the exact structure search, substructure search and also descriptor similarity. The use of a particular method depends upon the requirements of the researcher.

Keywords: biological activities, chemical similarity, clustering, natural products, secondary metabolites

3.1 Introduction

Chemical similarity could be defined in terms of the chemical entities (elements, molecules and compounds) with respect to their structural or functional qualities. This includes the effect of a compound on its immediate biological environment. The biological activity of a compound can be defined as its biological effects in a living system. Generally speaking, the biological function of a compound is related to its chemical structure [1].

This article has previously been published in the journal *Physical Sciences Reviews*. Please cite as: Ekaney, L. Y. E., Eni, D. B., Ntie-Kang, F. Chemical similarity methods for analyzing secondary metabolite structures. *Physical Sciences Reviews* [Online] 2021 DOI: 10.1515/psr-2018-0129

https://doi.org/10.1515/9783110668896-003

Chemical similarity could also be referred to as molecular similarity and plays an important role in the prediction of the properties of compounds, i.e. it is useful to design new chemicals with a predefined biological property, especially, in conducting drug design studies by identifying similar compounds from large screening databases, which contain available of potentially available chemical structures. The study of chemical similarity is based on the principle of chemical similarity property, which was originally proposed by Johnson and Maggiora and which states that: *similar compounds have similar properties* [2]. The effective application of similarity principle depends on the biological activity being investigated [3, 4]. In order to improve accuracy, several search methods should be applied wherever possible. This article concerns chemical similarity, as applied to the study of secondary metabolites (SMs), a distinct type of naturally occurring compounds, which are produced bacteria, fungi, and plants [5, 6]. These are compounds that can be transformed from primary metabolites in living cells, e.g. in plants, bacteria and fungi. Where present, they are not involved in the normal processes like growth, reproduction and development, but they rather primarily play maintenance and homeostatic roles within the organisms that produce them.

SMs are derived from SM in microbes, fungi and plants, often essential for the survival of the organism in the environment [5]. In general, mammals do not produce SMs, although they often contain these compounds in the form of xenobioitcs. Xenobiotics are defined as substances foreign to the cells containing them, i.e. substances found within an organism that is not naturally produced or expected to be present within the said organism. Metabolites may also be produced as by-products of the body's reaction to external substances, e.g. medications and antigens [7].

Natural products (NPs) are known to be a major source of compounds of pharmaceutical and industrial importance. Hence, it is important to quantify the similarity of NPs since their biological activities have been optimized through the process of natural selection. Plant SMs (often called phytochemicals), for example, are often very structurally complex rendering their synthesis very difficult. Besides, SMs are found as mixtures in different parts of plants (e.g. leaves, roots, shoots, and stem bark) often in different combinations at different growth stages (e.g. as seedlings, seeds, plantlets, shrubs or mature trees) and these combinations also differ according to the varying environmental conditions and plant families [8].

3.2 Different ways of expressing chemical similarity

The quantification of molecular similarity of chemical structures is vital and has diverse applications in drug discovery [5], e.g. in ligand-based virtual screening (VS), in toxicology predictions, in predictive pharmacology and in chemogenomics [3]. Various methods have been developed for efficiently representing chemical information and their application for determining how similar chemical structures are, with the goal of facilitating the calculation of chemical similarity. As a results of their potent biological activities and their use as lead compounds for drug discovery such similarity based approaches have been particularly applied to study NPs, e.g. to predict whether a given molecule belongs to known class of bioactive NPs, e.g. antimalarials, antiviral agents, etc.

3.2.1 (2D) Fingerprint-based similarity

The concept of diversity between compounds in a dataset is often linked to their physicochemical parameters and chemical fingerprints. Two-dimensional (2D) fingerprints are often used to compare how similar (or diverse) compounds are, using commonly known metrics like the Tanimoto coefficient [9]. However, due to the structural complexity of NP scaffolds, quantifying their similarity is often a major challenge [10, 11]. One of such methods applicable for comparing the similarity of NPs is the approach implemented in the Library for the Enumeration of MOdular Natural Structures (LEMONS) tool. The method outperforms conventional 2D fingerprints based methods and is based on an algorithm used to enumerate hypothetical modular NP structures [12]. A detailed discussion of the concept of chemical similarity is beyond the scope of this article.

In order to implement fingerprint-based similarity approaches, each compound structure is decomposed into its basic fragments. Then, each molecule's structural fragments or features are either turned ON (set or defined as 1 when present) or turned OFF (set defined as 0 when absent), as shown in Table 3.1. Each molecule is seen as a string which contains either 1s or 0s (called "bit strings"), not considering how often any given key occurs, i.e. each bit is only counted once [13]. Bits are set or defined on the basis of the presence or absence or given fragments, not on the entire structure of the molecule. The properties of the whole molecule are poorly "captured", since the structures are only compared based on their respective decomposed smaller fragments (or bits), based on the dictionary or set of all the prepared fragments, not on their connectivities (see Table 3.1).

Similarity calculation requires matching the fragments present in the individual NP with those in the dictionary of fragments. In our example above, the natural compound salicylic acid (**2**), first isolated from the willow tree (*Salix alba* L.), and it's acetylated derivative acetylsalicylic acid (**1**, or aspirin). If the fragment of one compound matches with that which is present in the dictionary, a value of "1" is attributed. When absent, a value of "0" is attributed. After preparing the bit- string fingerprint of each compound, any of the association coefficients for comparing the two molecules can be determined to compare the two molecules.

Table 3.1: Example for comparing the chemical structures and fragments of two compounds (**1** and **2**) [14].

	Acetyl salicylic acid (1)					Salicylic acid (2)			
Fragments (bit-strings or features)	benzene ring	OH	acetyl ester	isopropyl	OH	formyl ester	NH₂	ether O	thioether S
Compound **1**	1	1	1	0	0	1	1	0	0
Compound **2**	1	0	0	0	1	0	0	0	0
Feature numbers	1	2	3		4		5		

3.2.1.1 *Tanimoto* coefficient

This is one of the most widely used similarity-based on fingerprints (bit-string representations) methods. The approach is by comparing the bit-string fingerprints of two molecules, A and B, using eq. (1.1) [1]. If $N1$ is the number of features present in compound **1**, while $N2$ is the number of features present in compound **2**, and $N12$ is the number of features common to both compounds, the calculation proceeds as follows:

$$\tau = \frac{N12}{N1 + N2 - N12} \tag{3.1}$$

The *Tanimoto* similarity is only applicable for a binary variable and varies from 0 to 1 (with 0 = no similarity and 1 = the highest similarity, i.e. for two identical molecules, for example). A simple count of shared features (common fragment substructures) can therefore be regarded as a measure of chemical distance when used in some similarity coefficients. The value of τ is a simple and computationally efficient approach for the quantification of the extent structural similarity of two molecules [9].

In this case, there are altogether five features in compounds **1** and **2**, numbered 1 2 3 4 5 (two features in Table 3.1 are absent from both compounds), see Table 3.2:

Table 3.2: Features used in computing the *Tanimoto* coefficient between compounds **1** and **2**.

Feature numbers	1	2	3	4	5
Molecule **1**	1	1	0	1	1
Molecule **2**	1	1	1	0	0

The similarity assessment using the *Tanimoto* coefficient is given by:

$\tau = 2 /(4 + 3{-}2) = 0.4$, that is the sum of common bit-strings (or features) over the total set of features including the dictionary of all features in compounds **1** and **2**. Note that, although these compounds look so similar (one is derived synthetically from the other) and only the acetyl group is absent on compound 1, the similarity derived from the Tanimoto coefficient is so weak (less than 50% similarity!).

3.2.1.2 Jaccard coefficient

This was originally a statistical metric that measures the similarity between finite sample sets, defined in terms of the extent of intersection of the sample sets (or common features) divided by the size of the union set for the two sample sets:

$$J(A, B) = \frac{|A \cap B|}{|A \cup B|} = \frac{|A \cap B|}{|A| + |B| - |A \cap B|} \tag{3.2}$$

Like the Tanimoto coefficient, $0 \le J(A.B) \le 1$. In case A and B are both empty sets, $J(A,B) = 1$.

The *Tanimoto coefficient* is often related to the *Jaccard coefficient*, i.e. "1s" features in common divided by "1s" features not in common, where the 0's are regarded as "not significant", and the similarity coefficient is defined between completely dissimilar molecules [$J(A,B) = 0$] and completely identical ones [$J(A,B) = 1$]. A good cutoff of 0.7 or 0.8 is taken for two molecules with likely similar biologically activities.

3.2.1.3 Euclidean measures fingerprinting similarity
This is the Pythagorean distance-based similarity metric. For binary dimensions, defined as the square root of the Hamming distance, .i.e. the square root of the number of bits which are different. In this case, 0's are regarded as significant, with smaller values taken to mean more similar molecules [13].

3.2.2 Chemical similarity using physicochemical property descriptors

Similarity searches often make use of topological descriptors and this approach has proven to be extremely useful in large-scale screening for the identification of bioactive "hits" in compound databases. These types of chemical similarity are based on simple descriptors like molecular weight (MW), hydrogen bond donor (HBD) or acceptor (HBA), binding property class, predicted log P or experimental log P, partial atomic charges, etc [14]. The physicochemical property descriptors similarity is implemented using data mining techniques like those implemented in the Weka clustering tool [15].

3.2.3 Virtual screening based similarity search

The similarity-based VS approach assumes that for a given query compound, any compound in a database of compounds having a high degree of similarity should also have a similar biological activity as the compound in the query. VS often makes use of both information of 2D- and 3D similarity between the query compound and the compounds in the screening database, although the 2D-fingerprints (e.g. *Tanimoto* or *Jaccard* coefficients) are often more commonly used. When using such an approach, two compounds are often regarded as similar if the similarity metric is > 0.85 [9]. Many non-commercial databases are available to be screened via the web [16]. One which is widely used is the ZINC database, a compound library containing 13 million molecules (with predefined 3D structures). The chemical search graphical user interfaces for ZINC databases is shown in Figures 1.1 and 2 [17, 18]. Each molecule must have an assigned biologically-relevant protonation

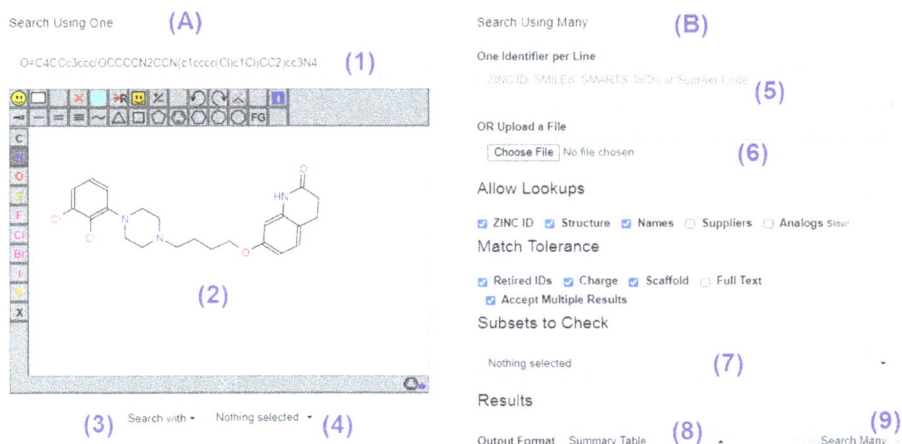

Figure 3.1: The graphical user interface for implementation of similarity searches in the ZINC database (https://zinc15.docking.org/substances/home/): (A) The user can search by keying the name of one known molecule using (1) the SMILES string (see next section), its ZINC IDs, InChI, InChI key, or its original catalog numbers are then inserted before initiating the search (2), the chemical structure could be drawn; (3) search options include "default", "substructure", or "smarts" could be used to searching; (4) the compound subset to be search is specified, e.g. FDA approved drugs, compounds in stock, etc. (B): Many molecules can be searched in a single operation; (5) by specifying one molecule at the start of each line in a file, which could be done by pasting the SMILES string or (6) by uploading the molecule file, (7) the compound subset to be search is specified, e.g. FDA approved drugs, compounds in stock, etc., then (8) specifying the format of results file. e.g. SMILES, SDF, MOL2, etc. before running the search my clicking the search button (9).

state and must be annotated with properties that can be screened using physico-chemical properties, e.g. MW, logP, HBD, HBA, NRB, the polar surface area (PSA), the net charge, and the polar desolvation, see Figure 3.2 [18].

3.2.4 SMILES based chemical similarity search

SMILES which stands for *simplified molecular-input line-entry system*, a (1D) line notation [14]. SMILES notations could be quite simple:

e.g. methane is simply C,

propane is CCC,

ethanol is CCO,

benzene is c1ccccc1 (lower case letters "c" representing aromatic carbon atoms and number 1 showing where the atoms are joint to form the ring).

However, more complex NPs like cephalostatin-1 (Figure 3.3) could be quite complex:

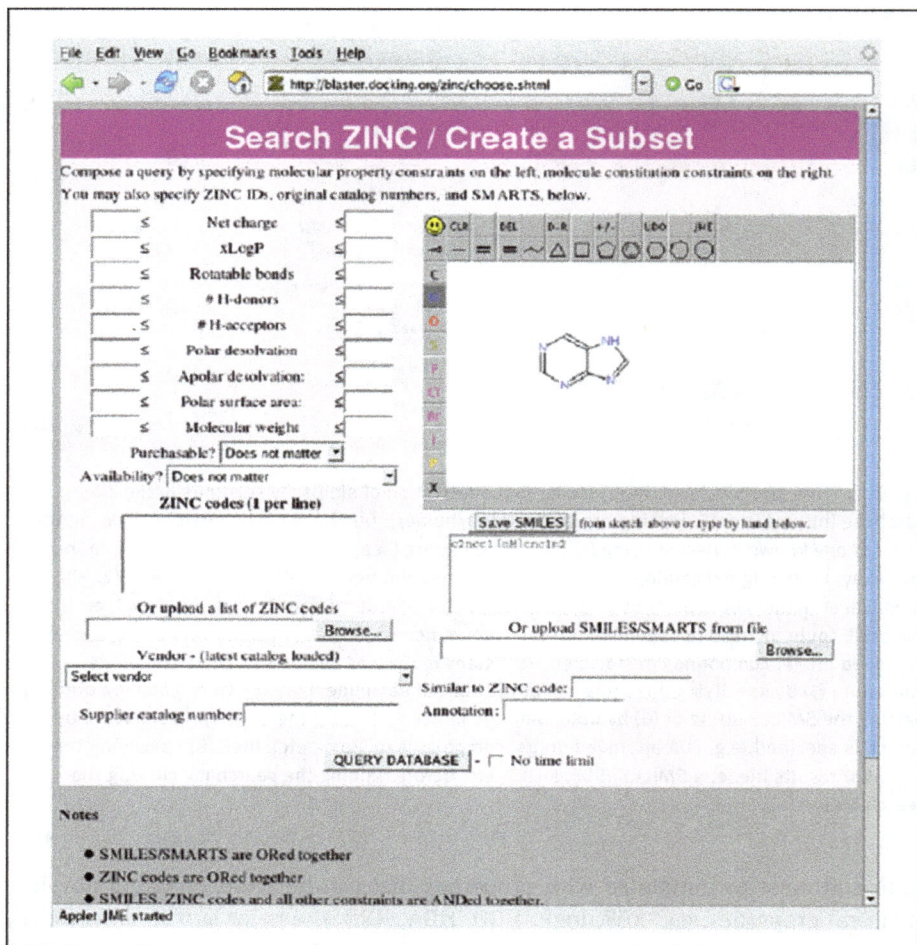

Figure 3.2: Searching the ZINC database using physicochemical property filters [18]. Figure reproduced by permission.

The SMILES string of cephalostatin-1, a complex NP:

CC(C)(O1)C[C@@H](O)[C@@]1(O2)[C@@H](C)[C@@H]3CC = C4[C@]3(C2)C(= O)C
[C@H]5[C@H]4CC[C@@H](C6)[C@]5(C)Cc(n7)c6nc(C[C@@]89(C))c7C[C@@H]8CC
[C@@H]%10[C@@H]9C[C@@H](O)[C@@]%11(C)C%10 = C[C@H](O%12)[C@]%11(O)
[C@H](C)[C@]%12(O%13)[C@H](O)C[C@@]%13(C)CO

A chemical similarity search is the ability to find chemicals based on their likeness to a specified structure, rather than an exact match. Similarity searches can be performed on SMILES or structures, by using the menu options to specify a similarity

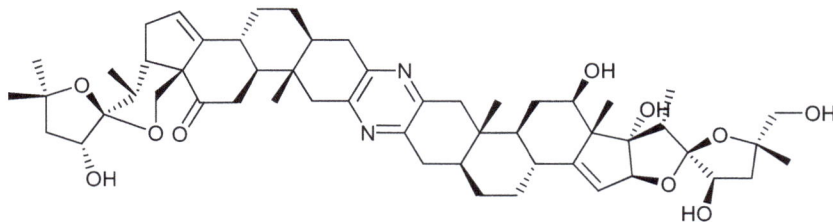

Figure 3.3: Chemical structure of cephalostatin-1 (a steroidic 13-ringed pyrazine) isolated from *Cephalodiscus gilchristi*, a worm collected from the Indian ocean [19].

cutoff instead of an exact match. First, a data set is prepared and then using Structured Query Language (SQL) chemical and descriptor similarity, the search is made. The general procedure involves the following:
- SMILES are stored as a string, e.g. in an Oracle database,
- The structure search is implemented using the EXACT function,
- The sub-structure search is implemented using the MATCHES function, and
- The similarity search implemented using the TANIMOTO and EUCLIDEAN functions.

3.2.5 Hypothetical enumeration method for comparing natural products

NPs are distinct from synthetic compounds in that their scaffolds are often larger and structurally more complex, in addition to their distinctive physical and chemical properties [16]. Hence, more advanced similarity search algorithms, e.g. LEMONS have been suggested for comparing NPs. This has been used to carry out similarity searches within the chemical space of NPs. LEMONS was designed with the goal to:
- enumerate the hypothetical modular NP structures.
- modify their monomer compositions,
- tailor their reactions, and
- compare the original and modified structures by use of 2D molecular fingerprints.

The approach requires the subsequent modification of each hypothetical structure by the substitution of one or more monomers, by the addition, removal, or replacement of the site of one or more tailoring reactions. The entire procedure implemented in LEMONS has been summarized in Figure 3.4 [17, 18].

3.3 Two-dimensional *versus* three-dimensional similarity methods

Both 2D and 3D descriptors have been developed and implemented in similarity search algorithms. Examples of 2D molecular fingerprint algorithms include the Tanimoto and Jaccard coefficients, whereby a chemical graph is decomposed into

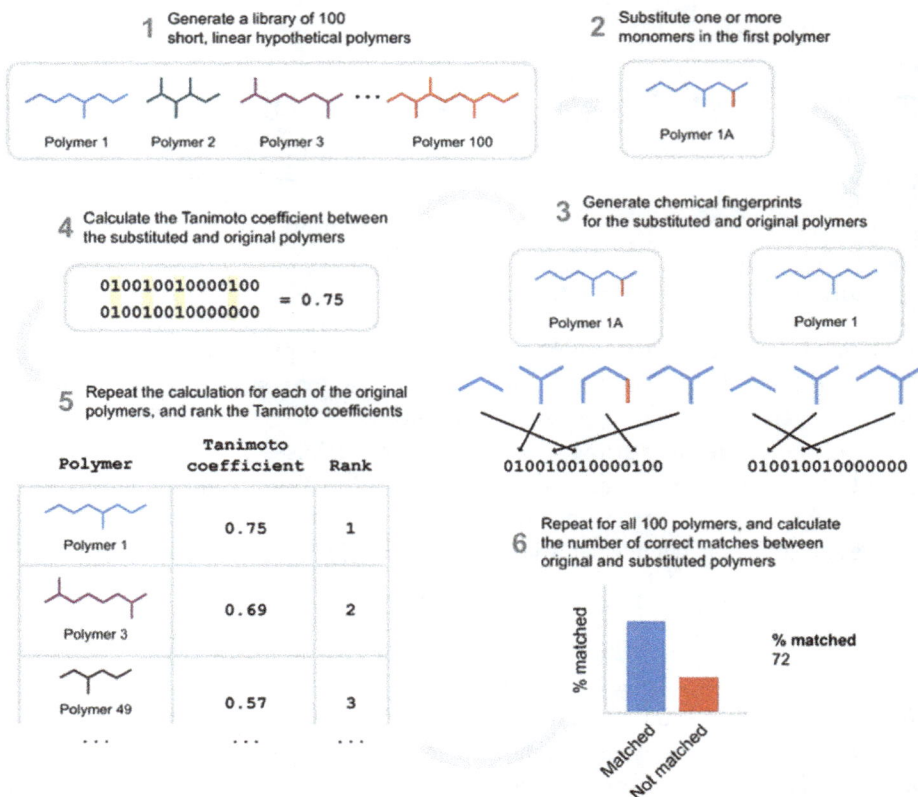

Figure 3.4: Application of the LEMONS algorithm for hypothetical modular NP structure enumeration [18]. The original figure was published under a Creative Commons License.

sets of bits, which are then used for assessing their similarity to each other [14]. Such 2D similarity search methods are used to identify "hits" or similar compounds to a known bioactive to be tested in experimental assays. This is known as virtual screening. In such approaches similar compounds having Tanimoto coefficients > 0.85 are generally accepted and are expected to exhibit similar biological activities. However, such approaches have faced several limitations, e.g. polyenes are known to have very high Tanimoto similarity scores but show quite vast differences in biological activities [20]. Besides, 2D fingerprints methods are limited by the available number of fingerprint, since it is impossible to assign one special bit for each pattern. This implies that two quite different molecules can have the same fingerprint. This may also be due to the fact that the position that corresponds to a given special pattern may explain why that particular pattern occurs. This also implies that when the same pattern occurs multiple times, it will result to the same fingerprint, e.g. molecules such as $C_{20}H_{22}$ or $C_{30}H_{32}$ without features would give the same fingerprint.

However, 2D fingerprint methods indicate whether a compound contains substructures present in another molecule or not [8, 21, 22].

3.3.1 Enantiomers with "high" 2D similarity but dissimilar activities

Another important consideration is the relationship between 2D Tanimoto and "biological activity" in relation to enantiomeric structures. Two enantiomers would be considered as "similar" molecules (with often high Tanimoto scores close to 1) but in real life enantiomers often have dissimilar activities. Experiments show that enantiomers often show different biological activities or the complete absence of activity in one enantiomers and the presence of activity in the other. This is often because enantiomers have distinct spatial (3D) arrangement of their atoms in space, which often affects the way each enantiomeric structure would interact with the macromolecular drug target at the site of action [23].

Since the 3D similarity between two molecules depends on the structural ensembles of the original 3D arrangement of atoms in space, this problem is often solved by the implementation of a conformer generation and clustering algorithms. As such, the parameters such as energy threshold for each compound (e.g. number of rotatable bonds), should be restricted in the 3D similarity approach. This implies that descriptors representing 3D information through conformational searches like docking and pharmacophore-based approaches help to overcome the limitations of 2D approaches [9, 14]. This implies that descriptors representing 3D information through conformational searches are advantageous over 2D methods.

3.3.2 3D-atom pair fingerprints

Several very fast similarity searching search approaches have been implemented for searching large databases, e.g. using various 2D fingerprints that make use of city-block distances as the similarity metric, e.g. 2D atom-pair fingerprints (APfp), extended atom-pair fingerprints (Xfp) [24]. Even though these approaches do not perceive stereochemistry, they do make use of molecular shape and pharmacophores. A 3D atom-pair fingerprints has been recently developed which enables the rapid stereoselective searche of large databases, e.g. ZINC (which contains ~ 23.2 million structures) [25]. This approach works by the design of molecular fingerprints which count atom-pairs at increasing through-space distance intervals. This was designed by the use of either all atoms (16-bit, 3DAPfp) or by use of different atom categories (80-bit, 3DXfp), see Figure 3.5. These 3D fingerprints are used to retrieve molecular shape and pharmacophore analogues and have been shown to have equal or better accuracy when compared with 2D-fingerprints (e.g. APfp and Xfp). Besides, the use of 3DXfp or 3DAPfp similarity searches in ligand-based virtual

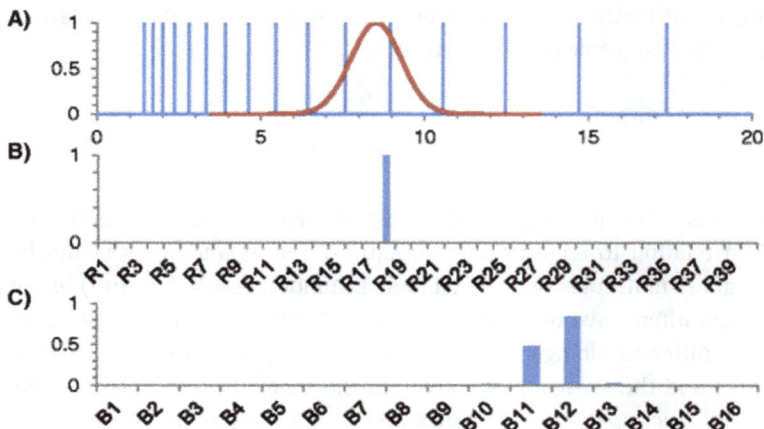

Figure 3.5: 3D-atom pair fingerprint design showing distance sampling for 3D-atom pair fingerprints illustrated for atom-pair distance of 8.51 Å. (A) A Gaussian curve (shown in red) with its maximum value centred at atom-pair distance of 8.51 Å and width as 18% of atom-pair distance. (B) Regular Binning: the atom-pair distance of 8.51 Å produced an increment of 1 in the R18 bin covering the range of 8.5–9 Å. (C) Bit values of B1-B16 for the atom pair at 8.51 Å from the Gaussian/exponential sampling principle in subfigure A. (D) The average bit value and standard deviation (SD) using the two methods applied on two datasets [25]. The original figure published under a Creative Commons License.

screening has proven to be stereoselective, giving very different analogues (as hits) when the starting ligands were diastereomers of the same chiral drug [25]. It must be mentioned that 3DAPfp and 3DXfp are useful when carrying out stereoselective searches for shape and pharmacophore analogs of query molecules contained in large databases, particularly in drug discovery projects. Similar approaches include the use of SMIfp (SMILES fingerprint) [26].

3.4 Similarity search algorithms for phylogenetic relationship between plants based on similarity of secondary metabolites

The metabolite contents of higher plants are mostly composed of structural diverse SMs. SMs are biologically active substances having diverse chemical structures ranging from small monoaromatics to complex tannins and saponins. As earlier mentioned, higher plants combine several of SMs to help them adapt to their environment, e.g. by defending themselves against herbivores and pathogenes, which is one way to depend on the ecological environment within which they thrive. This implies that similarity in metabolite contents of plant species could be useful in the assessment of the phylogenic similarity of higher plants. Structurally similar metabolites have been clustered using similarity search algorithms, e.g. network clustering

D)

R3DAPfp

3DAPfp

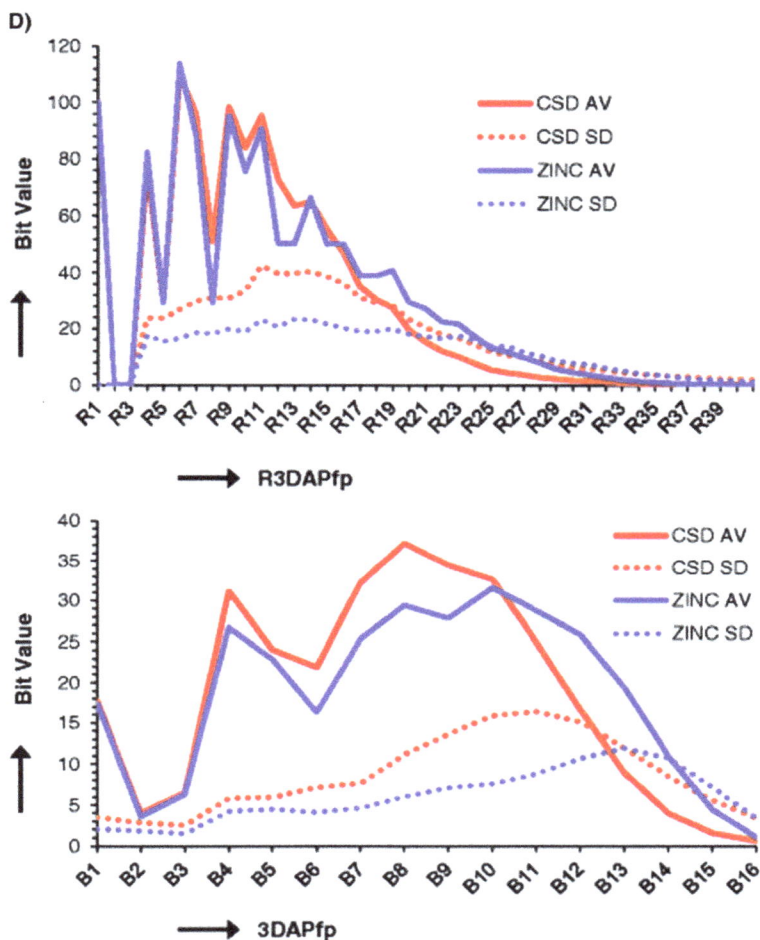

Figure 3.5 (continued)

algorithms like DPClus [27–29]. The DPClus algorithm has been applied to determine the structural similarity network of metabolites. It is a graph-clustering algorithm which divides metabolites into many groups. Each group is composed of compounds that are structurally similar or related. Thus, each group of compounds could be treated as a distinctive pattern of structure. Each plant would then be linked to a metabolite group, i.e. if the plant is related to any compound within a given metabolite group. The original plant-metabolite relationships are then transformed into plant to metabolite group relations [30].

This type of classification establishes the phylogenetic relationship between plants in terms of their potential bioactive properties, and to guide the prediction of medicinal properties of plants [27]. One useful tool is the KNApSAcK Core DB, which helps to establish extensive plant-metabolite relationships within its database and has been used in plant-metabolite research, e.g. to identify plant metabolites, construct integrated databases, and bioinformatics and systems biology [28].

The the DPClus algorithm determines structurally similar groups of metabolites by using binary vectors to link plants to metabolite groups instead of their individual metabolite contents [29]. The guiding principle is that structurally similar metabolites are often involved in or produced through the same metabolic pathway. The use of structurally similar compounds included in metabolite groups helps reduce the effect of missing data, as only a small proportion of plants have been exhaustively studied in terms of their metabolite contents. The similarity in metabolite content is computed based on the binary similarity coefficients, which are then transformed into metabolite-content distances used to compare the similarity between plant species. The classification results often reveal the phylogeny as well as relationships in biological activities among plants.

It must, however, be mentioned that a number of "plant metabolites" reported in the literature and which have been included in such algorithms, are in fact from endophytes/epiphytes and this renders the phylogenetic assignments of hierarchy in the classification of organisms by their chemistry quite complicated. In the marine invertebrate environment, for example, chemotaxonomy does not work. The trend is now being observed to occur in plants. The reader is referred to other packages for the computation of molecular fingerprints, e.g. the R package ChemmineR [30] and support vector machine (SVM) models for classification and regression analysis between plants and their metabolite contents [31, 32].

3.5 Towards the collapse of the concept of "chemotaxonomy"?

Similar SMs are often present within members of a family of related plant species and plants within a taxon often show similarity in metabolite content as well as in the biological properties of their metabolites. The metabolites present in a plant species can be used as "taxonomic markers" or used to distinguish higher plants and from organisms like bacteria, algae and fungi, although exceptions to the rule cannot often be excluded. It must, however, be mentioned that the concept of "chemotaxonomy" or "chemical taxonomy" [33], which consisted in classifying NP-producing organisms by the types of metabolites produced, is increasingly losing its validity. This is due to the realization that a number of previously known "plant metabolites" are actually known to be the products of endophytic and/or epiphytic microbes, as well as plant/microbial associations. In the area of plant SMs, for example, Maytansine (Figure 3.6) is an excellent example that is totally biosynthesized by rhizosphere microbes [34]. Although this antibiotic was isolated originally from the shrub *Maytenus serrata* from Ethiopian [35], it was recently shown that it's biosynthetic gene cluster is from the Actinomycete microorganism *Actinosynnema pretiosum* [36]. Maytansine is known to bind to tubulin at the rhizoxin binding site. The compound inhibits microtubule assembly, induces microtubule disassembly and disrupts mitosis. The compound also exhibits cytotoxic effects against a wide range of tumor cell lines and may even inhibit tumor growth *in vivo* [37]. In the area of marine invertebrate, chemical taxonomy has effectively been laid aside because most SMs are produced by associated microbes.

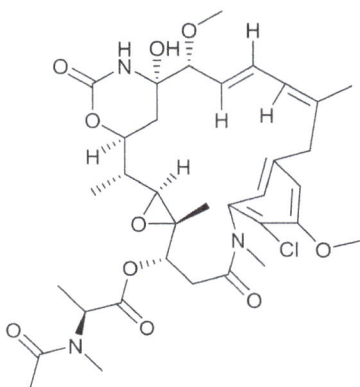

Figure 3.6: Chemical structure of Maytansine.

3.6 Conclusion

Metabolites are small molecules which are often products, bi-products or intermediates of metabolism in living organisms. In particular, the metabolites in plants, microbes and fungi are mostly SMs, which are often highly structurally diverse, and hence difficult to keep track of. Various approximation methods have been developed based on the similarity between chemical structures of descriptor values. These methods describe a compound by a set of numerical values which can be used as a means for comparing it with another compound. The similar property principle states that molecules that are structurally similar have a tendency to have similar properties, including exhibiting similar biological activities. Chemical similarity methods for identifying similar compounds, particularly SM structures have greatly eased the process of analyzing this group of compounds and have helped researchers to keep track of the structures and their properties for easy exploitation from nature. Nowadays, the approach of molecular networking systems (MNS), pioneered by the Dorrestein research group, is aimed at organizing mass spectrometric (MS/MS) data based on chemical similarity [38]. This powerfully complements traditional dereplication methods that require the MS/MS spectra of a NP mixture along with the MS/MS spectra of known standards (including synthetic compounds or compounds from well-characterized organisms). These are currently being organized into robust and useful databases to assist in NP dereplication [39–41]. Apart from helping to dereplicate known molecules from complex mixtures, MNS helps to capture related analogues, rending it more efficient than conventional dereplication strategies [42, 43]. The MNS approach now seems to "upset" metabolite similarity pathways.

Acknowledgements: FNK acknowledge funding from the European Structural and Investment Funds, through the OP RDE-funded project "ChemJets" (Award No. CZ.02.2.69/0.0/0.0/16_027/0008351). FNK also received an equipment donation from the Alexander von Humboldt Foundation, Germany. The technical support of Mme. Bokeng and Mr. Eseme are acknowledged. The reviewers are appreciated for their constructive comments to improve the final manuscript.

References

[1]   Nikolova N, Jaworska J. Approaches to measure chemical similarity - a review. QSAR Combi Sci. 2003;22:1006–26.

[2]   Johnson AM, Maggiora GM. Concepts and applications of molecular similarity. New York: John Willey & Sons, 1990. ISBN 978-0-471-62175–1.

[3]   Martin Y, Kofron J, Traphagen L. Do structurally similar molecules have similar biological activity. J Med Chem. 2002;45:4350.

[4]   Kubinyi H. Similarity and dissimilarity: a medicinal chemist's view. Perspect Drug Discovery Des. 1998;9:225.

[5]   Abegaz BM, Kinfe HH. Secondary metabolites, their structural diversity, bioactivity, and ecological functions: an overview. Phys Sci Rev. 2018. DOI:10.1515/psr-2018-0100.

[6]   Cragg G, Newman D. Natural products: a continuing source of novel drug leads. Biochim Biophys Acta. 2013;1830:3670.

[7]   Bennett R, Wallsgrove R. Secondary metabolites in plant defence mechanisms. New Phytol. 1994;127:617.

[8]   Liu K, Abdullah AA, Huang M, Nishioka T, Altaf-Ul-Amin M, Kanay S. Novel approach to classify plants based on metabolite-content similarity. BioMed Res Int. 2017;2017:296729.

[9]   Bajusz D, Rácz A, Héberger K. Why is Tanimoto index an appropriate choice for fingerprint-based similarity calculations? J Cheminform. 2015;7:20.

[10]   Lo YC, Senese S, Damoiseaux R, Torres JZ. 3D Chemical similarity networks for structure-based target prediction and scaffold hopping. ACS Chem Biol. 2016;11:2244–53.

[11]   Yan X, Liao C, Liu Z, Hagler AT, Gu Q1, Xu J. Chemical structure similarity search for ligand-based virtual screening: methods and computational resources. Curr Drug Targets. 2016;17:1580–5.

[12]   Skinnider MA, Dejong CA, Franczak BC, McNicholas PD, Magarvey NA. Comparative analysis of chemical similarity methods for modular natural products with a hypothetical structure enumeration algorithm. J Cheminform. 2017;9:46.

[13]   Schwartz J, Awale M, Reymond J-L. SMIfp (SMILES fingerprint) chemical space for virtual screening and visualization of large databases of organic molecules. J Chem Inf Model. 2013;538:1979–89.

[14]   Kumar A. Chemical similarity methods - a tutorial review. Chem Educator. 2011;16:1.

[15] Mackay D. Chapter 20, "An example inference task: clustering information theory, inference and learning algorithms. Cambridge University Press, 2003:284–92.

[16] Koulouridi E, Valli M, Ntie-Kang F, Bolzani VS. A primer on natural product-based virtual screening. Phys Sci Rev. 2018. DOI:10.1515/psr-2018-0105.

[17] Sterling T, Irwin JJ. ZINC 15 – ligand discovery for everyone. J Chem Inf Model. 2015;55:2324–37.

[18] Irwin JJ. Using ZINC to acquire a virtual screening library. In: Current protocols in bioinformatics (Suppl. 22) 14.6.1-14.6.23. Wiley Interscience John Wiley & Sons, Inc., 2008. DOI:10.1002/0471250953.bi1406s22.

[19] Atta-ur-rahmann CM. Chemistry and biology of steroidal alkaloids from marine organisms. Alkaloids. 1999;52:233.

[20] Kotler-Brajtburg J, Medoff G, Kobayashi GS, Boggs S, Schlessinger D, Pandey RC, et al. Classification of polyene antibiotics according to chemical structure and biological effects. Antimicrob Agents Chemother. 1979;15:716–22.

[21] Maggiora G, Vogt M, Stumpfe D, Bajorath J. Molecular similarity in medicinal chemistry. J Med Chem. 2014;57:3186–204.

[22] Bender A, Jenkins J, Scheiber J, Sukuru S, Glick M, Davies J. How similar are similarity searching methods? A principal component analysis of molecular descriptor space. J Chem Inf Model. 2009;49:108–19.

[23] Thimm M, Goede A, Hougardy S, Preibner R. Comparison of 2D similarity and 3D superposition. Application to searching a conformational drug database. J Chem Inf Computer Sci. 2004;44:1816–22.

[24] Awale M, Reymond JL. A multi-fingerprint browser for the ZINC database. Nucleic Acids Res. 2014;42:W234–39.

[25] Awale M, Jin X, Reymond J-L. Stereoselective virtual screening of the ZINC database using atom pair 3D-fingerprints. J Cheminform. 2015;7:3.

[26] Schwartz J, Awale M, Reymond JL. SMIfp (SMILES fingerprint) chemical space for virtual screening and visualization of large databases of organic molecules. J Chem Inf Model. 2013;53:1979–89.

[27] Wink M. Evolution of secondary metabolites from an ecological and molecular phylogenetic perspective. Phytochemistry. 2003;64:3–19.

[28] Nakamura Y, Afendi M, Parvin K. KNApSAcK metabolite activity database for retrieving the relationships between metabolites and biological activities. Plant Cell Physiol. 2014;55:e7.

[29] Altaf-Ul-Amin M, Tsuji H, Kurokawa H, Asahi H, Shinbo Y, Kanaya S. DPClus: a density-periphery based graph clustering software mainly focused on detection of protein complexes in interaction networks. J Comput-Aided Chem. 2006;7:150.

[30] Cao Y, Charisi L, Cheng C, Jiang T, Girke T. ChemmineR: a compound mining framework for R. Bioinformatics. 2008;24:1733–4.

[31] Cortes C, Vapnik V. Support-vector networks. Mach Learn. 1995;20:273–97.

[32] Durant JL, Leland BA, Henry DR, Nourse JD. Reoptimization of MDL keys for use in drug discovery. J Chem Inf Comput Sci. 2002;42:1273–80.

[33] Fox HM. Chemical taxonomy. Nature. 1946;157:511.

[34] Smith CR Jr, Powell RG. Plant sources of hepatotoxic pyrrolizidine alkaloids. In: Pelletier SW, editor. Alkaloids, vol. 2. NY: Wiley, 1984:149–204.

[35] Kupchan SM, Komoda Y, Court WA, Thomas GJ, Smith RM, Karim A, et al. Maytansine, a novel antileukemic ansa macrolide from *Maytenus ovatus*. J Am Chem Soc. 1972;94:1354–6.

[36] Yu T-W, Bai L, Clade D, Hoffmann D, Toelzer S, Trinh KQ, et al The biosynthetic gene cluster of the maytansinoid antitumor agent ansamitocin from *Actinosynnema pretiosum*. Proc Natl Acad Sci USA. 2002;99:7968–73.

[37] National Cancer Institute: Definition of Maytansine. https://www.cancer.gov/publications/ dictionaries/cancer-drug/def/maytansine?redirect=true. Accessed: 20 Aug 2019.

[38] Yang JY, Sanchez LM, Rath CM, Liu X, Boudreau PD, Bruns N, et al. Molecular networking as a dereplication strategy. J Nat Prod. 2013;769:1686–99.

[39] Aron AT, Gentry EC, McPhail KL, Nothias LF, Nothias-Esposito M, Bouslimani A, et al. Reproducible molecular networking of untargeted mass spectrometry data using GNPS. Nat Protoc. 2020;15:1954–91.

[40] Kang KB, Ernst M, Van Der Hooft JJ, Da Silva RR, Park J, Medema MH, et al. Comprehensive mass spectrometry-guided phenotyping of plant specialized metabolites reveals metabolic diversity in the cosmopolitan plant family Rhamnaceae. Plant J. 2019;98:1134–44.

[41] Nothias LF, Petras D, Schmid R, Dührkop K, Rainer J, Sarvepalli A, et al. Feature-based molecular networking in the GNPS analysis environment. Nat Methods. 2020;17:905–8.

[42] Gao YL, Wang YJ, Chung HH, Chen KC, Shen TL, Hsu CC. Molecular networking as a dereplication strategy for monitoring metabolites of natural product treated cancer cells. Rapid Commun Mass Spectrom. 2020;34:e8549.

[43] Kuo TH, Huang HC, Hsu CC. Mass spectrometry imaging guided molecular networking to expedite discovery and structural analysis of agarwood natural products. Anal Chim Acta. 2019;1080:95–103.

Part II: **Advanced Chemoinformatics Tools and Methods for Lead Compound Discovery and Development**

Abdulkarim Najjar, Abdurrahman Olğaç, Fidele Ntie-Kang
and Wolfgang Sippl

4 Fragment-based drug design of nature-inspired compounds

Abstract: Natural product (NP)-derived drugs can be extracts, biological macromolecules, or purified small molecule substances. Small molecule drugs can be originally purified from NPs, can represent semisynthetic molecules, natural fragments containing small molecules, or are fully synthetic molecules that mimic natural compounds. New semisynthetic NP-like drugs are entering the pharmaceutical market almost every year and reveal growing interests in the application of fragment-based approaches for NPs. Thus, several NP databases were constructed to be implemented in the fragment-based drug design (FBDD) workflows. FBDD has been established previously as an approach for hit identification and lead generation. Several biophysical and computational methods are used for fragment screening to identify potential hits. Once the fragments within the binding pocket of the protein are identified, they can be grown, linked, or merged to design more active compounds. This work discusses applications of NPs and NP scaffolds to FBDD. Moreover, it briefly reviews NP databases containing fragments and reports on case studies where the approach has been successfully applied for the design of antimalarial and anticancer drug candidates.

Keywords: fragment-based drug design, fragment optimization, molecular modeling, natural products

4.1 Introduction

Fragment-based drug design (FBDD) has been established in the last few decades as an approach for hit identification and lead generation [1, 2]. FBDD is based on screening small sized datasets (containing a few thousand compounds) in order to find low-affinity fragments within the range of micromolar or millimolar concentrations [3]. At lower chemical complexity, there is a higher probability of compounds fitting the receptor binding site even though it may be harder to detect such weak binders. More complex molecules are more likely to have more "clashes" and, thus, do not fit into the receptor binding site [4]. Therefore, the hit rate for screening smaller compounds should in principle be higher than for larger compounds. Starting with small compounds and selecting the most ligand efficient ones leads to the identification of

This article has previously been published in the journal *Physical Sciences Reviews*. Please cite as: Najjar, A., Olgaç, A., Ntie-Kang, F., Sippl, W. Fragment-based drug design of nature-inspired-compounds. *Physical Sciences Reviews* [Online] 2019 DOI: 10.1515/psr-2018-0110

https://doi.org/10.1515/9783110668896-004

more lead-like starting points that enhance the chances of successful optimization campaigns. Thus, by reducing the number of pharmacophores in the initial lead, only necessary interactions are built into the compound as it is being optimized. This should ensure the development of the resulting candidates [3]. Moreover, fragments can more efficiently sample the available chemical space at that level of complexity much more than is possible with more complex molecules [3]. For example, an estimation for the size of the chemical space below 160 Da is ~14 million compounds [5]. Thus, screening a set of fragments of 10,000 compounds captures substantially more chemical diversity than high-throughput screening (HTS), whose chemical space counts billions of molecules [3].

The most comprehensive definition of chemical fragments are reported by researchers at Astex through the "rule-of-3" as small molecules for which the molecular weight (MW) < 300, the number of hydrogen bond donors (HBD) ≤ 3, the number of hydrogen bond acceptors (HBAs) ≤ 3, and the clog P ≤ 3 [6]. Additionally, it was proposed that the number of rotatable bonds should not be higher than 3 and the polar surface area (PSA) should not be higher than 60 $Å^2$.

4.2 Methods for fragment screening

Fragment screening is more challenging than running typical high throughput (HTS) assays. This is due to the small sizes of the ligands and the resulting low binding affinities. Therefore, fragment screening is performed at high concentrations in functional screening assays [7]. Several biophysical methods have dominated the process of fragment screening, thus allowing identification of potential hits. However, non-biophysical methods contribute also to FBDD [8]. Table 4.1 displays the most relevant biophysical methods used for finding fragments, their performance regarding the size of the fragment library and the potency range of the identified hits.

Table 4.1: Biophysical methods for fragments screening.

Fragment screening methods	Description	Size of the library [8]	Affinity ranges in µM [9–11]
NMR [12]	It compares the two-dimensional NMR spectra of the protein in the presence and absence of the fragments. Thus, changes in protein chemical shifts indicate fragment binding. Moreover, the location of the binding site can be determined once the chemical shifts are assigned to specific protein residues.	1,000 compounds	10–2,500

Table 4.1 (continued)

Fragment screening methods	Description	Size of the library [8]	Affinity ranges in µM [9–11]
X-ray crystallography [13–15]	Crystallography can be applied to large proteins and provides very high-resolution data. X-ray crystallography is dominated on FBDD in many companies.	1,000–10,000 compounds	2–17,000
Surface plasmon resonance [16]	This is the most commonly known method often used as a primary screening technique. The protein is immobilized onto a metal-coated chip, then the ligands are allowed to flow past the chip. The bound ligand changes the reflectivity properties of the metals based on the mass of the ligand and the protein.	Up to 10,000 compounds	0.1–1,000
ITC [17]	ITC approach based on the measurement of the heat released when a ligand binds to a protein. This allows the calculation of the enthalpy and entropy of the binding. However, ITC is able to measure energetic parameters with accuracies higher than 0.1 kcal/mol. This technique is less commonly used in fragment screening.	n.d.	0.1–100
MS [18]	MS determines the identities of different complexes, as the mass of the individual molecule serves as the intrinsic detection label. In addition, the MS spectrum provides information about affinity and specificity [19]. MS can be used to detect the fragments that bind to a protein either covalently [20] or non-covalently [21]. This technique is less commonly used in fragment screening.	Large dataset	0.1–100
Fluorescence-based thermal shift assay	It uses an environmentally fluorescent dye to monitor protein unfolding. The affinity of a ligand to a protein can be assessed from the shift of the unfolding temperature obtained in the presence of ligands, compared to that obtained in the absence of the ligands [19]. This technique is less commonly used in fragment screening.	Small datasets in academia	0.1–1,000
Biochemical assays	This includes binding and functional assays [9, 10].	No limit	0.1–100
Interferometry	It relies on a shift in light due to changes in the refractive index and physical thickness of the layer of the protein upon binding to the fragments [22, 23].	Up to 10,000 compounds	0.1–1,000

NMR: Nuclear magnetic resonance; ITC: Isothermal titration calorimetry; MS: Mass Spectrometry.

A fast-growing area is computational FBDD, which represents a non-biophysical method. Computational approaches can contribute to FBDD at different stages. Focusing on fragments screening, computational approaches can be used to screen a large number of fragments. This requires a good degree of fragment diversity and increases the probability to find suitable fragments for a given binding pocket. However, virtual screening of a fragment library using either structure- or ligand-based approach is not able to give appealing binding energies or docking scores due to the relatively small size of the fragments. Therefore, further analysis of the known major protein-ligand interactions can help to detect suitable fragments. Once the suitable fragments have been identified, computational approaches e. g. structure-based design plays a major role to design lead compounds considering different strategies, to be discussed later. Furthermore, molecular docking can be used to predict binding modes of the proposed leads that can reveal protein-ligand interactions and provide insight into the mechanism of receptor activation or inhibition [19, 24].

4.3 Fragment optimization

Once the fragment hits are identified, the process of hit-to-lead generation is performed by applying several strategies, which aim to build additional interactions within a binding pocket. Several factors should be evaluated in order to design highly potent leads. For instance, a robust and efficient crystal system that provides high-resolution crystal structures ($R < 2.5$ Å), clear vectors available for fragments growing to improve the potency. Importantly, the metrics, which are available to aid fragment hit selection and optimization. The most common metrics are ligand efficiency (LE) and ligand lipophilicity efficiency (LLE) used in more of the 80 % and 40 % of the studies, respectively (Source: Practical Fragments October 3, 2016: http://practical fragments.blogspot.co.uk/2016/10/poll-results-affiliation-metrics-and.html).

– **LE**: LE is used to rank fragments and to monitor the progress of the optimization [25–27]. The fragments typically possess higher LE values than HTS-derived hits. LE is usually expressed in kcal/mol. The recommended range of LE for the fragment hits is ≥ 0.3 kcal/mol to be maintained or slightly decreased [25]. However, LE can be calculated using the following equation [28]:

$$\text{LE} = -\frac{\Delta G}{\text{HAC}} = -\frac{RT\ln(K_d)}{\text{HAC}}$$

where HAC is the heavy atoms count.
Simplified as [29, 30]

$$\text{LE} = \frac{1.4(-\log\text{IC}_{50})}{\text{HAC}}$$

where HAC is the heavy atoms count.

– **LLE**: LLE is a metric used to monitor the lipophilicity with respect to the *in vitro* potency of a molecule [25–27]. The goal of optimization is to improve potency without increasing lipophilicity. LLE does not consider the size of the ligand. Therefore, it might be used in the optimization process rather than selecting the top-ranked hits[30]. The recommended range of LLE for the fragment hits is ≥ 3 to be increased to 5–7 [25]. However, LLE is good for developing a working model. The chosen agent still has to penetrate a cell membrane or series of cell membranes. Sometimes active transport will help enable some non-lipophilic molecules to cross cell membranes. The following equation is used to estimate the LLE values [30]:

$$\text{LLE} = p\text{IC}_{50}(\text{or}pK_i) - c\text{Log}P(\text{or}\text{Log}D)$$

In order to prioritize the fragment hits, several attributes should be considered during the selection of the starting fragments for the optimization [31]. The hit rate of the screening methods and access to structural biology have a big impact on the selection process. Selecting hits with LE values of ≥ 0.3 kcal/mol and LLE ≥ 3 can increase the chance to design potent leads. Confidence in the binding mode obtained from crystallography is considered an additional criterion for selection. Chemical expansion vectors available are important to build additional interactions within the binding pocket and to grow the selected fragment. Finally, chemical diversity and synthetic tractability can influence the selection process [31].

4.4 Strategies for fragment optimization

The development of potent leads from fragments consumes more time than other methods since fragments involve additional development work to evolve into leads. However, three optimization strategies can aid to design potent leads. Once the fragments within the binding pocket of the protein are identified, they can be grown, linked, or merged to design potential ligands (Figure 4.1).

4.4.1 Growing

Fragment growing is the most commonly used approach for fragment optimization [25]. The identified fragment is grown up to have preferable interactions with a certain sub-pocket within the binding site. Therefore, several chemical substitutions can be added. The newly generated compounds are supposed to have a higher affinity toward the desired protein. Development of aurora kinase and phosphodiesterase inhibitors based on a fragment growing approach was recently published and represented examples for success stories in FBDD [32, 33].

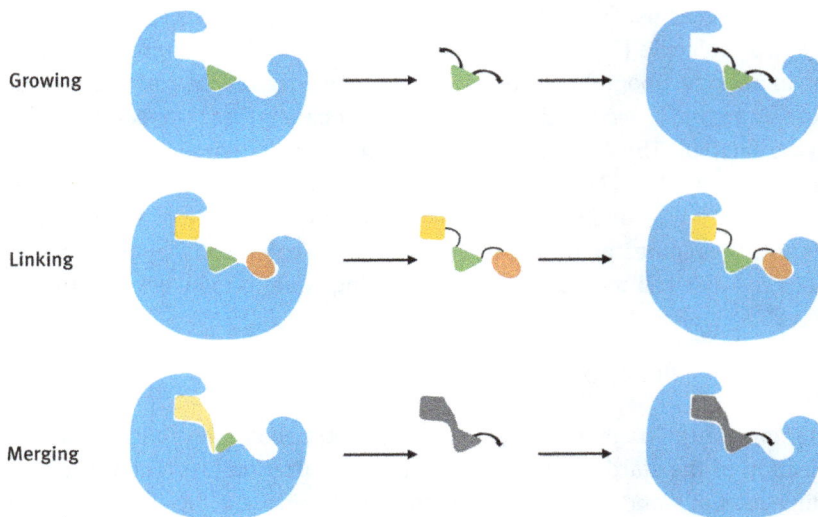

Figure 4.1: Strategies for fragment optimization.

4.4.2 Linking

This is an attractive approach for lead generation because of the rapid increases in the potency caused by super-additivity of fragment binding energies. Fragment linking can be applied to multiple fragments that occupy subregions simultaneously in a protein binding pocket [34]. However, linking two fragments is usually more difficult than to grow a fragment in the binding pocket. This is due to a necessity to maintain the binding modes of the fragments in the derived lead. The discovery of non-peptide stromelysin inhibitors was performed using fragment linking [35].

4.4.3 Merging

It combines information from multiple fragment hits found by screening. These fragments need to display complementary binding modes [10, 34]. Then, the new fragments can be designed that combine their key functional groups. Maintaining the binding modes of the parent fragments considered a big challenge for the fragment merging approach. A successful application of fragment growing was in the design of *Mycobacterium tuberculosis* P450 CYP121 inhibitors [36].

In a previous work, we discussed an application of computer modeling of NPs and NP scaffolds to drug discovery exemplified by a case study of PRK1 kinase inhibitors as potential drugs in prostate cancer treatment [37]. The present work discusses applications of NPs and NP scaffolds to FBDD. Moreover, it briefly reviews

NP databases containing fragments and case studies for two successful developments of antimalarial and anticancer drug candidates. Through them, we aim to demonstrate the performance of NPs with FBDD approaches to the development of new drugs.

4.5 Characterization of NP-related fragments

Natural active substances are useful products of a very long-term natural selection and evolution process that binds to their biological targets. Due to this long-term evolutionary filtering, the structural complexity and physicochemical nature of such molecules show a strong difference compared to synthetic ones. This was analyzed in various studies and the results show that NPs do not always show unwanted or chemically complex features and might be helpful to show the desired effect in drug discovery studies [38, 39].

Newman and Cragg collected new drugs which entered the pharmaceutical market from 1981 to 2017 [40, 41] and categorized them into different major categories; such as vaccines (V), biological macromolecules (B), and small molecules. Small molecules were categorized into original NPs, botanical mixture NPs, NP derivatives, and synthetic drugs (S) which may also contain NP pharmacophore or maybe the mimic of an NP. Here we merged the datasets published in different papers, covering different years. The results show that 31.76 % of all approved drugs contain original NPs (5.34 %) or NP derivatives (25.73 %), Table 4.2. In addition, it should be noted that synthetic drugs include the number of molecules which contain NP pharmacophore or NP-mimics. Such collections highlight the importance of NP-based drug discovery and development efforts. The results indicate that almost every year, NPs and NP-related molecules are entering the market.

Table 4.2: Analysis of the small molecules which entered the pharmaceutical market between 1981–2017 [40, 41].

	Original NPs	NP derivatives	Botanical mixture NPs	Sum of NPs	Synthetic drugs
Numbers	70	337	9	416	894
Percentage	5.34 %	25.73 %	0.69 %	31.76 %	68.24 %

NP-based molecules may have more complex structures and may be outliers of oral bioavailability rules such as Lipinski's Rule-of-5, but they still have the potential to show the desired pharmacokinetic profile, due to the natural selection process [42]. The same flexibility might be possible for the Astex's Rule-of-3 approach [6] (which was originally developed for the FBDD of synthetic molecules) in the development of NP-based drug candidates [43]. However, it is important to point out that Lipinski specifically exempted both NPs and actively transported agents from his Rule-of-5 [40].

4.5.1 Analysis of NP databases regarding the fragment-based definition

Chemical libraries of NPs represent important sources of molecules for both virtual and biological screening. They might be used directly in HTS or for the design of new molecules by fragment-based approaches based on combinatorial chemistry or other rational methods described in Figure 4.1. Various chemical data sets containing NP databases are available, which have been developed with different aims. These databases can be classified into two main categories: virtual databases and collections of readily obtainable NPs [44, 45] and both have free-of-charge or commercial ones inside these categories. Such databases can be used for virtual screening, target fishing, or molecular design of novel chemical NP-inspired mimic [40, 46]. Here several relevant NP databases are given, for detailed information, readers are invited to visit the NP database chapter of this book or the given references.

Chen *et al.* collected and clustered 25 virtual and 31 ready-to-obtain NP databases [44]. Each database was built with different aims such as structure-biological activity records, regional NP databases as botanical records, chemical substance records, natural source-related databases (e. g. plants, animals, or microorganisms), to be used for virtual screening, HTS or as building blocks [44, 45]. They found that over 250,000 unique NPs exist in virtual datasets and around 25,000 at commercial ones. In addition, their molecular similarity-based estimations indicate that around 10,000–30,000 of the compounds stored in chemical databases of the vendors are NP derivatives or analogs. A tenfold gap between the virtual NP libraries and commercial NP (or NP-like) libraries indicates that the commercial NP chemical space is not covering the chemical space of natural compounds. This is related to the difficulties of isolation of chemical substances from natural sources. To further investigate the physicochemical profile of both types of NP databases, the physicochemical properties of the known NPs were calculated and analyzed. The results indicate that the median MWs of the purchasable NPs are lower (267 Da) compared to virtual libraries (424 Da). The same trend can also be observed with the lower medians of cLogP values of the purchasable NPs (2.18) compared to the virtual libraries (2.92). These findings show that NPs are suitable sources for FBDD studies [44]. In a more recent study, researchers continued the analysis of the databases based on the major NP classes. They analyzed the chemical space of 208,166 unique NP compounds (excluding sugars and sugar like, because they rarely affect biological targets) and identified 50,366 unique Murcko scaffolds recorded in the curated virtual NP databases. They identified 25,524 unique compounds and 5,704 unique Murcko scaffolds in the obtainable NP databases. In this analysis, the referenced literature does not specify if the researchers excluded saccharidic antibiotics of the streptomycin class (aminoglycosides) from the detachment of sugars, as it is known that the presence of these sugar moieties is essential for the uptake of aminoglycosides [47]. The researchers continued the analysis of the physicochemical properties of the NP chemical space and they characterized the virtual NP libraries, obtainable NP libraries, and approved drugs. The results of the analysis indicate that the chemical spaces of the approved drugs and approved NP drugs (59 NPs and 320 NP derivatives are included in the dataset) are covered in the obtainable and purchasable NP databases. They also recognized that approved NP drugs contain less phenolic compounds compared to the obtainable NPs. Even though different databases share similar features between different clusters of the whole dataset, 58 % of the obtainable NPs are in fragment-scale MW (higher percentage coverage than virtual NP records [28 %] and drugs [41 %]). The $cLogP$ coverage of the known NPs indicates that virtual NPs are more lipophilic than the obtainable and approved ones. Fragment-featured subcluster of the databases can help the FBDD studies to derive different NP-related chemical compounds.

Known freely available downloadable chemical databases are merged in-house to be able to make further physicochemical analysis by fragment definition within the NP chemical space (with chemically stored molecules within the virtual databases) like those applied by Chen *et al.* AfroCancer [48], AfroDb [49], AfroMalariaDB [50], BIOFACQUIM [51], ConMedNP [52], HIM [53], HIT [54], NANPDB [55], NPs in PubChem [56], NPACT [57], NPASS [58], NPCARE [59], NuBBE [60], p-ANAPL [61], SANCDB [62], StreptomeDB [63], TCM-ID [64], TCMDatabase@Taiwan [65], TIPdb [66], UEFSNP, UNPD [67], and YaTCM [68] databases were used and resulted in 301,052 unique molecules, which are recorded in one database by keeping 2.5D molecular features, if recorded. It is well known that chirality and other stored features may change their biological activity of the final chemical product.

Analysis of the in-house database, based on the Astex fragment definition, showed that there are 63,461 molecules with equal or less MW than 300 Da; 166,239 molecules are found with equal or less than 3 $logP$ value; 82,062 of the molecules are found with less than 60 $Å^2$ of PSA; 206,208 of the molecules with equal or less than 3 HBD, 92,027 HBA; and 105,623 molecules with equal or less than 3 RBs. When the Astex fragment filters are applied together, 16,004 NPs are identified, which can be described as golden NP fragments. The detailed results of the analysis can be found in Figures 4.2 and 4.3.

4.5.2 Fragmentation of NPs

Koch *et al.* generated NP scaffold trees by structural classification of NPs to identify privileged NP scaffolds [69]. The identified fragments are also published and used in various NP-based FBDD campaigns [70]. The algorithm involves the removal of one ring system at each cycle to map frequently occurring NP scaffolds [71]. This led

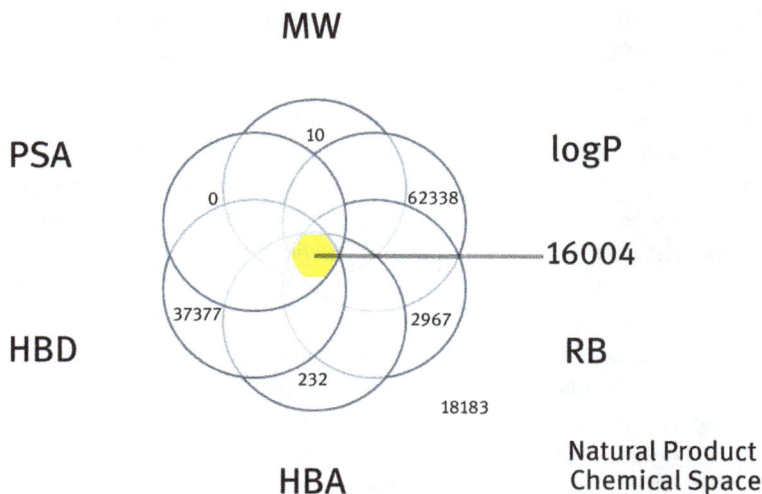

Figure 4.2: Analysis of the known and freely available NP chemical space based on the Astex fragment definitions [6].

to the rational design of new drug candidates carrying unique NP scaffolds unrelated to the volume of the NPs. Physicochemical analysis results can be used in FBDD-based studies. Careful analysis of NP databases may generate fruitful results by integrating with different computational FBDD approaches.

4.6 Case studies

Fragment libraries are increasingly becoming rich sources for the identification of small molecules which could eventually be optimized into lead compounds that constitute starting points for drug discovery [72, 73]. As previously mentioned, the identification of suitable fragments of generation for a focused fragment library is a highly crucial step in any such an endeavor [6]. Although only a few cases have been reported in the literature, recent success stories have been published on NP-based fragments for the discovery of lead compounds [74].

4.6.1 The identification of antimalarial NP fragments

Crowther *et al.* recently reported an extensive study involving the investigation of a small fragment-like NP library [28] of carefully selected compounds following the rules set by Congreve *et al.* [6]. This represented about 650 compounds from diverse collections (including the Dictionary of NPs, DNP, along with the chemical libraries of several compound suppliers). The majority of the derived fragments strictly

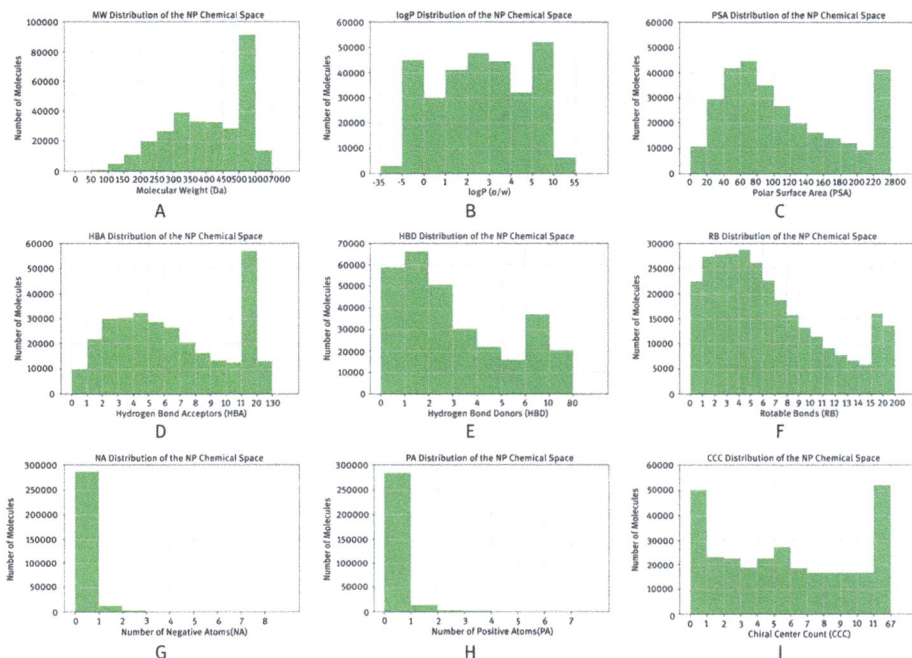

Figure 4.3: (A) Molecular weight (MW), (B) Logarithm of octanol/water partition coefficient (logP[o/w]), (C) polar surface area (PSA), (D) hydrogen bond acceptor (HBA), (E) hydrogen bond donor (HBD), (F) rotatable bonds (RB), (G) negative atom (NA), (H) positive atom (PA), (I) chiral centers count (CCC), (J) nitrogen atom (N), (K) oxygen atom (O), (L) heavy atom count (HAC), (M) ring count (RC), (N) aliphatic ring (ALR), (O) aromatic ring (ARR) distributions within the physicochemical properties of the NP chemical space.

complied by the aforementioned physicochemical property ranges, i. e. MW ≤ 250 Da, octanol-water partition coefficient ($ALogP$) < 4, HBD ≤ 4, HBAs ≤ 5, rotatable bonds (RB) ≤ 6, percent PSA < 45, with only a few fragments violating one of the criteria. The fragment scaffolds were used as queries for the substructure search within the molecular structures of interest. Each query was mapped against all of the molecules to be searched. For each molecule successfully mapped, the number and name of queries mapped were retrieved. These were investigated by HTS against 62 antimalarial drug targets, leading to the identification of about 100 fragment-like NPs binding to 32 of the investigated drug targets. Among these, four-fifths of the compounds inhibited the *Plasmodium falciparum* parasite at concentrations that warrant further investigations toward FBDD [28]. While the HTS results are freely available in ChEMBL [75], some selected fragment scaffolds among the antimalarial HTS hits, as well as the originating fragments and the structures of the active drug-sized molecules are shown in Table 4.3, i. e. molecules that could be developed through FBDD by fragment linking, growing, or merging from these fragments.

Table 4.3: Examples of fragment hit scaffolds present among antimalarial HTS hits [28].

Fragment scaffold	Original fragment	Antimalarial HTS hit molecule	Reference
		CHEMBL602817	[76]
		CHEMBL579557	[76]

[77]

[78]

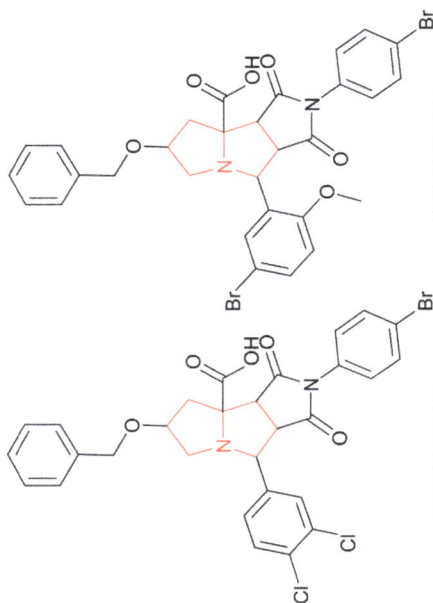

CHEMBL580814

CHEMBL534611

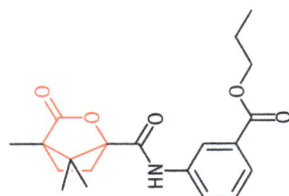

CHEMBL59366

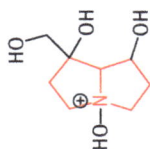

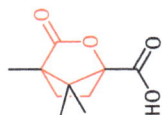

4.6.2 Fragment-based structural optimization of an NP as an anticancer lead compound

In another recent case itampolin A (Figure 4.4), a natural brominated tyrosine secondary metabolite (SM), initially isolated from the marine sponge *Iotrochota purpurea*, was structurally optimized using FBDD [79]. This study reported the synthesis of 45 brominated tyrosine derivatives with interesting p38α inhibitory activities for lung cancer.

The synthesis was guided by a docking/3D-QSAR study aimed at exploring the structural determinants responsible for the activity of brominated tyrosine skeleton p38α inhibitors (Figure 4.4). The lead compound was optimized by FBDD, then three series of brominated tyrosine derivatives were synthesized and evaluated for their inhibitory activities against p38α and tumor cells. The most potent nanomolar range inhibitor (with $IC_{50} = 0.66\,\mu M$) exhibited significant antitumor activity against nonsmall cell lung A549 cells (A549). This study demonstrated the feasibility of the FBDD method of structural optimization of a lead compound from the natural source.

4.7 Challenges and limitations of the NPs by using FBDD

There are several limitations to FBDD from NPs, which could be summarized as:
–   a limited number of NP fragments and samples,
–   challenges in NP synthesis procedures, and
–   very few examples to get inspiration from.

It must be mentioned that not so many samples of NPs are available from compound suppliers when compared with synthetic compounds [41]. This is because the process of isolating a bioactive compound from a natural source often only results to a few milligrams of compound, with limited possibilities of recovering the same compound after repeating the procedure [80]. Besides, a very little proportion of existing NP libraries include compounds that could be described as fragment like [81], following the aforementioned definition [6]. This could explain why some recent examples of NP FBDD employ fragments from the retrosynthetic combinatorial analysis procedure [82], resulting in the fragmentation of existing NP libraries like the DNP [28]. The point is that there is no guarantee of reconstituting an NP-like compound based on fragments from compounds currently included in NP libraries, as the synthetic accessibilities of NPs beginning with fragment libraries are often low [83, 84]. It must be mentioned that, with the presence of multiple chiralities and often fewer polar fragments, the total synthesis of naturally occurring compounds in the wet lab is quite a challenging process and goes way beyond linking molecular fragments together on a computer screen. Naturally occurring compounds have been biosynthesized by a battery of biological pathways, well regulated by biosynthetic gene cluster(s). This is

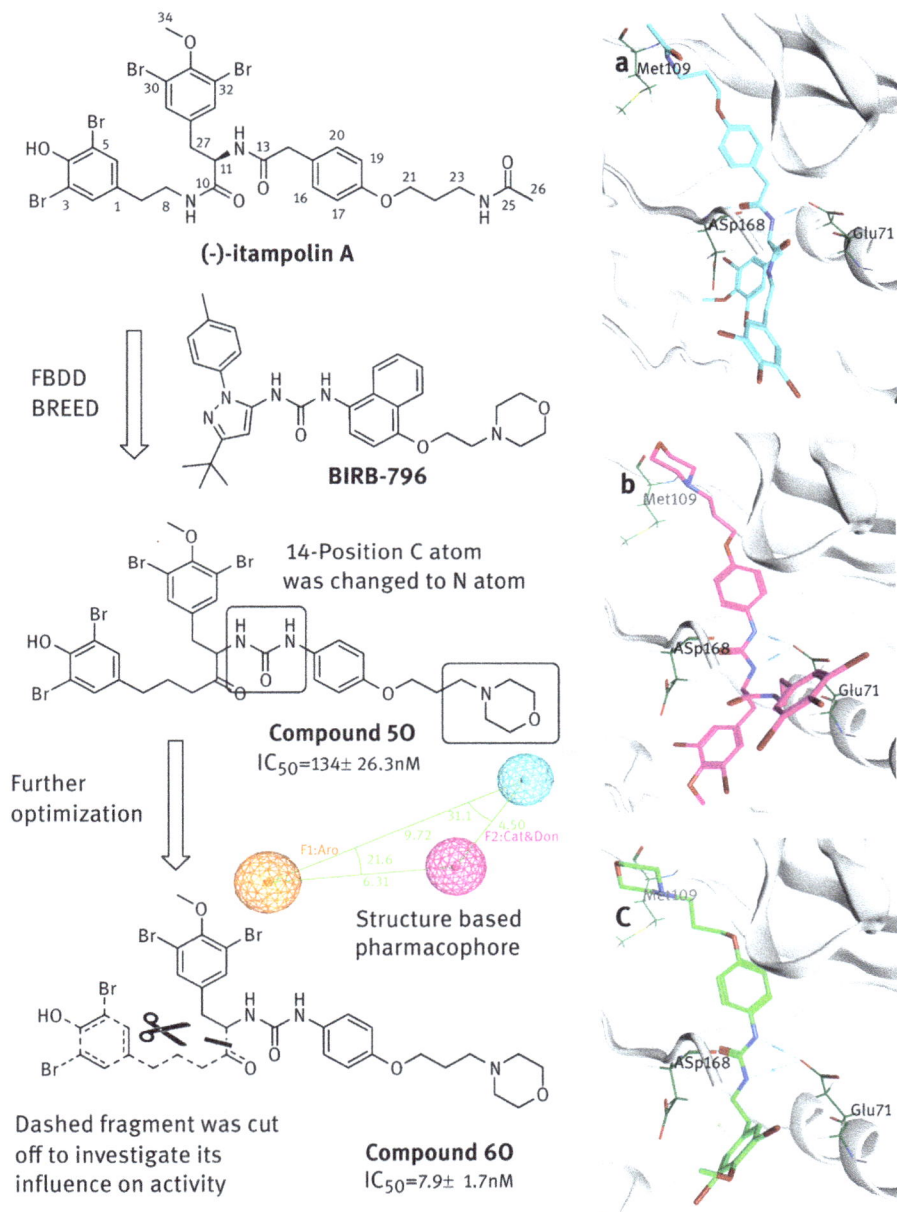

(-)-itampolin A

FBDD BREED

BIRB-796

14-Position C atom was changed to N atom

Compound 50
$IC_{50}=134\pm 26.3nM$

Further optimization

Structure based pharmacophore

Dashed fragment was cut off to investigate its influence on activity

Compound 60
$IC_{50}=7.9\pm 1.7nM$

Figure 4.4: Structural optimization of (–)-itampolin A; (a) the conformation of (–)-itampolin A docked to p38; (b) the conformation of compound 50 docked to p38; (c) the conformation of the most potent compound in the study docked to p38α [79].

often not the case in an organic synthesis laboratory. It is hoped that, with the advancement of metabolic engineering, the synthetic accessibility of NPs from bacterial and fungal sources, and hopefully from higher organisms like plants and marine organisms, would be improved. However, most current cases of NP-inspirational drug design projects rather involve the identification of the fragment including the bioactive pharmacophore in a naturally occurring metabolite and breaking down the (often bulky) metabolite into a structurally simpler compound, which could then undergo structural modifications toward optimization [85, 86].

4.8 Conclusion

It has been the aim of this chapter to vividly describe available tools for drug discovery from NPs using the FBDD approach. FBDD is increasingly becoming important in modern day drug discovery protocols. However, the implication of compounds of natural origin is much limited, when compared with synthetic compounds. It must be mentioned that the natural selection of more complex NP substances than synthetic ones can be helpful for designing novel lead-like or drug-like compounds with FBDD. This chapter has laid the foundation for such endeavors by briefly providing an overview of FBDD procedures, then describing NP databases including fragments, and providing recent successful cases or attempts of FBDD beginning with an entire metabolite (the case of itampolin A, the brominated SM from *Iotrochota purpurea*) and the identification of new antimalarial hits by HTS of small fragments of compounds included in the DNP against a broad array of antimalarial drug targets.

List of abbreviations

*c*Log*P*	The calculated logarithm of *n*-octanol/water partition coefficient
Da	Daltons
FBDD	Fragment-based drug design
HTS	High-throughput screening
LE	Ligand efficiency
LLE	Ligand lipophilicity efficiency
MW	Molecular weight
NP	Natural product
SM	Secondary metabolite

Acknowledgements: The work of AO was partially supported by the Scientific and Technological Research Council of Turkey (Technology and Innovation Funding Programmes Directorate Grant Number 7141231) and EU financial support, received through the cHiPSet COST Action IC1406. FNK acknowledges a Georg Forster return fellowship and equipment subsidy from the Alexander von Humboldt Foundation, Germany. Financial support is acknowledged from a ChemJets Fellowship awarded to FNK from the Ministry of Education, Youth and Sport, Czech Republic.

References

[1]   Baker M. Fragment-based lead discovery grows up. Nat Rev Drug Discov. 2012;12:5.
[2]   Hubbard RE, Murray JB. 20 - Experiences in fragment-based lead discovery. Fragm Based Drug Des Tools Pract Approaches Examples. 2011;493:509–31.
[3]   Hajduk PJ, Greer J. A decade of fragment-based drug design: strategic advances and lessons learned. Nat Rev Drug Discov. 2007;6:211–9.
[4]   Hann MM, Leach AR, Harper G. Molecular complexity and its impact on the probability of finding leads for drug discovery. J Chem Inf Comput Sci. 2001;41:856–64.
[5]   Fink T, Bruggesser H, Reymond JL. Virtual exploration of the small-molecule chemical universe below 160 daltons. Angew Chemie Int Ed. 2005;44:1504–8.
[6]   Congreve M, Carr R, Murray C, Jhoti H. A "Rule of Three" for fragment-based lead discovery? Drug Discov Today. 2003;8:876–7.
[7]   Maly DJ, Choong IC, Ellman JA. Combinatorial target-guided ligand assembly: identification of potent subtype-selective c-Src inhibitors. Proc Natl Acad Sci USA. 2000;97:2419–24.
[8]   Whittaker M, Law RJ, Ichihara O, Hesterkamp T, Hallett D. Fragments: past, present and future. Drug Discov Today. 2010;7:e163–71.
[9]   Bienstock RJ Overview: fragment-based drug design. ACS Symposium Series. American Chemical Society: Washington, DC, 2011.
[10] Lamoree B, Hubbard RE. Current perspectives in fragment-based lead discovery (FBLD). Essays Biochem. 2017;61:453–64.
[11] Alex A, Flocco M. Fragment-based drug discovery: what has it achieved so far? Curr Top Med Chem. 2007;7:1544–67.
[12] Shuker SB, Hajduk PJ, Meadows RP, Fesik SW. Discovering high-affinity ligands for proteins: SAR by NMR. Science. 1996;274:1531–4.
[13] Davies TG, Tickle IJ. Fragment screening using X-ray crystallography. Top Curr Chem. 2012;317:33–59.
[14] Bauman JD, Patel D, Arnold E. Fragment screening and HIV therapeutics. Top Curr Chem. 2012;317:181–200.
[15] Hennig M, Ruf A, Huber W. Combining biophysical screening and X-ray crystallography for fragment-based drug discovery. Top Curr Chem. 2012;317:115–43.
[16] Retra K, Irth H, Van Muijlwijk-Koezen JE. Technologies drug discovery surface plasmon resonance biosensor analysis as a useful tool in FBDD. Drug Discov Today. 2010;7:e181–7.
[17] Freire E. Technologies Drug Discovery Isothermal titration calorimetry: controlling binding forces in lead optimization. Drug Discov Today. 2004;1:295–9.
[18] Hofstadler SA, Sannes-Lowery KA. Applications of ESI-MS in drug discovery: interrogation of noncovalent complexes. Nat Rev Drug Discov. 2006;5:585–95.

[19] Gozalbes R, Carbajo RJ, Pineda-Lucena A. Contributions of computational chemistry and biophysical techniques to fragment-based drug discovery. Curr Med Chem. 2010;17:1769–94.

[20] Erlanson DA, Wells JA, Braisted AC. Tethering: fragment-based drug discovery. Annu Rev Biophys Biomol Struct. 2004;33:199–223.

[21] Vivat Hannah V, Atmanene C, Zeyer D, Van Dorsselaer A, Sanglier-Cianférani S. Native MS: an 'ESI, way to support structure- and fragment-based drug discovery. Future Med Chem. 2010;2:35–50.

[22] Concepcion J, Witte K, Wartchow C, Choo S, Yao D, Persson H, et al. Label-free detection of biomolecular interactions using biolayer interferometry for kinetic characterization. Comb Chem High Throughput Screen. 2009;12:791–800.

[23] Pröll F, Fechner P, Proll G. Direct optical detection in fragment-based screening. Anal Bioanal Chem. 2009;393:1557–62.

[24] Najjar A, Platzer C, Luft A, Aßmann CA, Elghazawy NH, Erdmann F, et al. Computer-aided design, synthesis and biological characterization of novel inhibitors for PKMYT1. Eur J Med Chem. 2019;161:479–92.

[25] Schultes S, De Graaf C, Haaksma EE, De Esch IJ, Leurs R, Krämer O. Ligand efficiency as a guide in fragment hit selection and optimization. Drug Discov Today. 2010;7:157–62.

[26] Bembenek SD, Tounge BA, Reynolds CH. Ligand efficiency and fragment-based drug discovery. Drug Discov Today. 2009;14:278–83.

[27] Hopkins AL, Groom CR, Alex A. Ligand efficiency: a useful metric for lead selection. Drug Discov Today. 2004;9:430–1.

[28] Crowther GJ, Rowley J, Nguyen B, Pouwer R, Pham NB, Andrews KT, et al. Fragment-based screening of a natural product library against 62 potential malaria drug targets employing native mass spectrometry. ACS Infect Dis. 2018;4:431–44.

[29] Shultz MD. Setting expectations in molecular optimizations: strengths and limitations of commonly used composite parameters. Bioorg Med Chem Lett. 2013;23:5980–91.

[30] Leeson PD, Springthorpe B. The influence of drug-like concepts on decision-making in medicinal chemistry. Nat Rev Drug Discov. 2007;6:881–90.

[31] Orita M, Ohno K, Niimi T. Two 'Golden Ratio' indices in fragment-based drug discovery. Drug Discov Today. 2009;14:321–8.

[32] Card GL, Blasdel L, England BP, Zhang C, Suzuki Y, Gillette S, et al. A family of phosphodiesterase inhibitors discovered by cocrystallography and scaffold-based drug design. Nat Biotechnol. 2005;23:201–7.

[33] Early TR, Carr MG, O'Reilly M, Navarro EF, Maman S, Boulstridge JA, et al. Fragment-based discovery of the pyrazol-4-yl urea (AT9283), a multitargeted kinase inhibitor with potent aurora kinase activity. J Med Chem. 2008;52:379–88.

[34] Bian Y, Xie X-Q. Computational fragment-based drug design: current trends, strategies, and applications. Aaps J. 2018;20:59.

[35] Hajduk PJ, Sheppard G, Nettesheim DG, Olejniczak ET, Shuker SB, Meadows RP, et al. Discovery of potent nonpeptide inhibitors of stromelysin using SAR by NMR. J Am Chem Soc. 1997;119:5818–27.

[36] Hudson SA, Surade S, Coyne AG, McLean KJ, Leys D, Munro AW, et al. Overcoming the limitations of fragment merging: rescuing a strained merged fragment series targeting *Mycobacterium tuberculosis* CYP121. ChemMedChem. 2013;8:1451–6.

[37] Najjar A, Ntie-Kang F, Sippl W. Application of computer modeling to drug discovery: case study of prk1 kinase inhibitors as potential drugs in prostate cancer treatment. In Unique Aspects of Anti-Cancer Drug Devevelopment (Jolanta Natalia Latosińska, Magdalena Latosińska). InTechOpen, 2017:18–49.

[38] Ertl P, Schuffenhauer A. Cheminformatics analysis of natural products: lessons from nature inspiring the design of new drugs. In: Petersen F, et al., editor. Natural compounds as drugs. Basel: Birkhäuser Basel, 2008:217–35.

[39] Harris MJ, Thompson JR, Evans JR, Whitt JA, Akee RK, Ewing TL, et al. NCI Program for natural product discovery: a publicly-accessible library of natural product fractions for high-throughput screening. ACS Chem Biol. 2018;13:2484–97.

[40] Newman DJ, Cragg GM. Natural products as sources of new drugs from 1981 to 2014. J Nat Prod. 2016;79:629–61.

[41] Newman DJ. From natural products to drugs. Phys Sci Rev. 2018;14:717–8.

[42] Lipinski CA. Rule of five in 2015 and beyond: target and ligand structural limitations, ligand chemistry structure and drug discovery project decisions. Adv Drug Deliv Rev. 2016;101:34–41.

[43] Laraia L, Waldmann H. Natural product inspired compound collections: evolutionary principle, chemical synthesis, phenotypic screening, and target identification. Drug Discov Today. 2017;23:75–82.

[44] Chen Y, De C, Kops B, Kirchmair J. Data resources for the computer-guided discovery of bioactive natural products. J Chem Inf Model. 2017;57:2099–111.

[45] Chen Y, Garcia De Lomana M, Friedrich NO, Kirchmair J. Characterization of the chemical space of known and readily obtainable natural products. J Chem Inf Model. 2018;58:1518–32.

[46] Olğaç A, Orhan IE, Banoglu E. The potential role of in silico approaches to identify novel bioactive molecules from natural resources. Future Med Chem. 2017;9:1663–84.

[47] Taber HW, Mueller JP, Miller PF, Arrow AS. Bacterial uptake of aminoglycoside antibiotics. Microbiol Rev. 1987;51:439–57.

[48] Ntie-Kang F, Nwodo JN, Ibezim A, Simoben CV, Karaman B, Ngwa VF, et al. Molecular modeling of potential anticancer agents from African medicinal plants. J Chem Inf Model. 2014;54:2433–50.

[49] Ntie-Kang F, Zofou D, Babiaka SB, Meudom R, Scharfe M, Lifongo LL, et al. AfroDb: a select highly potent and diverse natural product library from African medicinal plants. Plos One. 2013;8:e78085.

[50] Onguéné PA, Ntie-Kang F, Mbah JA, Lifongo LL, Ndom JC, Sippl W, et al. The potential of anti-malarial compounds derived from African medicinal plants, part III: an in silico evaluation of drug metabolism and pharmacokinetics profiling. Org Med Chem Lett. 2014;4:6.

[51] Pilón-Jiménez BA, Saldívar-González FI, Díaz-Eufracio BI, Medina-Franco JL. BIOFACQUIM: a Mexican compound database of natural products. Biomolecules. 2019;9:pii: E31.

[52] Ntie-Kang F, Onguéné PA, Scharfe M, Owono Owono LC, Megnassan E, Mbaze LM, et al. ConMedNP: A natural product library from Central African medicinal plants for drug discovery. RSC Adv. 2014;4:409–19.

[53] Kang H, Tang K, Liu Q, Sun Y, Huang Q, Zhu R, et al. HIM-herbal ingredients in-vivo metabolism database. J Cheminform. 2013;5:28.

[54] Ye H, Ye L, Kang H, Zhang D, Tao L, Tang K, et al. HIT: linking herbal active ingredients to targets. Nucleic Acids Res. 2011;39:1055–9.

[55] Ntie-Kang F, Telukunta KK, Döring K, Simoben CV, Moumbock AF, Malange YI, et al. NANPDB: A resource for natural products from Northern African Sources. J Nat Prod. 2017;80:2067–76.

[56] Ming H, Tiejun C, Yanli W, Stephen BH. Web search and data mining of natural products and their bioactivities in PubChem. Sci China Chem. 2013;56:10.

[57] Mangal M, Sagar P, Singh H, Raghava GP, Agarwal SM. NPACT: naturally occurring plant-based anti-cancer compound-activity-target database. Nucleic Acids Res. 2013;41:D1124–9

[58] Zeng X, Zhang P, He W, Qin C, Chen S, Tao L, et al. NPASS: natural product activity and species source database for natural product research, discovery and tool development. Nucleic Acids Res. 2018;46:D1217–22.

[59] Choi H, Cho SY, Pak HJ, Kim Y, Choi JY, Lee YJ, et al. NPCARE: database of natural products and fractional extracts for cancer regulation. J Cheminform. 2017;5:2.

[60] Pilon AC, Valli M, Dametto AC, Pinto ME, Freire RT, Castro-Gamboa I, et al. NuBBE $_{DB}$: an updated database to uncover chemical and biological information from Brazilian biodiversity. Sci Rep. 2017;7:7215.

[61] Ntie-Kang F, Onguéné PA, Fotso GW, Andrae-Marobela K, Bezabih M, Ndom JC, et al. Virtualizing the p-ANAPL library: a step towards drug discovery from african medicinal plants. PLoS One. 2014;9:e90655.

[62] Hatherley R, Brown DK, Musyoka TM, Penkler DL, Faya N, Lobb KA, et al. SANCDB: a South African natural compound database. J Cheminform. 2015;7:1–9.

[63] Klementz D, Döring K, Lucas X, Telukunta KK, Erxleben A, Deubel D, et al. StreptomeDB 2.0 - an extended resource of natural products produced by streptomycetes. Nucleic Acids Res. 2016;44:D509–14.

[64] Xue R, Fang Z, Zhang M, Yi Z, Wen C, Shi T. TCMID: traditional Chinese medicine integrative database for herb molecular mechanism analysis. Nucleic Acids Res. 2013;41:1089–95.

[65] Chen CY. TCM database@Taiwan: the world's largest traditional Chinese medicine database for drug screening in silico. PLoS One. 2011;6:e15939.

[66] Lin Y-C, Wang -C-C, Chen I-S, Jheng J-L, Li J-H, Tung C-W. TIPdb: A database of anticancer, antiplatelet, and antituberculosis phytochemicals from indigenous plants in Taiwan. Sci World J. 2013;2013:736386.

[67] Gu J, Gui Y, Chen L, Yuan G, Lu HZ, Xu X. Use of natural products as chemical library for drug discovery and network pharmacology. PLoS One. 2013;8:e62839.

[68] Li B, Ma C, Zhao X, Hu Z, Du T, Xu X, et al. YaTCM: yet another traditional chinese medicine database for drug discovery. Comput Struct Biotechnol J. 2018;16:600–10.

[69] Koch MA, Schuffenhauer A, Scheck M, Wetzel S, Casaulta M, Odermatt A, et al. Charting biologically relevant chemical space: A structural classification of natural products (SCONP). Proc Natl Acad Sci USA. 2005;102:17272–7.

[70] Prescher H, Koch G, Schuhmann T, Ertl P, Bussenault A, Glick M, et al. Construction of a 3D-shaped, natural product like fragment library by fragmentation and diversification of natural products. Bioorganic Med Chem. 2017;25:921–5.

[71] Schuffenhauer A, Ertl P, Roggo S, Wetzel S, Koch MA, Waldmann H. The scaffold tree - Visualization of the scaffold universe by hierarchical scaffold classification. J Chem Inf Model. 2007;47:47–58.

[72] Scott DE, Coyne AG, Hudson SA, Abell C. Fragment-based approaches in drug discovery and chemical biology. Biochemistry. 2012;51:4990–5003.

[73] Congreve M, Chessari G, Tisi D, Woodhead AJ. Recent developments in fragment-based drug discovery. J Med Chem. 2008;51:3661–80.

[74] Over B, Wetzel S, Grütter C, Nakai Y, Renner S, Rauh D, et al. Natural-product-derived fragments for fragment-based ligand discovery. Nat Chem. 2013;5:21–8.

[75] ChEMBL Neglected Tropical Diseases. http://www.ebi.ac.uk/chemblntd. Accessed: 18 Apr 2019.

[76] Plouffe D, Brinker A, McNamara C, Henson K, Kato N, Kuhen K, et al. *In silico* activity profiling reveals the mechanism of action of antimalarials discovered in a high-throughput screen. Proc Natl Acad Sci USA. 2008;105:9059–64.

[77] Guiguemde WA, Shelat AA, Bouck D, Duffy S, Crowther GJ, Davis PH, et al. Chemical genetics of *Plasmodium falciparum*. Nature. 2010;465:311–5.

[78] Gamo FJ, Sanz LM, Vidal J, De Cozar C, Alvarez E, Lavandera JL, et al. Thousands of chemical starting points for antimalarial lead identification. Nature. 2010;465:305–10.

[79] Liang J, Wang M, Wang S, Li X, Meng F. Fragment-based structural optimization of a natural product itampolin A as a p38α inhibitor for lung cancer. Mar Drugs. 2019;17:53.

[80] Bucar F, Wube A, Schmid M. Natural product isolation – how to get from biological material to pure compounds. Nat Prod Rep. 2013;30:525.

[81] Lucas X, Gru A, Bleher S, Gu S. The purchasable chemical space: a detailed picture. J Chem Inf Model. 2015;55:915–24.

[82] Carreira EM. Natural products synthesis: a personal retrospective and outlook. Isr J Chem. 2018;58:114–21.

[83] Lear MJ, Hirai K, Ogawa K, Yamashita S, Hirama M. A convergent total synthesis of the kedarcidin chromophore: 20-years in the making. J Antibiot. (Tokyo). 2019. DOI: 10.1038/s41429-019-0175-y.

[84] Lewell XQ, Judd DB, Watson SP, Hann MM. RECAP: retrosynthetic combinatorial analysis procedure : a powerful new technique for identifying privileged molecular fragments with useful applications in combinatorial chemistry. J Chem Inf Comput Sci. 1998;2338:511–22.

[85] Ertl P, Schuhmann T. A Systematic cheminformatics analysis of functional groups occurring in natural products. J Nat Prod. 2019. DOI: 10.1021/acs.jnatprod.8b01022.

[86] Zhu XL, Zhang R, Wu QY, Song YJ, Wang YX, Yang JF, et al. Natural product neopeltolide as a cytochrome bc 1 complex inhibitor: mechanism of action and structural modification. J Agric Food Chem. 2019;67:2774–81.

Fatima Baldo

5 Prediction of modes of action of components of traditional medicinal preparations

Abstract: Traditional medicine preparations are used to treat many ailments in multiple regions across the world. Despite their widespread use, the mode of action of these preparations and their constituents are not fully understood. Traditional methods of elucidating the modes of action of these natural products (NPs) can be expensive and time consuming, e. g. biochemical methods, bioactivity guided fractionation, etc. In this review, we discuss some methods for the prediction of the modes of action of traditional medicine preparations, both in mixtures and as isolated NPs. These methods are useful to predict targets of NPs before they are experimentally validated. Case studies of the applications of these methods are also provided herein.

Keywords: target prediction, mode of action, natural products

5.1 Introduction

Humans have used traditional medicines (TMs) since before written history began, ranging from the Neanderthals in the Palaeolithic era [1], the Sumerians in Mesopotamia [2], Ancient Egyptians [2], to India since 4000 BC [3] and China since 2000 BC. TM is defined as "the sum total of knowledge, skills, and practices based on the theories, beliefs, and experiences indigenous to different cultures that are used to maintain health, as well as to prevent, diagnose, improve or treat physical and mental illnesses" [1]. TMs consist of: (a) entire organisms e. g. plant or animal, (b) part of an organism e. g. leaf or gland, (c) extracts, (d) exudates, (e) pure compounds or (f) venoms and toxins [4].

Mode of action (MoA) analysis comprises the study of biochemical and physiological modes by which compounds or drugs elicit a response. This step is important for elucidating the MoA of a natural product (NP). The importance of elucidating the MoA of NPs is two-fold. First, knowing the target, and hence the pathway that these NPs modulate validates the use of the NPs by the herbalists and will inform relevant authorities in the regulation of their use. MoA information allows medicinal chemists to understand the side effects and toxicity of the NPs [5]. Here we will discuss the main approaches to target identification (and hence MoA prediction), namely genetic

This article has previously been published in the journal *Physical Sciences Reviews*. Please cite as: Baldo, F. Prediction of modes of action of components of traditional medicinal preparations. *Physical Sciences Reviews* [Online] 2020 DOI: 10.1515/psr-2018-0115

https://doi.org/10.1515/9783110668896-005

interaction methods and computational inference, including network methods. See Figure 5.1 for a summary of the different methods.

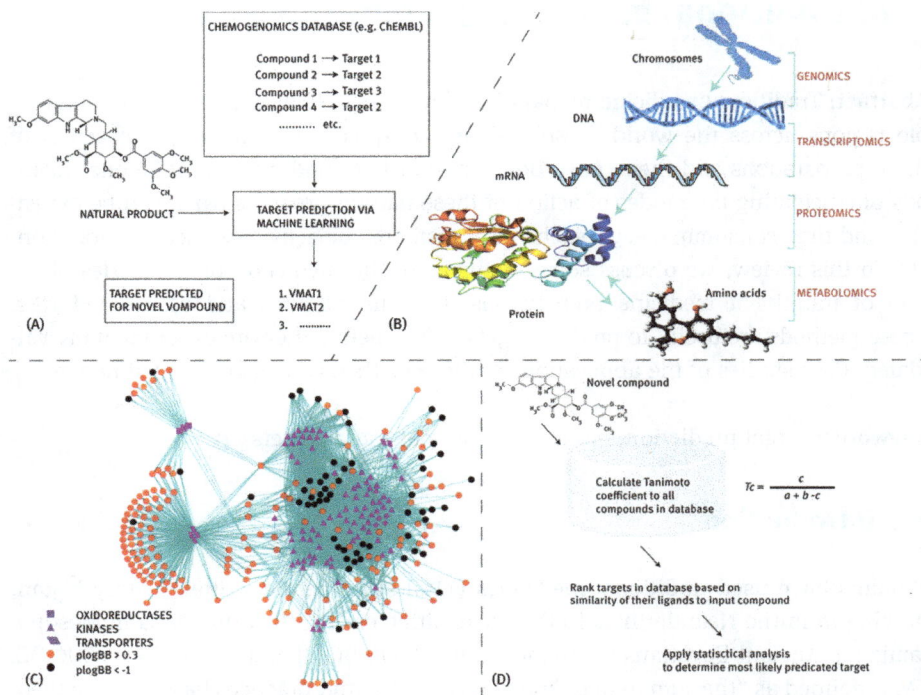

Figure 5.1: Determining the mode of action of natural product preparations. (A) An example of target prediction using data from a chemogenomic database. The machine-learning algorithm (e. g. SVM, Random Forest, etc.) is trained on both target and protein features to predict the most enriched targets. (B) A schematic illustrating the stages during which the various – omics technologies are used to predict targets. (C) A network of targets and compounds. Targets are shown in purple and compounds are shown in circles (red – for compounds that readily cross the BBB and back – compounds that do not cross the BBB). This is adapted from [5]. Networks can be used to visualise compound target interactions as well as predict the effect of targeting specific proteins. (D) A schematic of similarity searching to predict MoA. The Tanimoto coefficient is calculated between a novel compound and compounds in a chemogenomic dataset. Targets associated with the most similar compounds in the dataset are ranked to determine the most likely predicted target.

5.2 Genetic interaction methods

These methods involve altering the functions of putative targets by, for example, gene knockout, RNAi or small molecules [6, 7]. This allows for a target hypothesis to be generated. Furthermore, it is possible to predict activity using -omics methods, i. e.

proteomics [8], transcriptomics [9] and metabolomics [10]. A brief explanation of these methods along with examples of how they have been used to predict the activity of NPs is shown in Table 5.1.

Table 5.1: Omics technologies used in the prediction of the mode of action of natural products.

Technology	Definition	Example of use in MoA prediction
Genomics	Genomics analysis involves applying genetics and molecular biology techniques to the genetic mapping and DNA sequencing of sets of genes or the complete genomes of selected organisms.	Chemical genomics was used to identify Elongation factor 2 as the molecular target of girolline (isolated from the marine sponge *Stylissa aff. Carteri*), which inhibits signalling through both MyD88-dependent and – independent toll-like receptors and reduces cytokine production in human peripheral blood mononuclear cells and macrophages [11].
Proteomics	Proteomic analysis is used to quantify the expression of proteins involved in biological conditions under study. These methods can be used to determine post-translation modifications, cellular origins and location of activity within the system.	Proteomics study determined the MoA of Roemerine (an alkaloid from *Papaver rhoeas*) against *E. coli*, to be associated with the inhibition of membrane permeability and sugar transporter proteins [12]. Proteomic studies revealed anti-cancer MoA of marine anticancer compound rhizochalinin to be associated downregulation of isoforms of stathmin (regulates microtubule dynamics) and LASP1 (regulates cytoskeletal activities), in addition to the upregulation of Grp75, keratin 81, and precursor IL-1β [13].
Metabolomics	Metabolomics methods are used to study the effect of drugs, toxins, etc. on the global metabolite profile of the biological system under a given set of conditions.	Intracellular metabolic profile via HPLC ESI/MS of revealed that the MoA of the extract and main constituents of *Tinospora capillipes* treated *Staphylococcus aureus* to be similar to that of rifampicin and norfloxacin i. e. it is predicted to inhibit RNA polymerase, gyrase and topoisomerase IV [14].

Table 5.1 (continued)

Technology	Definition	Example of use in MoA prediction
		Comparing the metabolic profile of *Staphylococcus aureus* treated with Dihydrocucurbitacin F-25-*O*-acetate (from *Hemsleya pengxianensis*) and known antibiotics, revealed the antibacterial MoA of Dihydrocucurbitacin F-25-*O*-acetate to be via the inhibition of cell wall synthesis [15].
Transcriptomics	Transcriptomics (global gene expression profiling) is used to study the effect of drugs, small compounds, etc., on the gene expression levels of 1000s of genes in parallel. The underlying assumption is that genes involved in common processes are co-expressed and their patterns of expression correlate to the physiological status of the cell.	A microarray assay was used to evaluate the transcriptional profile of *S. aureus* treated with licochalcone A (chalcone isolated from *Glycyrrhiza inflate*. Genes related to autolysis, cell wall, pathogenic factors, protein synthesis, and capsule synthesis showed altered expression profiles [16]. Transcript profiling studies revealed the hypolipidemic and anti-apoptotic MoA of pterostilbene (naturally occurring dimethylether analogue of resveratrol) The expression of genes involved in methionine metabolism were downregulated, while the expression of genes involved in the induction of mitochondrial functions, drug detoxification, and transcription factor activity were significantly up-regulated [17].

5.3 Computational inference

Here, pattern recognition is used to compare the effects of the tested compounds to those with known and validated activities. Hypotheses are made about targets/pathways, but the results remain to be experimentally validated. Thus combining computational inference methods with direct measurements is a good approach for target convolution for MoA analysis [18].

It is important to remember that these methods detect interactions between ligands and targets and not the actual MoA of the ligands. Once a ligand is bound to a target/receptor it can act in a number of different ways to elicit a biological response, including but not limited to activating the receptor (agonist), blocking or reducing the biological response of the receptor (antagonist), or binding to the same receptor as an agonist but producing a pharmacological response opposite to that agonist (inverse agonist).

5.3.1 Ligand-based target prediction

There are several chemo-informatic approaches to investigating the potential targets of a NP. These can be broadly divided into three categories, based on information used, into single ligand-based, multiple ligand-based and ligand-target based. Single-ligand MoA studies include molecular similarity modelling and pharmacophore modelling. Multiple-ligand approaches include machine learning and quantitative structure-activity relationship (QSAR) modelling. Target–ligand approaches to MoA studies include proteochemometrics and docking studies. These methods are explored further in Table 5.2.

5.3.2 Machine learning in ligand-based target prediction

Machine-learning approaches are currently one of the most important in computer-aided drug discovery [29]. Machine-learning techniques use pattern recognition algorithms to detect mathematical relationships between empirical observations of small molecules [30]. The relationships are extrapolated to predict chemical, biological and physical properties of new compounds. Machine learning is also used to understand and exploit the relationship between chemical structures and their bioactivities [31]. Numerous target prediction approaches have been published, e. g. SEA (Similarity Ensemble Approach) [32] and PASS (Prediction of Activity Spectra for Substances) [33], that predict biological targets of a query ligand, including NPs. These methods rely on training models on bioactivity information of the ligands obtained from databases. Some of the open-source databases include PubChem [34], ChEMBL [35] and WOMBAT [36].

SEA is a molecular similarity method that quantitatively relates proteins to one another based on the chemical similarity of their bound ligands [32]. This method was used to predict the anti-malarial activity of the physalins B, D, F and G (isolated from *Physalis angulata*), with B, F, and G subsequently showing IC_{50} values of 2.8 μm, 2.2 μm and 6.7 μm (respectively) against *Plasmodium falciparum* [37].

Another tool that allows targets to be predicted based on the Tanimoto similarity of ligands to ligands associated with the target is TargetHunter [38]. TargetHunter uses the Targets Associated with its MOst Similar Counterparts (TAMOSIC) algorithm to predict the biological targets of query ligands. When a query compound is an input into TargetHunter, the TAMOSIC algorithm generates the fingerprints of the

Table 5.2: *In-silico* methods for NP target identification. Examples of their use to identify targets of NPs are shown in the notes column.

Method	Notes	Limitations
Molecular similarity	Molecular fingerprints of an input ligand are compared to those of ligands known to modulate the studied target. This method is based on the "similarity principle" where ligands with similar structures are predicted to bind similar targets [19].	– Does not take into account any information about the activities of known modulators of the target – Molecule descriptors describe different classes of compounds differently – Compounds with similar structures may have different activity (activity cliff) [20].
Pharmacophore modelling	Pharmacophores are the essential molecular features of ligands that are responsible for biological or pharmacological interactions. Pharmacophore models are built based on pharmacophore features of ligands known to bind a studied target. The method is discussed in detail in [21]. This method was used to identify the NP solanidine as a potent sigma-1 receptor ligand [22]. This method was also used to identify the targets of secondary metabolites from *Ruta graveolens* [23].	– Dependent on the pre-computed conformation database – Absence of good scoring metrics – No clear way to construct a pharmacophore query i. e. different pharmacophores can be created for similar targets
QSAR	This method involves calculating the correlation between the properties of the ligand, e. g. physicochemical properties and the experimentally validated biological activity of the various studied drug targets. The method is discussed in detail in [24]. This method has been used to identify the BH3-binding groove of the Bcl-2 protein to be the target of lonchocarpin, a chalcone isolated from *Pongamia pinnata* [25]. 3d-QSAR was used to identify the glycosomal GDPH of *Trypanosoma cruzi* as the target of the coumarin, chalepin from Rutaceae species [26].	– Success depends on selected molecular descriptors – The model must be trained on a dataset with enough activity data to be able to extract patterns

Proteo-chemometric Modelling (PCM)	PCM uses statistical modelling techniques to model ligand–target interactions. In PCM the ligand-descriptor matrix used for training the model in QSAR is extended to include protein descriptors [27]. There are currently no known studies using this method to identify targets of natural products.	– Success depends on selected molecular descriptors – The model must be trained on a dataset with enough activity data to be able to extract patterns
Docking	This method uses scoring functions to rank predictions of the predominant binding mode of a ligand with a target (protein) of known three-dimensional structure [28].	– The 3D structure of the target must be known – Scoring functions are not uniform

compound (chosen by the user, any of ECFP6, ECFP4, and ECFP2) and compares the Tanimoto similarity of this compound to compounds in a chemogenomics database, ChEMBL-11. Targets with the most similar compounds to the query compound are output as the predicted targets and ranked according to the similarity scores of the ligands to the input compound. TAMOSIC was trained on 117,535 unique compounds from ChEMBL and 794 targets. TargetHunter obtained 91.1 % prediction accuracy of the top three targets, i. e. 91.1 % of the compounds are assigned to their known targets in the top three predictions. TargetHunter was used to identify the MoA of compound CID46907796 from the PubChem database. This compound was reported in PubChem to display cellular apoptosis with AC_{50} values of 0.4136 and 4.908 µM, but the MoA of this compound was not known. TargetHunter predicted the nuclear factor erythroid 2-related factor 2 (Nrf2), which is known to have anti-apoptotic activity [39] as a likely target due to the Tanimoto similarity score of 0.78 and 0.63 to compounds in the dataset.

A machine-learning method, PASS, predicts the biological activity of a compound based on its structure [33]. The principle underlying this method is that biological activity equates to structure. This method has been used to predict the antioxidant and anti-microbial activity of the acetogenin alkaloid, neoannonin, from an extract of *Annona reticulata* [40]. It has also been used by Goel et al. [41] to predict the MoA of NPs in five plants. The authors generated a prediction coefficient, P, which calculates the number of activities predicted by PASS over the number of reported activities for the compound. An average prediction coefficient of 0.66 for the five compounds led to the conclusion by the authors that PASS can be applied to predicting the MOAs of NPs.

Another machine-learning target prediction method, used by Nidhi et al. [42], uses a Laplacian-modified Bayes model to predict biological targets of compounds in the MDDR Database. For *every* target class in the database, the authors built a Laplacian-modified Naïve Bayesian model. A query ligand is passed through each Laplacian-modified Naïve Bayesian model of each target class. The relative estimator score for each of the target classes is calculated. The most probable predicted target for that query compound is the target with the highest score. The model was trained on 103,735 compounds annotated to 964 target classes from World of Molecular BioAcTivity (WOMBAT) [36]. It was used to predict the top 3 most likely targets for compounds from the MDL Drug Database Report (MDDR) dataset [43]. The model predicted 77 % correct targets for compounds from 10 target classes in MDDR.

Self-organising maps (SOMs) have been used to predict ligands of targets as well as selective ligands of targets. Self-organising maps are a type of artificial neural network developed by Kohonen [44]. SOMs learn through unsupervised learning. They are essentially a clustering technique used to visualise similarity in the data where the geometric similarity between nodes indicates similarity. This method of SOM was used by Schneider et al. [45] to correctly predict prosanoid E receptor 3 as a target for the anti-cancer NP Doliculide. The authors trained the SOM model, on COBRA data, which contains 4,236 drugs and drug candidates [46]. They computed the p-values based on the background distribution of known ligands and drugs to rank the predicted targets [47].

Huang et al. [48] proposed the MOst-Similar ligand Target (MOST)-based approach to predict targets. This method incorporates the explicit bioactivity of the most similar ligand. The method is also able to remove false positive predictions due to incorporating p-values associated with explicit bioactivity information as an index. The method involved training a combination of different machine-learning algorithms including Naïve Bayes, Logistic Regression and Random Forest using compounds characterised by ECFP-4 Morgan-like fingerprints and FP-2 fingerprints. The model was trained on 61,937 compounds annotated with 173 targets from ChEMBL-19 and validated used seven-fold cross-validation. The dataset comprised 91.3 % active compounds and 8.7 % inactive compounds. The algorithm worked by calculating the Tanimoto similarity between the input compounds and annotated ligands of the targets. The Tanimoto similarities were then ranked and the most similar ligand was chosen. The Tanimoto similarity and pK_i ($-$log dissociation constant) of the most similar ligand were fed into the training model to generate the probability of how likely the input compound is to be inactive. If the probability of being active is greater than the probability of being inactive, then the query compound is classified as active and vice-versa. MOST was able to identify the MoA of aloe-emodin by predicting acetylcholinesterase as the target, which was validated *in vitro*. MOST was also able to predict novel targets for the drug Fluanisone (not in ChEMBL), where MOST correctly predicted adrenoreceptor alpha 1B and adrenoreceptor alpha 1D as the second and third most likely targets. These targets were validated by literature to be human targets of Fluanisone.

Compound activity mapping [49] is a method used to predict the MoA of NPs from complex NP libraries. This cytological profiling tool integrates data from image-based screening platforms [50–53] and metabolic data from an extract library to compare the bioactivity landscape of NPs to those in a training set with known MoAs [54, 55]. By correlating the mass signals with specific phenotypes from a high-content cell-based screen the authors were able to identify four new quinocinnolinomycins and predict their MoA to induce endoplasmic reticulum stress and protein unfolding.

Case study: Fruits of the plant *Psorospermum aurantiacum*, Family Hypericaceae are used in Cameroon and other parts of Africa for the treatment of cancer as well as gastrointestinal and urinary tract infections, skin infections, venereal diseases, gastrointestinal disorder, infertility, epilepsy and microbial infections [56]. PIDGINv2 [57], a random forest target prediction algorithm, was used to predict the targets of the compounds extracted from the plant [5]. Results for the target predictions for the five NPs isolated and characterised from *P. aurantiacum* [58] are shown in Figure 5.2.

Compound 1 is predicted to bind to protein kinase C gamma, and it has previously been shown to be linked to keratosis. Neutrophilic cutaneous infiltrates are

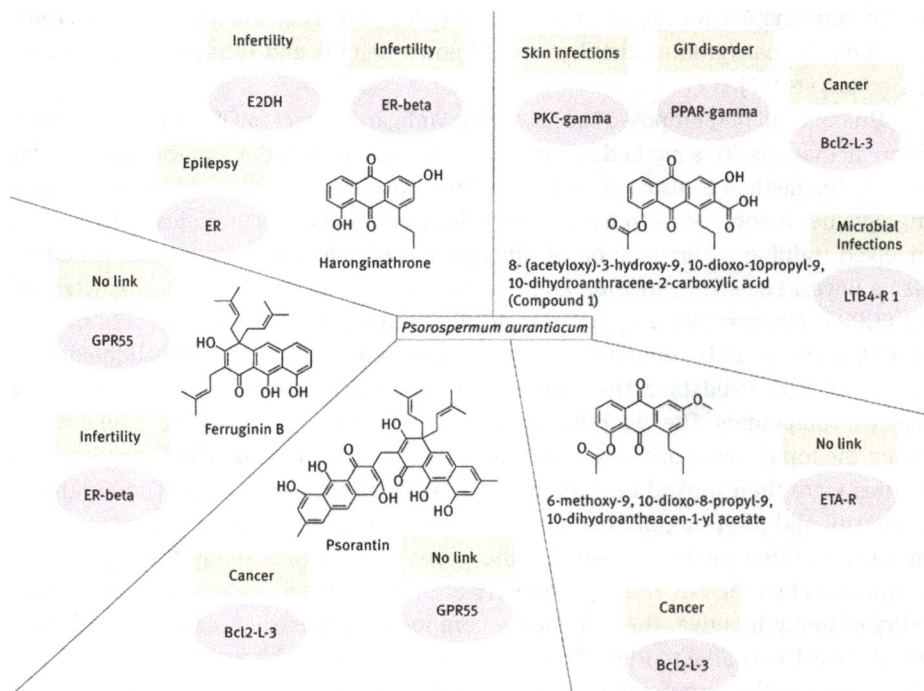

Figure 5.2: The targets were predicted for compounds isolated from *Psorospermum aurantiacum*. Targets predicted for each compound are shown in purple ellipses. Disorders for which the plant is traditionally used and are associated with the target are shown in the orange boxes.

produced as a result of activating Protein kinase C, and they act to prevent relapse of the tumour by mediating antibody-dependent cellular toxicity against the tumour cells [59, 60]. It thus appears that the activity of Compound 1 against protein kinase C gamma may be responsible for its utilization for treating skin infections. At the same time, Compound 1 is also predicted to bind to Leukotriene B4 receptor 1 and Induced myeloid leukaemia cell differentiation protein Mcl-1. Elevated levels of Leukotriene B4 receptor 1 have been found in a number of inflammatory diseases [61], so inhibition may explain the anti-inflammatory activity associated with this plant. Apart from annotated indications, an experimental leukotriene B4 inhibitor was found to inhibit proliferation and induce apoptosis in pancreatic cells [62], and suppression of induced myeloid leukaemia cell differentiation protein Mcl-1 is known to induce apoptosis [63], so this compound may have the potential to also act as an anti-tumour agent. The targets predicted for Haronginanthrone can explain the seemingly unrelated annotated bioactivity of this plant in treating epilepsy and infertility. This compound was predicted to bind to the oestrogen receptor and oestrogen receptor beta, which play an important role in cancer [64, 65] and catamenial epilepsy [66]. More proof that oestrogen receptors play a role in epilepsy is shown by the fact that reproductive dysfunction is associated with epilepsy [67] as well as anti-epileptic therapy [68]. A literature review revealed that oestrogen receptor knockout mice have been shown to exhibit infertility as well as reduced fertility [69]. The fact that *P. aurantiacum* is used to treat infertility indicates that the compounds may bind to oestrogen receptor β and act as agonists. Looking at the bioactivity profiles of the compounds in *P. aurantiacum* and similar compounds in the ChEMBL database the authors found that these compounds are predicted to have activity in a variety of cell lines and targets e. g. Ferruginin C and Vismin (> 80 % similar to compounds in *P. aurantiacum*) have reported activities against cancer lines including MCF-7 [70]. Target prediction allowed the authors to develop a plausible MoA hypothesis for this plant, despite the rather dissimilar indications for which it is being used.

5.3.2.1 Applicability domain of target prediction models to NP chemistry

It is important to note that the accuracy of predictions of a model is only as good as the training set, i. e. the accuracy of the model beyond the training set (outside the applicability domain) cannot be guaranteed. Several limitations have been highlighted by Kalliokoski et al. [71] and these are as follows:
1. Chemical structure related errors;
2. Transcription errors;
3. Inaccurate and insufficient target annotations;
4. Ineffective and incomplete archiving of original data;
5. Redundancy.
6. Different lab conditions with different protocols for measurement.

When extrapolating the predictions of a QSAR model to compounds outside of the training set, it is possible to get good predictions for compounds that are relatively similar to the training set [72]. Predictions may fail for those compounds that are very different from those in the training set [72]. This concept is known as the applicability domain of a model and is usually defined using the similarity of molecular structures or a similarity measure based on descriptors of the compounds [73]. Several chemoinformatic analyses comparing the properties of different sets of NPs and synthetic compounds [74–76] found that NPs differ considerably from synthetic compounds in several molecular properties. They found that NPs on average tend to have: more oxygen atoms, fewer nitrogen atoms, more stereogenic centres, more fused rings, but fewer aromatic rings and fewer rotatable bonds. These models also use fingerprint similarity to compare training sets to the test compounds. This is not the best representation since fingerprints lead to a loss of atom order as well as the fact that they do not capture many aspects important for NP activity, e. g. repeat units, stereochemistry, etc., as mentioned above. This leads to one of the limitations of utilising ligand-based target prediction trained on ChEMBL, WOMBAT, etc. e. g. PIDGINv2. NPs generally do not share the same chemical space as the training space of the algorithm.

One way to overcome this problem is to fragment the NPs into smaller entities and predict their targets by comparing the smaller entities to synthetic drugs with known targets [77]. Another way to do this is to include NPs with experimentally validated results in the training set [78].

5.3.2.2 Network analysis

A commonly used approach of deciphering molecular modes of individual NPs and mixtures of NPs is to construct and analyse drug–gene–disease networks [79, 80].

1. Specifically, for drugs or compounds of interest or a type of disease of interests, a drug–gene–disease network is constructed by integrating the experimentally validated and predicted disease target interactions with gene–disease associations. The gene–disease associations are usually collected from databases such as CTD [81], DisGeNET [82] and Online Mendelian Inheritance in Man (OMIM) [83].
2. Network tools such as Cytoscape [84] or Gephi [85], etc., can be used to visualise the network.
3. Bioinformatics enrichment tools [86] are then employed to analyse the functions of the genes in the network.

Network analysis is carried out on the network and several metrics are used to predict targets, effects of knockdown, important proteins, driver genes, etc. Several functions scores can be calculated, (see Figure 5.3) including the "Betweenness Centrality", which shows which nodes (in this case targets) are more likely to be in communication paths between other nodes (targets). This metric is useful in determining which targets can be modulated where the protein network would break apart i. e. modulating a target with high "Betweenness Centrality" will cause a larger downstream effect. This measure demonstrates how likely the target is to be the most direct route between two targets in the network. Other useful measures that can be calculated include "Eigenvectors", which determine how well a node (target) is connected to other well-connected nodes (targets). Eigenvector centrality is, therefore, a measure of the connectivity of the nodes (targets) that are connected to the node (target) of interest. "Degree Centrality" is a metric that shows how many nodes (targets) a particular node (target) interacts with directly. Integrating the systematic analysis results with previously published data and literature in pharmacology and clinics, it may be possible to understand molecular modes of NPs in the NP–gene–disease networks.

5.4 Pathway annotations

When attempting to understand MoA of compounds, studying individual target/gene information does not give insight into the underlying mechanistic action. Following

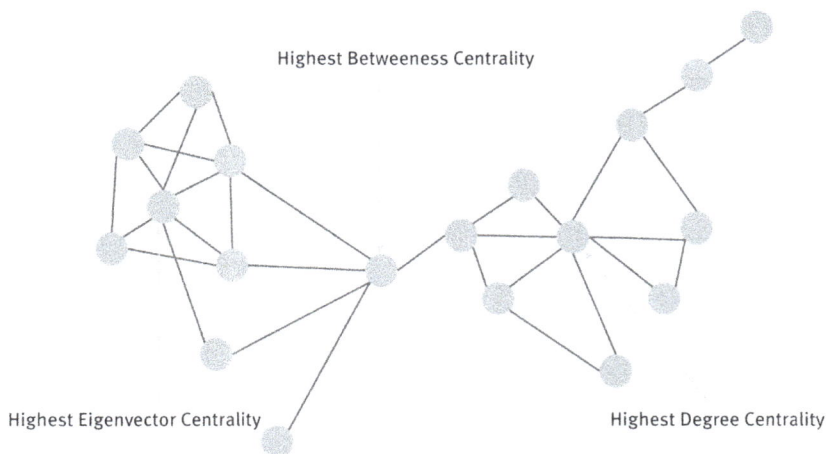

Figure 5.3: Network showing the most common metrics calculated in a network. Modulating the target with the "Highest Betweenness Centrality" is predicted to have great influence as it lies on a high proportion of paths between other targets in the network. The target with the "Highest Degree Centrality" is the most central target in the network i. e. it is an important target involved in a large number of interactions. Targets with high degree centrality are more likely to be essential for survival and growth than targets with low-degree centrality values [87]. The target with the "Highest Eigenvector Centrality" is an important target because it is connected to important neighbouring targets.

target prediction, it is common to annotate predicted targets with pathways, with the aim of understanding their effects of modulating particular proteins in the human body. These pathways consist of genes, proteins and small molecules interacting with each other in a cellular setting to elicit cellular change or creating products. The most well-known types of biological pathways are signalling pathways, which move a signal from outside a cell to its interior; metabolic pathways, responsible for chemical reactions within the body; and gene-regulatory pathways that control the activation and deactivation of genes. In (thesis) use pathway annotations to put the isolated targets into biological context [88]. This is carried out by combining information from databases with statistical testing [88]. Enrichment calculations are carried out on the predicted targets and pathways to determine those that are statistically relevant to the phenotype of study. This facilitates both the interpretation of isolated target information and the generation of a hypothesis of MoA. Pathway annotation can be used to identify the biological roles of candidate genes or targets [89]. It has been applied to predicted targets of NPs and it improved the mechanistic understanding of the MoA of the studied NPs [90]. Pathway annotation has also been used to identify important targets to be modulated in order to elicit a required response, e. g. stop a particular function or inhibit a particular mode [91].

Previous studies have utilised pathway annotations to understand the MoA of TCM and Indian TMs [90, 92] as well as African NPs [5] as target predictions alone do not provide information on downstream biological effects.

Several databases are available that provide signalling, metabolic and gene-regulatory pathway annotations when provided with gene lists, e. g. Reactome [93], KEGG [94, 95], gene ontology (GO) [96, 97], PANTHER [98] and Comparative Toxicogenomics Database (CTD) [99].

Definitions and abbreviations

NP	Chemical compound or substance produced by a living organism that is found in nature
TAM	Traditional African medicine
TCM	Traditional Chinese medicine
Ayurveda	Traditional Indian medicine

References

[1]   Solecki RS. Shanidar IV, a Neanderthal flower burial in Northern Iraq. Science. 1975;190:880–1.
[2]   Sumner J. The natural history of medicinal plants. Portland, Oregon: Timber Press, 2000.
[3]   Ebbell B. The Ebers Papyrus. The Greatest Egyptian medical document, vol. 17. London: H. Milford and Oxford University Press, 1937:123.
[4]   Elufioye TO, Badal S. Chapter 1 - Background to pharmacognosy. In: Badal S, Delgoda R, editor(s). Pharmacognosy. ed Boston: Academic Press, 2017:3–13.
[5]   Baldo F. Integrating chemical, biological and phylogenetic spaces of African natural products to understand their therapeutic activity. Doctor of Philosophy, Cambridge, United Kingdom: Chemistry, University of Cambridge, 2019.
[6]   Zheng XS, Chan TF, Zhou HH. Genetic and genomic approaches to identify and study the targets of bioactive small molecules. Chem Biol. 2004;11:609–18.
[7]   Boutros M, Ahringer J. The art and design of genetic screens: RNA interference. Nat Rev Genet. 2008;9:554–66.
[8]   Wright MH, Sieber SA. Chemical proteomics approaches for identifying the cellular targets of natural products. Nat Prod Rep. 2016;33:681–708.
[9]   Amos GC, Awakawa T, Tuttle RN, Letzel A-C, Kim MC, Kudo Y, et al. Comparative transcriptomics as a guide to natural product discovery and biosynthetic gene cluster functionality. Proc Nat Acad Sci. 2017;114:E11121.
[10] Yuliana ND, Khatib A, Choi YH, Verpoorte R. Metabolomics for bioactivity assessment of natural products. Phytother Res. 2011;25:157–69.
[11] Fung S-Y, Sofiyev V, Schneiderman J, Hirschfeld AF, Victor RE, Woods K, et al. Unbiased screening of marine sponge extracts for anti-inflammatory agents combined with chemical genomics identifies girolline as an inhibitor of protein synthesis. ACS Chem Biol. 2014;9:247–57.

[12] Gokgoz NB, Akbulut BS. Proteomics evidence for the activity of the putative antibacterial plant alkaloid (-)-roemerine: Mainstreaming omics-guided drug discovery. Omics. 2015;19:478–89.

[13] Dyshlovoy SA, Otte K, Venz S, Hauschild J, Junker H, Makarieva TN, et al. Proteomic-based investigations on the mode of action of the marine anticancer compound rhizochalinin. Proteomics. 2017;17:1700048

[14] Yu Y, Yi Z-B, Liang Y-Z. Main antimicrobial components of Tinospora capillipes, and their mode of action against Staphylococcus aureus. FEBS Lett. 2007;581:4179–83.

[15] Biao-Yi Z, Yu Y, Zeng-Liang Y. Investigation of antimicrobial model of Hemsleya pengxianensis W.J. Chang and its main active component by metabolomics technique. J Ethnopharmacol. 2008;116:89–95.

[16] Shen F, Tang X, Wang Y, Yang Z, Shi X, Wang C, et al. Phenotype and expression profile analysis of Staphylococcus aureus biofilms and planktonic cells in response to licochalcone A. Appl Microbiol Biotechnol. 2015;99:359–73.

[17] Pan Z, Agarwal AK, Xu T, Feng Q, Baerson SR, Duke SO, et al. Identification of molecular pathways affected by pterostilbene, a natural dimethylether analog of resveratrol. BMC Med Genomics. 2008;1:7–7.

[18] Schenone M, Dancik V, Wagner B, Clemons P. Target identification and mechanism of action in chemical biology and drug discovery. Nat Chem Biol. 2013;9:232–40.

[19] Klopmand G. Concepts and applications of molecular similarity, by Mark A. Johnson and Gerald M. Maggiora, eds., John Wiley & Sons, New York, 1990, 393 pp. Price: $65.00. J Comput Chem. 1992;13:539–40.

[20] Martin YC, Kofron JL, Traphagen LM. Do structurally similar molecules have similar biological activity? J Med Chem. 2002;45:4350–8.

[21] Lin S-K. Pharmacophore perception, development and use in drug design. Edited by Osman F. Güner. Molecules. 2000;5:987.

[22] Laggner C, Schieferer C, Fiechtner B, Poles G, Hoffmann RD, Glossmann H, et al. Discovery of high-affinity ligands of sigma1 receptor, ERG2, and emopamil binding protein by pharmacophore modeling and virtual screening. J Med Chem. 2005;48:4754–64.

[23] Rollinger JM, Schuster D, Danzl B, Schwaiger S, Markt P, Schmidtke M, et al. In silico target fishing for rationalized ligand discovery exemplified on constituents of Ruta graveolens. Planta Med. 2009;75:195–204.

[24] Jitender V, Vijay MK, Evans CC. 3D-QSAR in drug design - A review. Curr Top Med Chem. 2010;10:95–115.

[25] Chen G, Zhou D, Li X-Z, Jiang Z, Tan C, Wei X-Y, et al. A natural chalcone induces apoptosis in lung cancer cells: 3D-QSAR, docking and an in vivo/vitro assay. Sci Rep. 2017;7:10729

[26] Menezes IR, Lopes JC, Montanari CA, Oliva G, Pavao F, Castilho MS, et al. 3D QSAR studies on binding affinities of coumarin natural products for glycosomal GAPDH of Trypanosoma cruzi. J Comput Aided Mol Des. 2003;17:277–90.

[27] Lapinsh M, Prusis P, Gutcaits A, Lundstedt T, Wikberg JE. Development of proteo-chemometrics: a novel technology for the analysis of drug-receptor interactions. Biochim Biophys Acta. 2001;1525:180–90.

[28] Morris GM, Lim-Wilby M. Molecular docking. Methods Mol Biol. 2008;443:365–82.

[29] Varnek A, Baskin I. Machine learning methods for property prediction in chemoinformatics: Quo Vadis? J Chem Inf Model. 2012;52:1413–37.

[30] Lo Y-C, Rensi SE, Torng W, Altman RB. Machine learning in chemoinformatics and drug discovery. Drug Discov Today. 2018;23:1538–46

[31] Ali SM, Hoemann MZ, Aube J, Georg GI, Mitscher LA, Jayasinghe LR. Butitaxel analogues: synthesis and structure-activity relationships. J Med Chem. 1997;40:236–41.

[32] Keiser M, Roth B, Armbruster B, Ernsberger P, Irwin J, Shoichet B. Relating protein pharmacology by ligand chemistry. Nat Biotechnol. 2007;25:197–206.

[33] Lagunin A, Stepanchikova A, Filimonov D, Poroikov V. PASS: prediction of activity spectra for biologically active substances. Bioinformatics. 2000;16:747–8.

[34] Kim S, Thiessen PA, Bolton EE, Chen J, Fu G, Gindulyte A, et al. PubChem substance and compound databases. Nucleic Acids Res. 2016;44:D1202–13

[35] Bento AP, Gaulton A, Hersey A, Bellis LJ, Chambers J, Davies M, et al. The ChEMBL bioactivity database: an update. Nucleic Acids Res. 2014;42:D1083–90.

[36] Marius Olah MM, Ostopovici L, Rad R, Bora A, Hadaruga N, Olah I, et al. WOMBAT: world of molecular bioactivity. In: Mannhold R, Kubinyi H, Folkers G, Oprea TI, editor(s). Chemoinformatics in drug discovery Methods and Principles in Medicinal Chemistry. Mannheim: Wiley-VCH. 2005 May 199783527307531

[37] Sá MS, de Menezes MN, Krettli AU, Ribeiro IM, Tomassini TC, Ribeiro Dos Santos R, et al. Antimalarial activity of physalins B, D, F, and G. J Nat Prod. 2011;74:2269–72.

[38] Wang L, Ma C, Wipf P, Liu H, Su W, Xie X-Q. Target Hunter: an in silico target identification tool for predicting therapeutic potential of small organic molecules based on chemogenomic database. Aaps J. 2013;15:395–406.

[39] Tan KP, Yang M, Ito S. Activation of nuclear factor (erythroid-2 like) factor 2 by toxic bile acids provokes adaptive defense responses to enhance cell survival at the emergence of oxidative stress. Mol Pharmacol. 2007;72:1380–90.

[40] Jamkhande PG, Wattamwar AS, Pekamwar SS, Chandak PG. Antioxidant, antimicrobial activity and in silico PASS prediction of Annona reticulata Linn. root extract. Beni-Suef Univ J Basic Appl Sci. 2014;3:140–8.

[41] Goel RK, Singh D, Lagunin A, Poroikov V. PASS-assisted exploration of new therapeutic potential of natural products. Med Chem Res. 2011;20:1509–14.

[42] Nidhi MG, Davies J, Jenkins J. Prediction of biological targets for compounds using multiple-category Bayesian models trained on chemogenomics databases. J Chem Inf Model. 2006;46:1124–33.

[43] L. MDDR licensed by Molecular Design. MDL Drug Data Report (MDDR). San Leandro, CA. http://www.mdli.com., ed.

[44] Kohonen T. Self-organized formation of topologically correct feature maps. Biol Cybern. 1982;43:59–69.

[45] Schneider G, Reker D, Chen T, Hauenstein K, Schneider P, Altmann KH. Deorphaning the macromolecular targets of the natural anticancer compound doliculide. Angew Chem Int Ed. 2016;55:12408–11.

[46] Schneider P, Schneider G. Collection of bioactive reference compounds for focused library design. QSAR Comb Sci. 2003;22:713–18.

[47] Reker D, Rodrigues T, Schneider P, Schneider G. Identifying the macromolecular targets of de novo-designed chemical entities through self-organizing map consensus. Proc Nat Acad Sci. 2014;111:4067.

[48] Huang T, Mi H, C.-y L, Zhao L, Zhong LL, Liu F-B, et al. MOST: most-similar ligand based approach to target prediction. BMC Bioinf. 2017;18:165.

[49] Kurita KL, Glassey E, Linington RG. Integration of high-content screening and untargeted metabolomics for comprehensive functional annotation of natural product libraries. Proc Natl Acad Sci USA. 2015;112:11999–2004.

[50] Navarro G, Cheng AT, Peach KC, Bray WM, Bernan VS, Yildiz FH, et al. Image-based 384-well high-throughput screening method for the discovery of skyllamycins A to C as biofilm inhibitors and inducers of biofilm detachment in Pseudomonas aeruginosa. Antimicrob Agents Chemother. 2014;58:1092–9.

[51] Peach KC, Bray WM, Winslow D, Linington PF, Linington RG. Mechanism of action-based classification of antibiotics using high-content bacterial image analysis. Mol Biosyst. 2013;9:1837–48.

[52] Schulze CJ, Bray WM, Woerhmann MH, Stuart J, Lokey RS, Linington RG. "Function-first" lead discovery: mode of action profiling of natural product libraries using image-based screening. Chem Biol. 2013;20:285–95.

[53] Wong WR, Oliver AG, Linington RG. Development of antibiotic activity profile screening for the classification and discovery of natural product antibiotics. Chem Biol. 2012;19:1483–95.

[54] Perlman ZE, Slack MD, Feng Y, Mitchison TJ, Wu LF, Altschuler SJ. Multidimensional drug profiling by automated microscopy. Science. 2004;306:1194–8.

[55] Woehrmann MH, Bray WM, Durbin JK, Nisam SC, Michael AK, Glassey E, et al. Large-scale cytological profiling for functional analysis of bioactive compounds. Mol Biosyst. 2013;9:2604–17.

[56] Ntie-Kang F, Onguene PA, Scharfe M, Owono LC, Megnassan E, Mbaze LM, et al. ConMedNP: a natural product library from Central African medicinal plants for drug discovery. RSC Adv. 2014;4:409–19.

[57] Lewis AM, Mervin H, Bulusu KC, Kalash L, Svensson F, Firth MA, et al. Orthologue chemical space and its influence on target prediction. Bioinformatics. 2018;34:72–9

[58] Kouam SF, Njonkou YL, Kuigoua GM, Ngadjui BT, Hussain H, Green IR, et al. Psorantin, a unique methylene linked dimer of vismin and kenganthranol E, two anthranoid derivatives from the fruits of Psorospermum aurantiacum (Hypericaceae). Phytochem Lett. 2010;3:185–9.

[59] Le TT, Gardner J, Hoang-Le D, Schmidt CW, MacDonald KP, Lambley E, et al. Immunostimulatory cancer chemotherapy using local ingenol-3-angelate and synergy with immunotherapies. Vaccine. 2009;27:3053–62.

[60] Challacombe JM, Suhrbier A, Parsons PG, Jones B, Hampson P, Kavanagh D, et al. Neutrophils are a key component of the antitumor efficacy of topical chemotherapy with ingenol-3-angelate. J Immunol. 2006;177:8123–32.

[61] Crooks SW, Stockley RA. Leukotriene B4. Int J Biochem Cell Biol. 1998;30:173–8.

[62] Tong W-G, Ding X-Z, Hennig R, Witt RC, Standop J, Pour PM, et al. Leukotriene B4 receptor antagonist LY293111 inhibits proliferation and induces apoptosis in human pancreatic cancer cells. Clin Cancer Res. 2002;8:3232–42.

[63] Bingle CD, Craig RW, Swales BM, Singleton V, Zhou P, Whyte MK. Exon skipping in Mcl-1 results in a Bcl-2 homology domain 3 only gene product that promotes cell death. J Bio Chem. 2000;275:22136–46.

[64] Nelson AW, Tilley WD, Neal DE, Carroll JS. Estrogen receptor beta in prostate cancer: friend or foe? Endocr Relat Cancer. 2014;21:T219–34.

[65] Palmieri C, Cheng GJ, Saji S, Zelada-Hedman M, Warri A, Weihua Z, et al. Estrogen receptor beta in breast cancer. Endocr Relat Cancer. 2002;9:1–13.

[66] Fucic A, Miškov S, Želježić D, Bogdanovic N, Katić J, Gjergja R, et al. Is the role of estrogens and estrogen receptors in epilepsy still underestimated? Med Hypotheses. 2009;73:703–5.

[67] Bauer J, Isojarvi JI, Herzog AG, Reuber M, Polson D, Tauboll E, et al. Reproductive dysfunction in women with epilepsy: recommendations for evaluation and management. J Neurol Neurosurg Psychiatry. 2002;73:121–5.

[68] Calabro RS, Marino S, Bramanti P. Sexual and reproductive dysfunction associated with antiepileptic drug use in men with epilepsy. Expert Rev Neurother. 2011;11:887–95.

[69] Dupont S, Krust A, Gansmuller A, Dierich A, Chambon P, Mark M. Effect of single and compound knockouts of estrogen receptors alpha (ERalpha) and beta (ERbeta) on mouse reproductive phenotypes. Development. 2000;127:4277–91.

[70] Hussein AA, Bozzi B, Correa M, Capson TL, Kursar TA, Coley PD, et al. Bioactive constituents from three vismia species. J Nat Prod. 2003;66:858–60.

[71] Kalliokoski T, Kramer C, Vulpetti A. Quality Issues with Public Domain Chemogenomics Data. Mol Inform. 2013;32:898–905.

[72] Gedeck P, Kramer C, Ertl P. Computational analysis of structure-activity relationships. In: Lawton G, Witty DR, editor(s). Progress in medicinal chemistry Vol. 49. Amsterdam: Elsevier. ed: 2010:113–60. http://www.sciencedirect.com/science/article/pii/S0079646810490049.

[73] Tetko IV, Sushko I, Pandey AK, Zhu H, Tropsha A, Papa E, et al. Critical assessment of QSAR models of environmental toxicity against Tetrahymena pyriformis: focusing on applicability domain and overfitting by variable selection. J Chem Inf Model. 2008;48:1733–46.

[74] Kristina G, Gisbert S. Properties and architecture of drugs and natural products revisited. Curr Chem Biol. 2007;1:115–27.

[75] Henkel. Statistical investigation into the structural complementarity of natural products and synthetic compounds. Angew Chem-Int Ed. 1999;38:643–7.

[76] Feher M, Schmidt JM. Property distributions: Differences between drugs, natural products, and molecules from combinatorial chemistry. J Chem Inf Comput Sci. 2003;43:218–27.

[77] Reker D, Perna AM, Rodrigues T, Schneider P, Reutlinger M, Mönch B, et al. Revealing the macromolecular targets of complex natural products. Nat Chem. 2014;6:1072–8.

[78] Ehrman TM, Barlow DJ, Hylands PJ. Virtual screening of Chinese herbs with Random Forest. J Chem Inf Model. 2007;47:264–78.

[79] Chamberlin SR, Blucher A, Wu G, Shinto L, Choonoo G, Kulesz-Martin M, et al. Natural product target network reveals potential for cancer combination therapies. Front Pharmacol. 2019;10:557.

[80] Fang J, Wu Z, Cai C, Wang Q, Tang Y, Cheng F. Quantitative and systems pharmacology. 1. In silico prediction of drug-target interactions of natural products enables new targeted cancer therapy. J Chem Inf Model. 2017;57:2657–71.

[81] Davis AP, Grondin CJ, Johnson RJ, Sciaky D, McMorran R, Wiegers J, et al. The Comparative Toxicogenomics Database: update 2019. Nucleic Acids Res. 2019;47:D948–54.

[82] Piñero J, Queralt-Rosinach N, Bravo À, Deu-Pons J, Bauer-Mehren A, Baron M, et al. DisGeNET: a discovery platform for the dynamical exploration of human diseases and their genes. Database: J Bio Database Curation. 2015;2015:bav028–bav028.

[83] Amberger JS, Hamosh A. Searching online mendelian inheritance in man (OMIM): A knowledgebase of human genes and genetic phenotypes. Curr Protoc Bioinf. 2017;58:1.2.1–1.2.12.

[84] Shannon P, Markiel A, Ozier O, Baliga NS, Wang JT, Ramage D, et al. Cytoscape: a software environment for integrated models of biomolecular interaction networks. Genome Res. 2003;13:2498–504.

[85] Bastian M, Heymann S, Jacomy M. Gephi: An open source software for exploring and manipulating networks DOI: Association for the Advancement of Artificial Intelligence (www.aaai.org). 2009;8.

[86] Huang Da W, Sherman BT, Lempicki RA. Bioinformatics enrichment tools: paths toward the comprehensive functional analysis of large gene lists. Nucleic Acids Res. 2009;37:1–13.

[87] Melak T, Gakkhar S. Comparative genome and network centrality analysis to identify drug targets of mycobacterium tuberculosis H37Rv. Biomed Res Int. 2015;2015:212061–212061.

[88] García-Campos MA, Espinal-Enríquez J, Hernández-Lemus E. Pathway analysis: state of the art. Front Physiol. 2015;6:383.

[89] Folger O, Jerby L, Frezza C, Gottlieb E, Ruppin E, Shlomi T. Predicting selective drug targets in cancer through metabolic networks. Mol Syst Biol. 2011;7:501.

[90] Mohamad Zobir SZ, Mohd Fauzi F, Liggi S, Drakakis G, Fu X, Fan T-P, et al. Global mapping of traditional Chinese medicine into bioactivity space and pathways annotation improves mechanistic understanding and discovers relationships between therapeutic action (Sub) classes. Evidence-Based Complement Altern Med. 2016;2016:25.

[91] Liggi S, Drakakis G, Koutsoukas A, Cortes-Ciriano I, Martinez-Alonso P, Malliavin T, et al. Extending in silico mechanism-of-action analysis by annotating targets with pathways: application to cellular cytotoxicity readouts. Future Med Chem. 2014;6:2029–56.

[92] Lagunin AA, Goel RK, Gawande DY, Pahwa P, Gloriozova TA, Dmitriev AV, et al. Chemo- and bioinformatics resources for in silico drug discovery from medicinal plants beyond their traditional use: a critical review. Nat Prod Rep. 2014;31:1585–611.

[93] Croft D, Mundo AF, Haw R, Milacic M, Weiser J, Wu G, et al. The Reactome pathway knowledgebase. Nucleic Acids Res. 2014;42:D472–7.

[94] Kanehisa M, Goto S. KEGG: kyoto encyclopedia of genes and genomes. Nucleic Acids Res. 2000;28:27–30.

[95] Kanehisa M. The KEGG database. Novartis Found Symp. 2002;247:91–101.

[96] Ashburner M, Ball C, Blake J, Botstein D, Butler H, Cherry J, et al. Gene ontology: tool for the unification of biology. The Gene Ontology Consortium. Nat Genet. 2000;25:25–9.

[97] Consortium TG. Expansion of the Gene Ontology knowledgebase and resources. Nucleic Acids Res. 2017;45:D331–8.

[98] Mi H, Huang X, Muruganujan A, Tang H, Mills C, Kang D, et al. PANTHER version 11: expanded annotation data from Gene Ontology and Reactome pathways, and data analysis tool enhancements. Nucleic Acids Res. 2017;45:D183–9.

[99] Davis AP, Grondin CJ, Johnson RJ, Sciaky D, King BL, McMorran R, et al. The Comparative Toxicogenomics Database: update 2017. Nucleic Acids Res. 2017;45:D972–8.

Fidele Ntie-Kang

6 Mechanistic role of plant-based bitter principles and bitterness prediction for natural product studies I: Database and methods

Abstract: This chapter discusses the rationale behind the bitter sensation elicited by chemical compounds, focusing on natural products. Emphasis has been placed on a brief presentation of BitterDB (the database of bitter compounds), along with available methods for the prediction of bitterness in compounds. The fundamental basis for explaining bitterness has been provided, based on the structural features of human bitter taste receptors and have been used to shed light on the mechanistic role of a few out of the 25 known human taste receptors to provide the foundation for understanding how bitter compounds interact with their receptors. Some case studies of ligand-based prediction models based on 2D fingerprints and 3D pharmacophores, along with machine learning methods have been provided. The chapter closes with an attempt to establish the relationship between bitterness and toxicity.

Keywords: bitterness, chemoinformatics prediction, drug discovery, natural products

6.1 Introduction

6.1.1 Natural products and bitterness

Natural products (NPs), also known as secondary metabolites, are compounds biosynthesized by living organisms (e. g. plants, fungi and microbes in order to protect them and help to adapt to the environment). NPs elicit a wide variety of tastes, including bitter [1]. Previous studies have attempted to explain taste on a molecular basis, and have attributed the bitterness sensation in humans, for example, to molecules interacting with 25 known bitter taste receptors (hTas2Rs). These are the G-protein coupled receptors (GPCRs), expressed both orally and extra-orally [2], the 25 receptors constituting the 2nd largest group of chemosensory GPCRs in humans. These perform a variety of functions, e. g. bitter taste perception, detection of toxins, hormonal secretion, etc. Tas2Rs, in general, have been proposed as novel targets for several indications [3]. Bitterness is a property of both natural and synthetic

This article has previously been published in the journal *Physical Sciences Reviews*. Please cite as: Ntie-Kang, F. Mechanistic role of plant-based bitter principles and bitterness prediction for natural product studies I: Database and methods. *Physical Sciences Reviews* [Online] 2019 DOI: 10.1515/psr-2018-0117

https://doi.org/10.1515/9783110668896-006

compounds, many therapeutic drugs in clinical use, as well as toxins [4]. It should be noted that hTas2Rs are able to recognize several hundred structurally diverse agonists and each agonist is able to be recognized by more than one hTas2R. One hypothesis is that plants, for example, make secondary metabolites specifically in order to dissuade predators. A second hypothesis is that bitterness was inherently a property in toxic compounds in order to protect humans and herbivores. However, we must note that:

- many toxins are not bitter
- many important bitter NPs have been used as rather curative agents (not toxins), e. g. the antimalarial drug Quinine (Figure 6.1), discovered from the bark of the cinchona (quina-quina) tree [5], has been used for four centuries
- some known bitter chemicals do not deter herbivores [6].

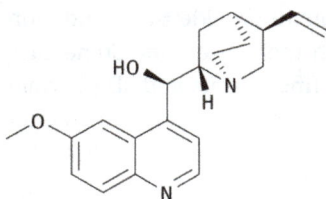

Figure 6.1: Chemical structure of the bitter antimalarial principle, Quinine, discovered since the 1600s.

6.1.2 Distinguishing between sweet, salty, sour, umami and bitter

Taste is a sensation that can be categorized into five; sweetness, sourness, saltiness, bitterness, and umami (or "savory" taste characteristic of foods with high protein content, a fifth recently established taste [7]). This distinction is made by taste buds by detecting the interaction with different molecules or ions, e. g. sweet, umami, and bitter tastes are triggered when the tasted molecule binds to GPCRs located on the cell membranes of taste buds. On the contrary, saltiness and sourness are perceived when alkali metal or hydrogen ions enter taste buds, respectively [8]. Like all taste senses, bitterness could be both harmful and beneficial [9], e. g. while sweetness helps to identify energy-rich foods, bitterness warns against poisons [10].

6.1.3 Advantage and disadvantage of bitterness

While bitterness would be advantageous if it helps dissuade against intoxication, it would clearly be a disadvantage for orally administered drugs, particularly with children [11]. On the contrary, extensive research has been conducted on the health benefits of consuming plant-based bitter foods, suggesting that their metabolites might play a role in protection against aging, heart disease, and cancer [12–17].

6.1.4 Rationale of this work

In this chapter, an attempt is made to provide an overview of computer-based methods for the prediction of bitterness, including the database of bitter compounds [19] and an attempt to establish a relationship between bitterness and toxicity has been provided.

6.2 BitterDB -the database of bitter compounds

This collection made by the Niv Group, was first published as several hundred compounds [18], but has recently been updated to include more than 1,000 known bitter compounds collected from literature sources [19]. BitterDB is unique as the largest collection of data on bitter compounds and currently includes data curated from ~ 100 sources (journal articles and Merck index). Figure 6.2 shows the web interface of BitterDB and its possible uses.

Special features of the BitterDB include:
- searching compounds through browsing by name search, searching for similar molecules, or by keying search terms contained in the compound name
- otherwise the BitterDB receptor ID can be used to browse for ligands binding to a selected receptor
- linking a given compound with the 25 known Tas2Rs
- providing different ways to search for bitter compounds by using different criteria, e. g. querying for molecules with structures, browsing through bitter receptors and their potential binders, etc.
- advanced search options include searching for compounds within given ranges of physico-chemical properties, e. g. molecular weight (MW), AlogP, the logarithm of the n-octanol/water partition coefficient using Crippen's method [20]
- acute oral LD_{50} in rats and data related to association with specific receptors
- other tools accessible through the database include several utilities, e. g. BitterPredict (a tool that predicts whether a compound could be bitter or not, based on its chemical structure) [21], BitterToxic (a tool which accesses the bitterness-toxicity relationship of a compound) [22],

– a page where users could upload new data and
– some very useful help pages and tutorial videos to assist new users and visitors.

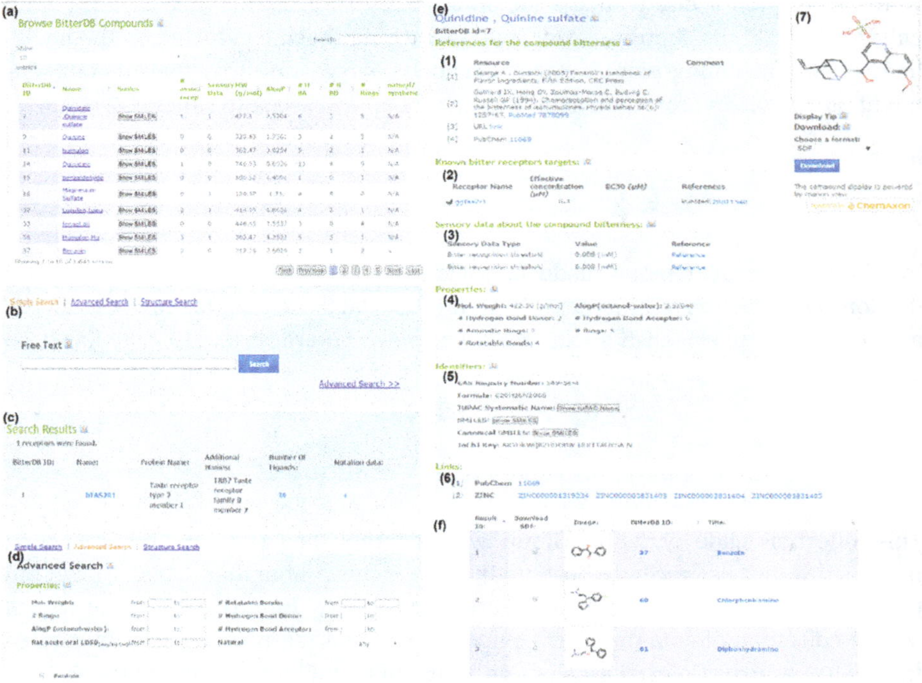

Figure 6.2: Several snapshots of the BitterDB interface: (a) compounds search, e. g. by compound name, (b) by keyword search, and (c) by receptor name; (d) compound advanced search with options to search for compounds within a range of properties including MW, AlogP, data on rat acute oral LD_{50} with respect to associated receptors, etc.; (e) properties of a selected compound in the query output, e. g. quinine sulfate: (1) references in the literature to confirm bitterness in the output compound, (2) experimental data, e. g. *in vitro* studies of associated receptor in chosen species models. (3) quantitative sensory data (4) molecular properties and toxicity data (e. g. rat acute LD_{50}); (5) CAS number, IUPAC name, SMILES, InChI Key, etc.; (6) links to PubChem and ZINC; (7) chemical structure of query compound; (f): a typical output for a query search showing compounds that fit the criteria.

The current version of BitterDB [19] contains data on additional species bitter receptors and ligands, e. g. chicken, mouse and cat, as well as homology models of the bitter receptors, available in the receptors pages.

The importance of this database within the scientific community is highlighted by the number of users visiting the website over the years since its publication in 2012 [18]. There are currently ~ 20,000 recorded users, meanwhile more than four fifths of former users returned to it (Figure 6.3). The average user spent about 5 minutes on the database, returning on average at least once.

(a)

(b)

Users	New Users	Sessions
19,659	19,614	37,479

Number of Sessions per User	Pageviews	Pages/ Session
1.91	184,243	4.92

Avg.Session Duration	Bounce Rate
00:05:09	45.43%

(c)

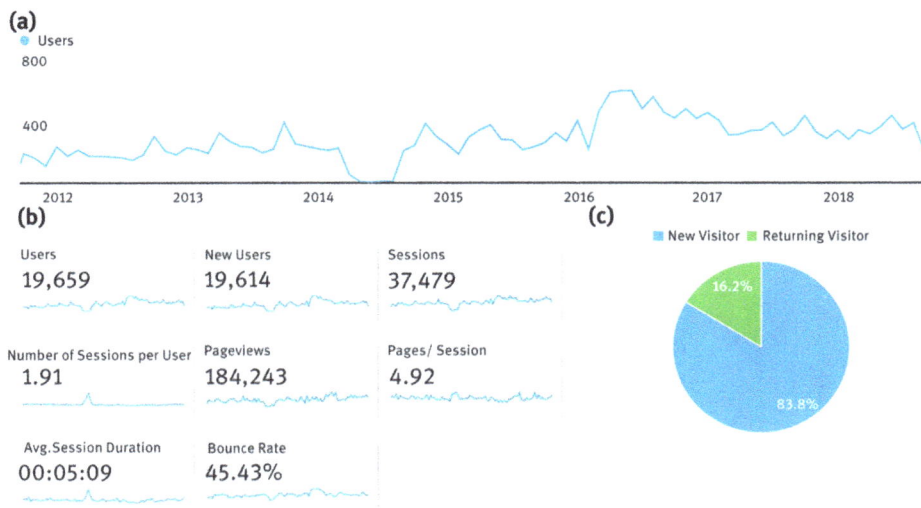

Figure 6.3: Overview of users of BitterDB since publication; (a) fluctuation curve of number of users over the years, (b) users and work sessions, (c) most current users are returnees. Image adapted from information of the BitterDB website [23].

6.3 In silico methods for predicting bitterness

6.3.1 Structure-based methods

Even though docking GPCRs has proven to be challenging, a community-wide GPCR modeling and docking (GPCR Dock) assessment was able to achieve close-to-experimental accuracy for small rigid orthosteric ligands using homology models, although predictions of long loops and GPCR activation states remained problematic [24].

6.3.1.1 Structural features of hTa2SR

The general structure of hTAS2Rs has seven transmembrane (TM) domains, which are composed of helices, e. g. Figure 6.4 shows the various TM domains in hTas2R46 along with binding of the NP strychnine. In an attempt to understand the structure to function relationship of the 25 hTas2R receptors and eventually develop specific antagonists required for human bitter taste perception, Brockhoff et al. investigated the structure of hTas2R binding pockets, by focusing on a sub-family of closely related hTas2Rs which had shown close amino acid (AA) sequence identities but which each bind to unique agonists [25].

The investigation led to the identification of receptor regions and showed the AA residues important for the activation of hTas2R46 by strychnine, for example (Figure 6.5). An investigation consisted in exchanging two residues located in the

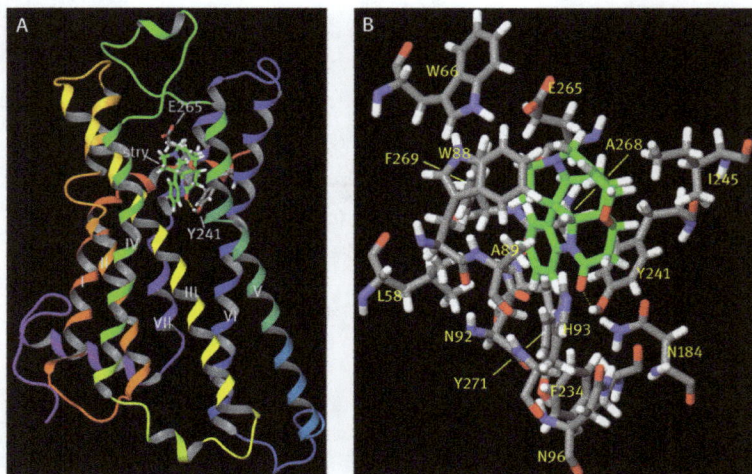

Figure 6.4: (A) TM domains (numbered in Roman numerals) shown in the homology model of hTas2R46 with the putative mode with its agonist strychnine, (B) binding interactions of strychnine docked into the proposed binding pocket of hTas2R46 [25]. Figure reproduced by permission.

transmembrane domain VII between hTas2R46, activated by strychnine, and hTas2R31, activated by aristolochic acid. This led to the inversion of agonist selectivity [25]. It was further shown that the transfer of functionally relevant AAs in hTas2R46 to the corresponding positions of hTas2R43 and hTas2R31 resulted in pharmacological properties indistinguishable from the parental hTas2R46.

Like some members of the subclass (known as class A GPCRs), hTas2R46 has been shown by molecular (MD) simulations to have 2 stable binding pockets (Figure 6.6), an orthosteric agonist-binding site but also a more extracellular or "vestibular" site, involved in the binding process, the two sites being capable of hosting the agonist strychnine [26]. The MD method implements a hybrid molecular mechanics/coarse-grained (MM/CG) molecular dynamics approach, which has been described in details in the literature [27–29].

6.3.1.2 Insight from MM/CG to explain the difference between specific and promiscuous hTa2SRs

Several extensive MM/CG simulations have been conducted on hTas2Rs in order to investigate specific *versus* promiscuous binding to agonists [26, 29, 30]. As an example, hTas2R38, in contrast with hTas2R46, did not capture the vestibular site [31]. This could perhaps be due to the fact that the latter is not a requirement for a more selective receptor like hTas2R38, which is a lot less promiscuous than hTas2R46, for example. In hTas2R46, the existence of a second binding site may be vital for recognizing a wider variety of agonists. Evidence from MD and experimental validation showed that this

```
                              ↓1.50                            ↓2.50
hTAS2R43   1 MITFLPIIFSSLVVVTFVIGNFANGFIALVNSIESFKRQKISFADQILTALAVSRVGLLW  60
hTAS2R31   1 MTTFIPIIFSSVVVVLFVIGNFANGFIALVNSIERVKRQKISFADQILTALAVSRVGLLW  60
hTAS2R46   1 MITFLPIIFSILIVVTFVIGNFANGFIALVNSIEWFKRQKISFADQILTALAVSRVGLLW  60

                                      ↓3.50
hTAS2R43  61 VLLLNWYSTVLNPAFNSVEVRTTAYNIWAVINHFSNWLATTLSIFYLLKIANFSNFIFLH 120
hTAS2R31  61 VLLLNWYSTVFNPAFYSVEVRTTAYNVWAVTGHFSNWLATSLSIFYLLKIANFSNLIFLH 120
hTAS2R46  61 VLVLNWYATELNPAFNSIEVRITAYNVWAVINHFSNWLATSLSIFYLLKIANFSNLIFLH 120
                +    * *       *          *        * *
             ↓4.50
hTAS2R43 121 LKRRVKSVILVMLLGPLLFLACHLFVINMNEIVRTKEFEGNMTWKIKLKSAMYFSNMTVT 180
hTAS2R31 121 LKRRVKSVILVMLLGPLLFLACQLFVINMKEIVRTKEYEGNMTWKIKLRSAVYLSDATVT 180
hTAS2R46 121 LKRRVKSVVLVILLGPLLFLVCHLFVINMNQIIWTKEYEGNMTWKIKLRSAMYLSNTTVT 180
                                         * *  *                     *
         ↓5.50
hTAS2R43 181 MVANLVPFTLTLLSFMLLICSLCKHLKKMQLRGKGSQDPSTKVHIKALQTVISFLLLCAI 240
hTAS2R31 181 TLGNLVPFTLTLLCFLLLICSLCKHLKKMQLHGKGSQDPSTKVHIKALQTVIFFLLLCAV 240
hTAS2R46 181 ILANLVPFTLTLISFLLLICSLCKHLKKMQLHGKGSQDPSMKVHIKALQTVTSFLLLCAI 240

             ↓6.50                          ↓7.50
hTAS2R43 241 YFLSIMISVWSFGSLENKPVFMFCKAIRFSYPSIHPFILIWGNKKLKQTFLSVFWQMRYW 300
hTAS2R31 241 YFLSIMISVWSFGSLENKPVFMFCKAIRFSYPSIHPFILIWGNKKLKQTFLSVLRQVRYW 300
hTAS2R46 241 YFLSIIMSVWNSFESLENKPVFMFCEAIAFSYPSTHPFILIWGNKKLKQTFLSVLWHVRYW 300
                +                          *    * +

hTAS2R43 301 VKGEKTSSP 309
hTAS2R31 301 VKGEKPSSP 309
hTAS2R46 301 VKGEKPSSS 309
```

Figure 6.5: Alignment of the amino acid sequences of the receptors hTAS2R43, hTAS2R31, and hTAS2R46. Identical positions are highlighted in gray. The transmembrane regions are indicated by solid lines [25]. Figure reproduced by permission.

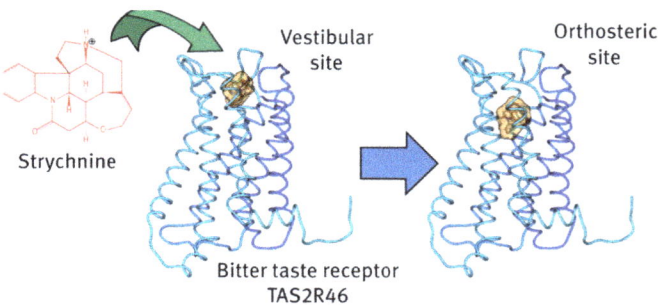

Figure 6.6: Vestibular and orthosteric binding sites of the agonist strychnine to the bitter taster receptor hTas2R43 [26]. Figure reproduced by permission.

receptor has a two-step authentication mechanism, which could explain this promiscuity [32].

6.3.1.3 Using structure-based pharmacophore models

Among the human bitter taste receptors hTas2R39 is known to be activated by NPs, particularly phenolics like flavonoids and isoflavonoids [33–35], commonly present in, and associated with the healthiness (along with the bitterness) of fruits and vegetables [34, 35]. Roland et al. built a structure-based pharmacophore model of the hTas2R39 binding pocket using two sets of compounds:

– an (iso)flavonoid-dedicated set, and
– a more generic, structurally diverse set.

Both datasets contained actives and inactives [36]. It was observed that agonists showed a linear binding geometry and were deeply bound in the hTas2R39 pocket. These also showed mapping to key pharmacophore hydrogen donor features (Figure 6.7).

Apart from their crooked geometry, it was observed that blockers lacked the hydrogen-bond donors enabling contact to the receptor. This enabled them to sterically hinder movement of the TM domains upon receptor activation. This study revealed the characteristics of hTas2R39 agonists and bitter blocker binding and is vital for the development of blockers suitable to counter the bitterness of dietary hTas2R39 agonists in food applications.

6.3.1.4 Case of rational design of bitter agonists

The hTas2R14 is one of the most broadly tuned human bitter taste receptors. This receptor recognizes several chemically diverse synthetic and natural bitter compounds, including many currently used drugs [37]. This is because the receptor's ligand binding pocket is easily accessible. This was proven by the fact that several diverse chemically modified agonists have shown receptor responses in functional assays [38]. Based on this, Karaman et al. modified several hTas2R14 agonists by adding large functional groups and showed that the derived compounds retained agonistic activity [39]. The results could be explained by modeling the newly synthesized agonist derivatives in the binding site of the receptor, comparing with the parent compounds and rationalization the *in vitro* activities of the compound series. From this work, it could be shown that chemical modification of the receptor agonists permits the exploration of the spatial capacity of the receptor binding pocket, i. e. the receptor has the capacity to accommodate agonists with diverse sizes, which is a general characteristic of multi-specific GPCRs [40]. This investigation also indicated that agonist-receptor contact points do not envelop the ligand tightly, proving that promiscuity in human bitter taste receptors correlates with the

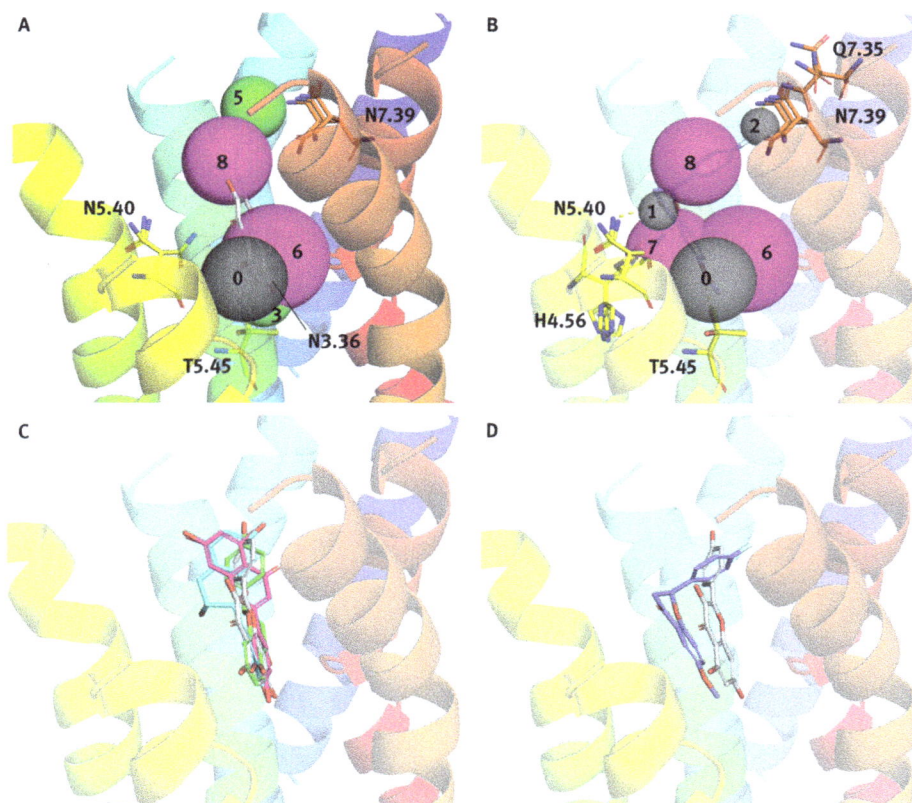

Figure 6.7: Structure-based pharmacophore modeling of hTas2R39. A) Fitting of agonist kaempferol (gray) into the pharmacophore features 0, 3, 5, 6, 8. Residues, which make hydrogen bonds to the agonist, are shown as sticks. B) Fitting of the blocker 4'-fluoro-6-methoxyflavanone (S-enantiomer, blue) into the pharmacophore features 0, 1, 2, 7, 8. Residues, which make hydrogen bonds (yellow dashes) to the blocker, are shown as sticks. C) Fitting of kaempferol (gray), luteolin (pink), naringenin (green), and epicatechin (cyan). D) Fitting of the kaempferol (gray) and 4'-fluoro-6-methoxyflavanone (blue) [36]. Material reproduced originally published under Creative Commons (CC BY) License.

binding site surface [41] and stands as the molecular basis for the promiscuity of this receptor apart from its apparent spacious ligand binding site.

6.3.2 Ligand-based methods

Since flavonoids and isoflavonoids are known for their capacity to activate bitter taste receptors, Roland et al. undertook a study aimed at detecting the essential molecular fingerprints and pharmacophore features for activating the bitter receptors hTas2R14 and hTas2R39 [34]. The training sets consisted of 68 hTas2R14 activators and 70 hTas2R39 activators, among which 58 ligands were overlapping. This was with the intention to predict their structural requirements to activate these receptors.

6.3.2.1 Molecular fingerprints for detecting SAR

In order to identify key molecular features involved in bitter receptor activation, the authors established "good" and "bad" fingerprint fragments (see Figure 6.8 for illustrative fragments for the hTas2R14 fingerprint model). It was observed that the flavanones and isoflavones were likely to trigger hTas2R14. Although flavones could activate this receptor as well, some fragments (e. g. methoxy and glycosyl groups) within some flavones assigned to the "bad" fragments, implying that some flavones caused less or no activation.

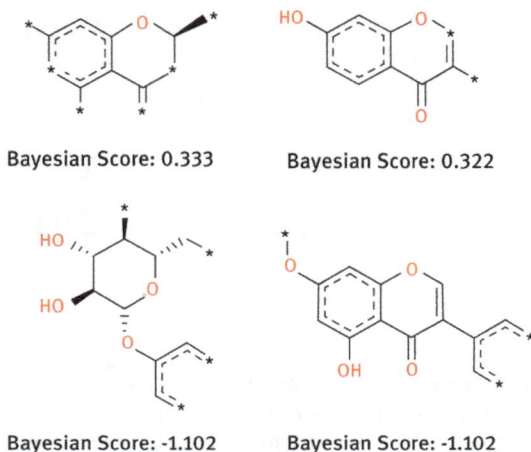

Bayesian Score: 0.333   Bayesian Score: 0.322

Bayesian Score: -1.102   Bayesian Score: -1.102

Figure 6.8: Selected "good" (positive Bayesian score) and "bad" (negative Bayesian score) fingerprint fragments for hTas2R14 [34]. Figure reproduced by permission.

As per the hTas2R39 receptor, a high probability of activation by flavanones and flavonols was observed in addition to a similar behavior toward flavones, when compared with hTas2R14. The presence of methoxy (and obviously glycosyl) fragments were not beneficial for the activation of this receptor by isoflavonoids. The fingerprint fragments helped shed light on favorable and unfavorable molecular features for receptor activation, showing that the 2D models could represent excellent predictive tools for identification of bitter (iso)flavonoids activating the hTas2R14 or hTas2R39 receptors. However, this could not provide sufficient explanations for understanding of the general molecular signature involved in bitter receptor activation of (iso)flavonoids. This required 3D modeling of pharmacophores to investigate chemical characteristics that influence bitter receptor interaction.

6.3.2.2 Insight from 3D pharmacophore models

The structural requirements for (iso)flavonoids to activate the two investigated receptors have been shown in Figure 6.9. As shown, activation of hTas2R14 was best described by a five-feature pharmacophore (Figure 6.9(a)); two hydrogen donor features, one hydrogen acceptor feature, one "hydrophobic aromatic" feature, and one "ring aromatic" feature, while modeling the activation of hTAS2R39 by (iso)flavonoids was best described with a six-feature pharmacophore (Figure 6.9(b)); three hydrogen donor features, one hydrogen acceptor feature, one "hydrophobic aromatic" feature, and one "ring aromatic" feature.

ROC analysis (with area under the curve, AUC = 0.751) showed that the hTas2R14 model was able to correctly predict the activation or absence of activation of two-thirds of the ligands and performed better for highly active than for moderately active compounds. Of the highly active compounds, 81 % were predicted correctly, but for the moderately active compounds this was 52 %. In addition, two-thirds of the inactive compounds were predicted correctly. As per the hTas2R39 model, the best results for the ROC plot AUC value was 0.873. The model returned nine false positives and three false negatives, out of 73 tested compounds. It was additionally able to map 84 % of the compounds correctly. This included predicting 94 % of the highly active compounds, 75 % of the moderately active compounds, and 79 % of the inactive compounds correctly.

6.3.3 Machine learning approaches

Machine learning, generally speaking, is a discipline in computer science, which involves the development algorithms that are trained to find patterns within data. Based on the fact that these algorithms "learn" from data (training set) in a supervised or an unsupervised manner, we derive the term "machine learning". There are several machine learning methods, including artificial neural networks, Naïve Bayes classification, Support Vector Machine (SVM), random forest, etc. The principle behind the algorithm when used to classify molecules is that a set of molecules with a known (and/or contrasting) property, e. g. bitter and non-bitter, bitter and sweet, active or not active against a given receptor are included in the training set. The algorithm then learns in an iterative manner how to classify these molecules as either positive or negative within the training set, and then validated on test set(s), which are molecules with and without the given property, which were not included in the initial training set for building the model. The performance of each machine learning model would typically depend on the training set and the type of algorithm used.

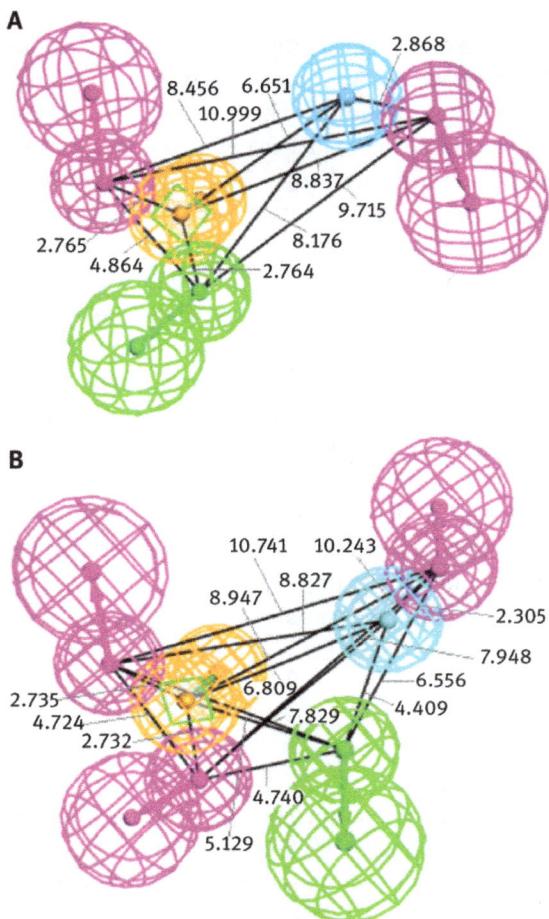

Figure 6.9: Pharmacophores for hTAS2R14 (A) and hTAS2R39 (B). The colors of the spheres represent the following features: pink, hydrogen donors; green, hydrogen acceptors; blue, hydrophobic aromatic; and yellow, ring aromatic. The green rectangle represents a plane within a ring structure. The small spheres indicate the presence of the feature on the ligand, and the large spheres indicate the possible position of the amino acids on the receptor, interacting with this feature. The direction of interaction is shown with an arrow. The numbers represent the distance between the features in Ångströms [34]. Figure reproduced by permission.

6.3.3.1 Using Bayesian classifiers

This was the first reported comprehensive classification method for bitterness of small molecules, which was based on the 2D structures. A dataset of 649 bitter and 13,530 randomly selected molecules was employed in the study [42]. The structures were analyzed by circular fingerprints (MOLPRINT 2D) and information-gain feature selection (i. e. sub-structural features which are statistically correlated to bitterness).

A Naïve Bayes classifier was used to discriminate between bitter and random compounds, and was able to predict 72.1 % of the bitter compounds correctly.

6.3.3.2 Case study: an "AdaBoost-based" machine learning classifier for bitterness and non-bitterness prediction

Adaptive Boosting (AdaBoost), a machine learning classifier, is based on decision
trees algorithm implemented in the BitterPredict tool (more on this in the next chapter) [21]. This is used to classify a compound is bitter or not, based on its chemical
structure. The algorithm was trained using BitterDB and used as the positive set,
while randomly selected non-bitter molecules from the literature were used as the
negative set. Based on computed physicochemical and ADME/Tox descriptors, it was
able to classify over 80 % of the compounds in the hold-out test set, and 70–90 % of
the compounds in three independent external sets and in sensory test validation. In
addition, the tool suggested ~ 40 % of a set of randomly selected molecules as bitter,
along with a large portion of NPs (~77 %) and clinical drugs (~66 %). A summary of
the aforementioned bitterness prediction methods has been provided in Table 6.1.

6.4 Bitterness-toxicity relationship

Nissim et al. have attempted to establish a relationship between toxicity and bitterness
by predicting the toxicity profiles of a dataset of known bitter compounds (BitterDB
[19]) and by using the BitterPredict tool [21] to predict the bitterness of a set of known
toxic compounds [22]. The results showed that about 60 % of the bitter compounds
have documented toxicity and only 56 % of the toxic compounds are known or predicted to be bitter. The toxicity data represented as the lethal dose at 50 % concentration (LD_{50} value) distributions suggested that most of the bitter compounds were not
very toxic. Flavonoids and alpha acids were noted to be more common in the bitter
dataset, when compared with the toxic dataset. On the contrary, alkaloids were noted
to be more common in the toxic datasets when compared to the bitter dataset. There
was no trend linking LD_{50} values with the number of activated bitter taste receptors
(Tas2Rs) subtypes, suggesting that the bitter taste is not a very reliable marker for toxicity, and is likely to have other physiological roles. This will be further discussed in
the next chapter within the context of toxicity *versus* bitterness of plant-based secondary metabolites aimed at protecting plants from pests, diseases and predators (including humans).

Table 6.1: Summary of chemoinformatics bitterness prediction methods.

Method	Description	Examples	References
Structure-based	Derives insight from the structural features of the receptor binding site or features required for activating the receptor and trigger off the bitterness sensation.	-Strychnine binding to both the vestibular and orthosteric binding sites of the bitter taster receptor hTas2R43. -Pharmacophore models derived from the homology models of the hTas2R39 receptor could explain the activation and non-activation of this receptor by a set of (iso)flavonoids. - Several hTas2R14 agonists were derived by adding large fragments to existing agonists, while retaining their activation property.	[26, 37, 39]
Ligand-based	Insight is derived from the structural features of known activators of taste receptors.	-Essential 2D molecular fingerprints and pharmacophore features for activating the bitter receptors hTas2R14 and hTas2R39 showed that methoxy and glycosyl groups are non-favorable for activating the receptor. -3D pharmacophore models derived from known (iso)flavonoids that activate these receptors show high performance in ROC analysis and could be successfully used to identify bitter principles that activate these receptors.	[34]
Machine learning	An algorithm that "learns" to classify bitter and non-bitter, bitter and sweet, active or not active against a given receptor are included in a training set.	-A Naïve Bayes classifier was used to discriminate between bitter and random compounds, and was able to predict 72.1 % of the bitter compounds correctly. -AdaBoost, a decision trees classifier was able to correctly classify over 80 % of the compounds in the hold-out test set.	[21, 42]

6.5 Conclusions

Taste is a very important physiological function, among which bitterness. In this chapter, an attempt has been made to provide an explanation for the bitter taste sensation at the molecular level by covering an overview of the various prediction methods, along with the database of bitter compounds. In the next chapter, software tools to assist in the prediction of bitterness in compounds will be described and an attempt will be made to explain why some medicinal plants are bitter. An exploration of the role of bitter natural products in the defense mechanism of plants will also be provided, with case studies focused on phenolic derivatives.

Acknowledgements: The author acknowledges a return fellowship and an equipment subsidy from the Alexander von Humboldt Foundation, Germany. Financial support for this work is acknowledged from the Ministry of Education, Youth and Sports of the Czech Republic.

References

[1] Quora. What is the biochemistry behind bitter plants being medicinal? (https://www.quora.com?What-is-the-biochemistry-behind-bitter-plants-being-medicinal.html). Assessed on 03 Januray 2019.

[2] Shaik FA, Singh N, Arakawa M, Duan K, Bhullar RP, Chelikani P. Bitter taste receptors: extraoral roles in pathophysiology. Int J Biochem Cell Biol. 2016;77:197–204.

[3] Bahia MS, Nissim I, Niv MY. Bitterness prediction *in-silico*: a step towards better drugs. Int J Pharm. 2017;536:526–9.

[4] Mennella JA, Spector AC, Reed DR, Coldwell SE. The bad taste of medicines: overview of basic research on bitter taste. Clin Ther. 2013;35:1225–46.

[5] Achan J, Talisuna AO, Erhart A, Yeka A, Tibenderana JK, Baliraine FN, et al. Quinine, an old anti-malarial drug in a modern world: role in the treatment of malaria. Malar J. 2011;10:144.

[6] Nolte DL, Russell Mason J, Lewis SL. Tolerance of bitter compounds by an herbivore, *Cavia porcellus*. J Chem Ecol. 1994;20:303–8.

[7] Ikeda K. New seasonings. Chem Senses. 2002;27:847–9.

[8] Gowthamarajan K, Kulkarni GT, Kumar MN. Pop the pills without bitterness. Resonance. 2004;9:25–32.

[9] Davis JD. Drug Cosmet. India: Encyclopedia of Pharmaceutical Technology, II, 2000.

[10] Li SP, Kowarski CR, Feld KM, Grim WM. Recent advances in microencapsulation technology and equipment. Drug Development and Industrial Pharmacy. 1988;14:353–76.

[11] Mennella JA, Beauchamp GK. Optimizing oral medications for children. Clin Ther. 2008;30:2120–32.

[12] Vital Plan: The Impressive Health Benefits of Bitter Foods (https://vitalplan.com/blog/the-impressive-health-benefits-of-bitter-foods). Accessed on 03 January 2019.

[13] Barratt-Fornell A, Drewnowski A. The taste of health: nature's bitter gifts. Nutr Today. 2002;37:144–50.

[14] Craig WJ. Phytochemicals: guardians of our health. J Am Diet Assoc. 1997;97:S199–204.

[15] Drewnowski A, Gomez-Carneros C. Bitter taste, phytonutrients and the consumer: a review. Am J Clin Nutr. 2000;72:1424–35.

[16] Rouseff RL. Bitterness in food products: an overview. In: Rouseff RL, editor. Bitterness in foods and beverages. Developments in food science. Vol. 25. Amsterdam: Elsevier; 1990: 1–14.

[17] Drewnowski A. Taste preferences and food intake. Ann Rev Nutr. 1997;17:237–53.

[18] Wiener A, Shudler M, Levit A, Niv MY. BitterDB: a database of bitter compounds. Nucleic Acids Res. 2012;40:D413–9.

[19] Dagan-Wiener A, Di Pizio A, Nissim I, Bahia MS, Dubovski N, Margulis E, et al. BitterDB: taste ligands and receptors database in 2019. Nucleic Acids Res. 2019;47:D1179–85. DOI:.

[20] Wildman SA, Crippen GM. Prediction of physicochemical parameters by atomic contributions. J Chem Inf Comput Sci. 1999;39:868–73.

[21] Wiener AD, Nissim I, Abu NB, Borgonovo G, Bassoli A, Niv MY. Bitter or not? BitterPredict, a tool for predicting taste from chemical structure. Sci Rep. 2017;7:12074.

[22] Nissim I, Wiener AD, Niv MY. The taste of toxicity: a quantitative analysis of bitter and toxic molecules. IUBMB Life. 2017;69:938–46.

[23] Bitter DB, Institute of biochemistry, food science and nutrition, faculty of agriculture, The Hebrew University of Jerusalem (http://bitterdb.agri.huji.ac.il/dbbitter.php). Accessed on 15 January 2019.

[24] Kufareva I, Katritch V, Rc S, Abagyan R. Advances in GPCR modeling evaluated by the GPCR Dock 2013 assessment: meeting new challenges. Structure. 2014;22:1120–39.

[25] Brockhoff A, Behrens M, Niv MY, Meyerhof W. Structural requirements of bitter taste receptor activation. Proc Natl Acad Sci USA. 2010;107:11110–15.

[26] Sandal M, Behrens M, Brockhoff A, Musiani F, Giorgetti A, Carloni P, et al. Evidence for a transient additional ligand binding site in the TAS2R46 bitter taste receptor. J Chem Theory Comput. 2015;11:4439–49.

[27] Neri M, Anselmi C, Cascella M, Maritan A, Carloni P. Coarse-grained model of proteins incorporating atomistic detail of the active site. Phys Rev Lett. 2005;95:218102.

[28] Leguèbe M, Nguyen C, Capece L, Hoang Z, Giorgetti A, Carloni P. Hybrid molecular mechanics/coarse-grained simulations for structural prediction of G-protein coupled receptor/ligand complexes. PLoS One. 2012;7:e47332.

[29] Marchiori A, Capece L, Giorgetti A, Gasparini P, Behrens M, Carloni P, et al. Coarse-grained/molecular mechanics of the TAS2R38 bitter taste receptor: experimentally-validated detailed structural prediction of agonist binding. PLoS One. 2013;8:e64675.

[30] Biarnes X, Marchiori A, Giorgetti A, Lanzara C, Gasparini P, Carloni P, et al. Insights into the binding of phenyltiocarbamide (PTC) agonist to its target human TAS2R38 bitter receptor. PLoS One. 2010;5:e12394.

[31] Behrens M, Meyerhof W. Vertebrate bitter taste receptors: keys for survival in changing environments. J Agric Food Chem. 2018;66:2204–13.

[32] Suku E, Fierro F, Giorgetti A, Alfonso-Prieto M, Carloni P. Multi-scale simulations of membrane proteins: the case of bitter taste receptors. J Sci Adv Mater Devices. 2017;2:15–21.

[33] Roland WS, Vincken J-P, Gouka RJ, van Buren L, Gruppen H, Smit G. Soy isoflavones and other isoflavonoids activate the human bitter taste receptors hTAS2R14 and hTAS2R39. J Agric Food Chem. 2011;59:11764–71.

[34] Roland WS, van Buren L, Gruppen H, Driesse M, Gouka RJ, Smit G, et al. Bitter taste receptor activation by flavonoids and isoflavonoids: modeled structural requirements for activation of hTAS2R14 and hTAS2R39. J Agric Food Chem. 2013;61:10454–66.

[35] Roland WS, Gouka RJ, Gruppen H, Driesse M, van Buren L, Smit G, et al. 6-Methoxyflavanones as bitter taste receptor blockers for hTAS2R39. PLoS One. 2014;9:e94451.

[36] Roland WS, Sanders MP, van Buren L, Gouka RJ, Gruppen H, Vincken JP, et al. Snooker structure-based pharmacophore model explains differences in agonist and blocker binding to bitter receptor hTAS2R39. PLoS One. 2015;10:e0118200.

[37] Meyerhof W, Batram C, Kuhn C, Brockhoff A, Chudoba E, Bufe B, et al. The molecular receptive ranges of human TAS2R bitter taste receptors. Chem Senses. 2010;35:157–70.

[38] Levit A, Nowak S, Peters M, Wiener A, Meyerhof W, Behrens M, et al. The bitter pill: clinical drugs that activate the human bitter taste receptor TAS2R14. FASEB J. 2014;28:1181–97.

[39] Karaman R, Nowak S, Di Pizio A, Kitaneh H, Abu-Jaish A, Meyerhof W, et al. Probing the binding pocket of the broadly tuned human bitter taste receptor TAS2R14 by chemical modification of cognate agonists. Chem Biol Drug Des. 2016;88:66–75.

[40] Levit A, Beuming T, Krilov G, Sherman W, Niv MY. Predicting GPCR promiscuity using binding site features. J Chem Inf Model. 2014;54:184–94.

[41] Di Pizio A, Niv MY. Promiscuity and selectivity of bitter molecules and their receptors. Bioorg Med Chem. 2015;23:4082–91.

[42] Rodgers S, Glen RC, Bender A. Characterizing bitterness: identification of key structural features and development of a classification model. J Chem Inf Model. 2006;46:569–76.

Fidele Ntie-Kang

7 Mechanistic role of plant-based bitter principles and bitterness prediction for natural product studies II: prediction tools and case studies

Abstract: The first part of this chapter provides an overview of computer-based tools (algorithms, web servers, and software) for the prediction of bitterness in compounds. These tools all implement machine learning (ML) methods and are all freely accessible. For each tool, a brief description of the implemented method is provided, along with the training sets and the benchmarking results. In the second part, an attempt has been made to explain at the mechanistic level why some medicinal plants are bitter and how plants use bitter natural compounds, obtained through the biosynthetic process as important ingredients for adapting to the environment. A further exploration is made on the role of bitter natural products in the defense mechanism of plants against insect pest, herbivores, and other invaders. Case studies have focused on alkaloids, terpenoids, cyanogenic glucosides and phenolic derivatives.

Keywords: bitterness, computer-based tools, drug discovery, machine learning, medicinal plants, natural products

7.1 Introduction

Many natural products (NPs) are bitter, including many drugs in the market [1]. In the previous chapter [2], an attempt was made to explain taste on a molecular basis in humans, based on molecules interacting with bitter taste receptors (hTas2Rs) [3]. NPs or secondary metabolites (SMs) are compounds of natural origin but which do not appear to have any direct functions in growth and development, i. e. they have no generally recognized roles in the process of photosynthesis, respiration, solute transport, translocation, nutrient assimilation, and differentiation. NPs and SMs play a significant role in the direct defense against herbivores by impairing performance by one of two general mechanisms:
- they may reduce the nutritional value of plant food, or
- they may act as feeding deterrents or toxins.

This article has previously been published in the journal *Physical Sciences Reviews*. Please cite as: Ntie-Kang, F. Mechanistic role of plant-based bitter principles and bitterness prediction for natural product studies II: prediction tools and case studies. *Physical Sciences Reviews* [Online] 2019 DOI: 10.1515/psr-2019-007

https://doi.org/10.1515/9783110668896-007

There has been considerable debate as to which of these two strategies is more important for host plant selection and herbivore resistance, particularly related to questions as to what extent variation in the levels of primary and secondary metabolites has evolved as a plant defense [4].

Medicinal plants are known to make several metabolites for the purpose of setting up their defense mechanism and to protect themselves against predators. Bitterness is an inherent property of many toxic chemicals, protecting humans and animals from self-poisoning. It must, however, be recalled that many toxins are not bitter, that many important bitter NPs have been used as rather curative agents (not toxins) [5], and that some known bitter chemicals do not deter herbivores [6]. Apart from the fact that several bitter principles are known to be therapeutic [5], a number of *in silico* models have been developed to predict the toxicity of chemicals based on chemical structure [7].

The focus was on structure-based, ligand-based, and machine learning (ML) methods. Bitterness is a deterrent factor for orally administered drugs. Due to the expensive and laborious experimental screening for determining if a compound (in foods or drugs) is bitter, *in silico* models are urgently needed. Besides, bitter compounds are quite diverse in chemical structure and are currently known to bind 21 out of the 25 known hTas2Rs, making bitterness prediction quite a difficult task.

In this chapter, an attempt is made to provide a mechanistic view of the role of bitter principles in the defense mechanism of plants. The first part, however, provides an overview of computer-based tools for the prediction of bitterness. These are based on ML models, e. g. decision trees (DTs), random forest (RF), support vector machines (SVMs), etc., built from known datasets of bitter compounds (e. g. BitterDB [8–10]) *versus* known sweet compounds and/or known non-bitter compounds.

7.2 Available tools for predicting bitterness

Several tools are currently available for predicting bitterness in chemical compounds, all of which are based on ML methods (Table 7.1). The description is presented based on the methods implemented, the datasets used in the training and test sets, the main results obtained, etc. The tools in Table 7.1 are arranged according to the order of publication.

7.2.1 A note on machine learning methods

Full coverage of ML is beyond the scope of this chapter. Basically, ML methods are algorithms that are trained to find patterns within data and could be classified as supervised (e. g. deep neural networks, support vector machines, etc.) and unsupervised (e. g. random forests). Let us first define a couple of terms recurrent in ML:

Table 7.1: Summary of software/tools for bitterness prediction.

Tools	Methods implemented/useful links	Advantages	References
BitterX	Support vector machines (SVMs) classifier. http://mdl.shsmu.edu.cn/BitterX	– Freely accessible web server. – Includes a user-friendly web interface.	[11]
BitterPredict	Decision trees-based tool for predicting bitterness. https://github.com/Niv-Lab/BitterPredict1	– The Matlab code is accessible.	[12]
BitterSweetForest	Random forest classifier for distinguishing bitter from sweet compounds. http://bioinformatics.charite.de/sweet/	– Open access tool. – Runs on a fully available KNIME workflow.	[13]
e-Bitter	Uses consensus votes from multiple machine learning methods (e. g. deep learning etc.). https://winpython.github.io/ https://www.dropbox.com/sh/3sebvza3qzmazda/AADgpCRXJtHAJzS8DK_P-q0ka?dl=0	– Adopts a consensus model for the prediction. – Has a free stand-alone software.	[14]
BitterSweet	Combines random forest and adaptive boosting (AB) to enhance classifier performance in order to predict both bitterness and sweetness. https://github.com/cosylabiiit/bittersweet/ https://github.com/cosylabiiit/bittersweet/data/	– The software is freely available for non-commercial use. – The prediction server is equipped with a user-friendly interface. – Data used in constructing the models are freely available for download. – Training sets cover a wider chemical space than previous models.	[15]

7.2.1.1 Training set

The dataset of compounds used to build the model. Like the test set, this is often composed of compounds with a certain characteristic and those that do not have, e. g. bitter compounds *versus* non-bitter compounds.

7.2.1.2 Test set
The dataset of compounds used to prove or validate the model.

7.2.1.3 Decision trees
A DT uses a tree-like model of decisions and their possible consequences. They are used commonly used in operations research to arrive at decisions but are also popular in ML. Typically, a DT is a flowchart-like structure in which each internal node represents a "test" on an attribute (e. g. either an event occurs or not), each branch represents the outcome of the test, and each leaf node represents a class label (decision taken after computing all attributes). The paths from the root to the leaf represent classification rules. DTs can easily become unstable, i. e. a small change in the data could lead to a large change in the structure of the optimal DT. As a result, predictors derived by other ML methods would perform better with similar data. One way to solve this problem could be by replacing a single DT with a random forest (RF) of DTs. However, RF is not often easy to interpret when compared with a single DT.

7.2.1.4 Artificial neural network
Artificial neural networks (ANN) are inspired by the neural networks in the biological system that form animal brains since the original goal of the ANN approach was to solve problems as the human brain would do. However, the neural network is not an algorithm in itself, but a framework where several different ML algorithms work together in order to treat the data inputs. As such, an ANN is based on a collection of connected units (nodes) called artificial neurons (AN), with each connection functioning like the synapses in the brain, processing and transmitting signals from one AN to another. At a connection, the signal between ANs is a real number, and the output is computed by some non-linear function of the sum of its inputs. The connections between ANs are called "edges", ANs and edges typically having a weight that adjusts as learning proceeds. The weight increases or decreases the strength of the signal at a connection. ANs may have a threshold such that the signal is only sent if the aggregate signal crosses that threshold. Typically, ANs are aggregated into layers. Different layers may perform different kinds of transformations on their inputs. Signals travel from the first layer (the input layer) to the last layer (the output layer), possibly after traversing the layers many times.

7.2.1.5 Deep neuron network
Deep neuron network or deep neural network (DNN) is a neural network with more than one hidden layer between the input and output layers. In DNN, thousands of neurons in each layer can be extensively applied to the dataset with thousands of features, and more advanced regularization technique such as the dropout can be

used to prevent the overfitting problem. Nevertheless, DNN requires users to adjust a variety of parameters.

7.2.1.6 k-nearest neighbors
The k-nearest neighbors (k-NN) algorithm is among the simplest of all ML algorithms. Both for classification and regression, a weight is assigned to the contributions of the neighbors, so that the nearer neighbors contribute more to the average than the more distant ones. For example, a common weighting scheme consists in giving each neighbor a weight of $1/d$, where d is the distance to the neighbor. This can be thought of as the training set for the algorithm, though no explicit training step is required.

7.2.1.7 Random forest
An RF algorithm is a type of ensemble learning method that constructs a large number of decisions trees (usually greater than 100), and outputs predictions based on a collection of the votes of the individual trees. A subset of the training dataset is chosen to grow individual trees, while the remaining samples are used to estimate the optimal fit. The constructed trees are grown by splitting the training set (subset) at each node according to the value of the random variable, which is sampled independently from a subset of variables.

7.2.1.8 Support vector machine
Support vector machine (SVM) or support vector network is a popular supervised ML technique that is used for classification and regression. Given a set of training examples, each marked as belonging to one of two categories, an SVM training algorithm builds a model that predicts whether a new example falls into one category or the other. The algorithm performs the classification by constructing the hyperplanes in the multi-dimensional space that separates the different classes.

7.2.1.9 Validation or performance assessment
The performance of statistical learning methods like ML is often measured by the number of true positives (*TP*), true negatives (*TN*), false positives (*FP*), and false negatives (*FN*). In this scenario, *TP*, *TN*, *FP*, and *FN* would represent true bitterant, true non-bitterant, false bitterant, and false non-bitterant compounds, respectively.

Precision (P_{re}) is a measure of accuracy for a specific, predicted class.

$$P_{re} = \frac{TP}{TP + FP} \tag{7.1}$$

Accuracy (A_{cc}) is another frequently used index for the overall classification performance, but it may be misleading due to the highly unbalanced class distribution in the used datasets.

$$A_{cc} = \frac{TP + TN}{TP + TN + FP + FN} \tag{7.2}$$

Sensitivity (S_e) or *recall* and *specificity* (S_p) can assess a model's ability to correctly identify *TP*s and *TN*s, respectively. These two parameters are usually interpreted in combination with each other.

$$S_e = \frac{TP}{TP + FN} \tag{7.3}$$

$$S_p = \frac{TN}{TN + FP} \tag{7.4}$$

The indices in eqs. (7.1) to (7.4) are often used for model validation and comparison.

In addition, the *F1 measure* (or *F1-score*) and *non-error rate* (*NER*) are defined, respectively, as:

$$F1\ measure = \frac{2 \cdot P_{re} \cdot S_e}{P_{re} + S_e} = \frac{2 \times TP}{2 \times TP + FP + FN} \tag{7.5}$$

$$NER = \frac{S_e \cdot S_p}{2} \tag{7.6}$$

F1-score (cross-validation) is evaluated on the internal validation dataset during the cross-validation. Meanwhile, *F1- score* (test test) is when *F1-score* is assessed on the test set. ΔF1-score is the absolute value of the difference between *F1- score* (*cross-validation*) and *F1-score* (*test set*), i. e.:

$$\Delta F1 - score = |F1 - score\ (cross - validation) - F1 - score\ (test\ test)| \tag{7.7}$$

ΔF1-score is used to monitor the potential overfitting (or underfitting), i. e. if ΔF1-score is small, it means that the model performances are similar on the internal-validation dataset and test set.

7.2.1.10 Area under the curve (AUC)

This prediction metric is derived from a Receiver Operator Characteristics (*ROC*) plot. The ROC curve is a plot of the *FP* rate (1− S_p) on the *y*-axis against the *TP* rate (S_e) on the *x*-axis while varying the decision threshold. The area under the curve (*AUC*) of the *ROC* plot provides a convenient way of comparing classifiers. An *AUC* value of 0.5 represents a random classifier, while an ideal classifier has an area of 1.0.

7.2.1.11 Matthews correlation coefficient

F1-score and Matthews correlation coefficient (*MCC*), eq. (7.8), are commonly used to measure the quality of binary classifications.

$$MCC = \frac{(TP \times TN - FP \times FN)}{\sqrt{(TP + FP)(TP + FN)(TN + FP)(TN + FN)}} \qquad (7.8)$$

7.2.1.12 Y-randomization

This is a tool used in the validation of statistical models, whereby the performance of the original model in data description (r^2) is compared to that of models built for permuted (randomly shuffled) responses, based on the original descriptor pool and the original model building procedure [16].

7.2.1.13 Principal component analysis (PCA)

PCA is a mathematical technique that captures the linear interactions between the underlying attributes in a dataset. Every principal component can be expressed as a combination of one or more existing variables. All principal components are orthogonal to each other, and each one captures some amount of variance in the data.

7.2.2 BitterX – a support vector machines bitterness predictor

BitterX was the first web server tool that could be used to predict the human bitter taste receptors that bind certain small molecules [11]. It is available at http://mdl. shsmu.edu.cn/BitterX, with a web interface, Figure 7.1.

This tool serves two functions:
- identifying if a compound is a bitterant (or bitter taste receptor activator) and
- predicting its possible bitter taste receptors (Tas2Rs).

The SVMs model was built using a training set manually curated from the literature using PubMed and BitterDB [8]. This included 540 bitterants, i. e. 260 positive and 2379 negative bitterant-Tas2R interactions. Data on the bitterant and bitterant-Tas2R interactions were collected manually from the literature in order to be used for identifying bitterant-Tas2R interactions. The molecular structure file of each bitter compound obtained from PubChem [17] was input into a program Checker and ChemAxon's Standardizer for predicting the interactions with Tas2Rs. The benchmark evaluations showed that the models for bitterant determination and receptor recognition could accurately predict the activities of the test dataset [11]. Besides, BitterX could accurately predict the known Tas2Rs of several experimentally proven bitterants.

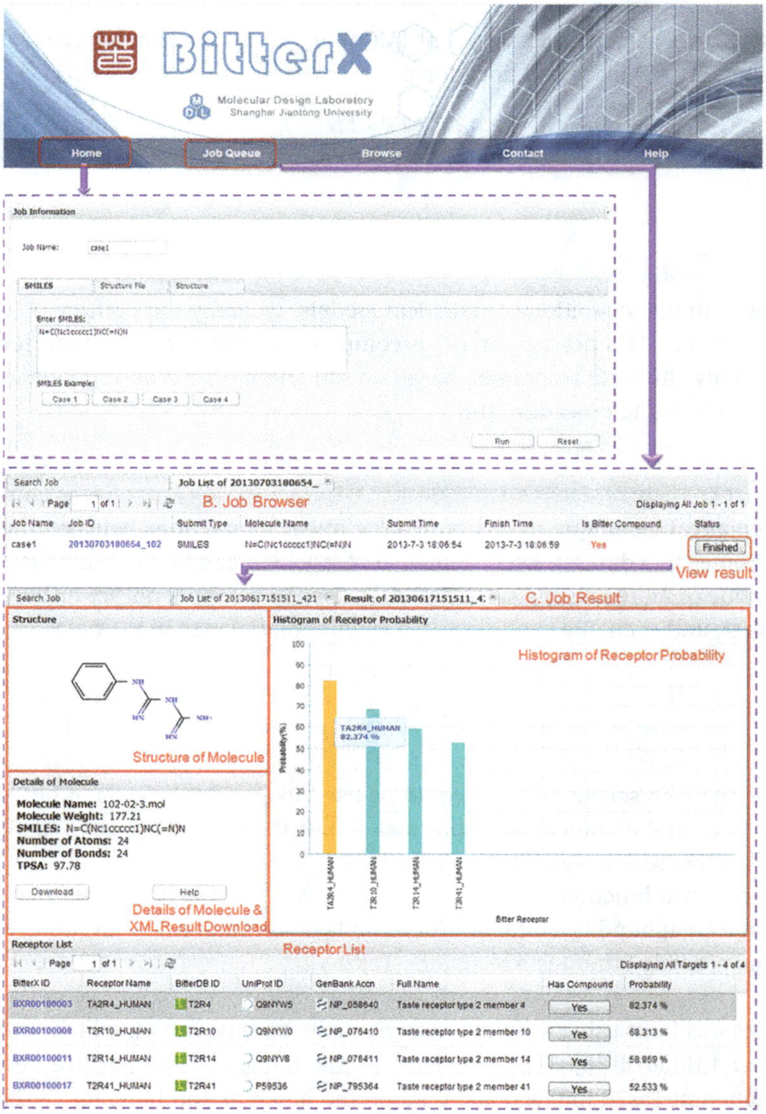

Figure 7.1: Web interface and output of BitterX [11]; (A) Query input in homepage (B) Bitterant-Tas2R interaction entries in "Browse" page. (C) An example of an output page after submitting a chemical molecule. A confidence score in probability is displayed along with the associated Tas2R in both the "Receptor List" and the Column Chart, which can be retrieved by clicking "Show Receptor Histogram". Material reproduced from data originally published under a Creative Commons (CC BY) License.

7.2.3 BitterPredict – a decision trees-based tool for predicting taste from chemical structure

BitterPredict predicts whether a compound is bitter or not and is built on the adaptive boosting (AdaBoost) DTs classifier [12]. It implements an algorithm in which the DTs are built sequentially by learning from mis-classified samples of the former DT. The positive training set includes 632 molecules from BitterDB [8], while about 2,000 non-bitter molecules were gathered from literature to create the negative set. The non-bitter set was composed into three subsets: non-bitter flavors, sweet molecules, and tasteless molecules. The classifier was based on physicochemical and ADME/Tox descriptors. BitterPredict was able to correctly classify >80% of the compounds in the hold-out test set, and 70–90% of the compounds in three independent external sets and in sensory test validation. This implies that BitterPredict is a quick and reliable tool for classifying large sets of compounds into bitter and non-bitter groups. In addition, the tool suggested ~ 40% of random molecules, and a large portion (66%) of clinical and experimental drugs, and of NPs (77%) to be bitter. The Matlab code for BitterPredict is provided *via* BitterDB http://bitterdb.agri.huji.ac.il/dbbitter.php#Bitter-Predict and *via* GitHub repository https://github.com/Niv-Lab/BitterPredict1.

7.2.4 BitterSweetForest – a random forest open access tool

BitterSweetForest [13] uses a random forest (RF) classifier, based on molecular fingerprints that were used to discriminate between sweet- and bitter-tasting molecules. It is an open access model and is implemented on a KNIME workflow [18] that provides a platform for predicting if a compound would be bitter or sweet. A training set 1,202 of compounds, i.e. 517 artificial and natural sweeteners from the SuperSweet [19] against 685 bitter compounds from the BitterDB [8], was used to construct the model. The original model yielded an accuracy of 95% and an area under the curve (*AUC*) of 0.98 in cross-validation. The model was validated using an independent test set with an accuracy of 96% and an AUC of 0.98 for bitter and sweet taste prediction. This was then applied for the prediction of bitterness and sweetness in NPs from the Super natural II dataset [20], approved drugs from Drugbank [21], and known toxic compounds (with experimentally proven acute oral toxicity) from the Protox web server [22]. The BitterSweetForest tool predicted that up to 70% and 10% of the NPs from the Super natural II dataset, to be bitter and sweet, respectively, with a confidence score of 0.60 and above. In the same way, 77% and 2% of the approved drugs were predicted as bitter and as sweet, respectively, with a confidence score of 0.75 and above. Moreover, 75% of the toxic compounds were predicted only as bitter with a minimum confidence score of 0.75. This model, thus, suggested that toxic compounds, NPs, and approved drugs are mostly bitter.

7.2.5 e-Bitter – a free software for bitterness prediction

The e-Bitter tool [13] is based on harnessed consensus votes from the multiple machine-learning methods (e. g. deep learning), combined with molecular fingerprints, to build classification models of compounds into either bitter or bitterless (non-bitter). The training set is composed of 707 experimentally proven bitterants (a majority from BitterDB [8]) and 592 non-bitterants (including sweet compounds downloaded from the SuperSweet dataset [19] and SweetenersDB [23], along with 132 tasteless and 17 non-bitter compounds retrieved from the literature). The extended-connectivity fingerprint (ECFP) [24] was adopted as the molecular descriptor to build the bitter/bitterless classification models. Five algorithms – k-NN, SVM, RF, gradient boosting machine (GBM), and DNN – were used to train the models via the Scikit-learn, Keras, and TensorFlow python libraries. The model was validated with a five-fold cross-validation.

Through an exhaustive parameter exploration with the five-fold cross-validation, all the models are carefully scrutinized by the Y-randomization test to ensure their reliability, and subsequently nine consensus models are constructed based on the individual or average models, which differ in term of accuracy, speed, and diversity of models. One of the best consensus models showed that accuracy, precision, specificity, sensitivity, $F1$-$score$, and Matthews correlation coefficient (MCC) gave respective values of 0.929, 0.918, 0.898, 0.954, 0.936, and 0.856 on the test set. It was additionally demonstrated that e-Bitter outperforms BitterX on three test sets, while showing better results than BitterPredict for two test sets.

A graphical user interface (Figure 7.2) was developed for the convenience of users. The tool is unique in that it adopts a consensus model for bitterness prediction and was the first free stand-alone software for bitterness prediction. Another advantage is that the entire training dataset is publicly available from the e-Bitter program and users can view the 3D structure of each compound and its corresponding classification as bitter or bitterless (Y: bitterant or N: non-bitterant).

7.2.6 BitterSweet – a freely available state-of-the-art software for bitter *versus* sweet taste prediction

This is the most recently published tool for bitterness (and sweetness) prediction, which combines random forest and adaptive boosting to enhance classifier performance [15]. The dimensionality of the molecular descriptors was reduced using principal component analysis (PCA) and t-distributed stochastic neighbor embedding (t-SNE) [25]. The effort was motivated by the inconsistencies observed in the curation process that led to the training datasets used to develop the models implemented in the previously described tools [11–14]. These include possible incorrect predictions that could result from including molecules with unverified taste information

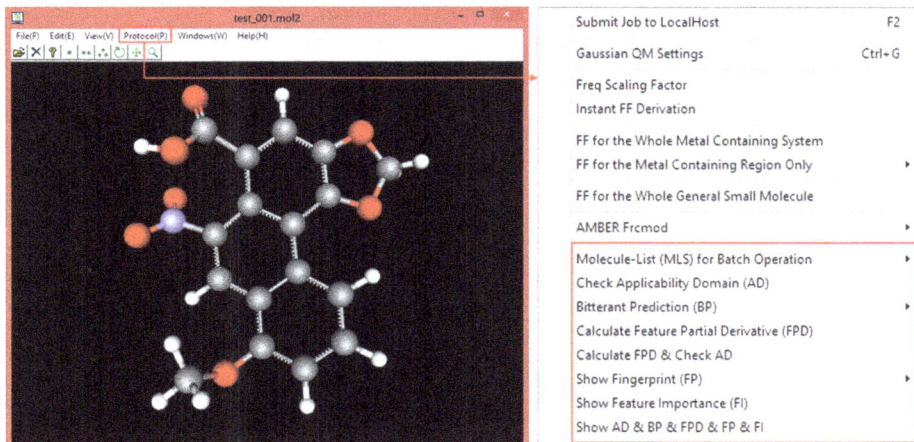

Figure 7.2: The basic functions in the e-Bitter program, which is highlighted by the red rectangle [14]. Material reproduced from data originally published under a Creative Commons (CC BY) License.

or incomplete representation of chemical space in the training set. For example, the training sets for constructing BitterX [11] and BitterPredict [12] included compounds with unverified non-bitterness (corresponding to 50 % and 55.6 % of non-bitter compounds, respectively), while BitterSweetForest [13] and e-Bitter [14] only used experimentally verified data. This considerably reduced the size of the training sets (and eventually) the possible bitter-sweet chemical space representation in the models.

BitterSweet is built on an exhaustive compilation of bitter, non-bitter, sweet, and non-sweet compounds from the literature, aimed at spanning the chemical space while not compromising the accuracy of taste information of the molecules. Its training set includes 918 bitter and 1510 non-bitter molecules as well as 1205 sweet and 1171 non-sweet molecules resulting from the careful curation of data from a wide variety of sources, ranging from scientific publications to books. Tasteless molecules were included as important controls for both bitter and sweet taste prediction. The datasets were separated into training and test sets, with the test set taken from the external validation/test sets obtained for the BitterPredict models [12].

The bitter-sweet taste prediction models were trained and evaluated using a wide spectrum of molecular descriptors, e. g. Dragon 2D/3D quantitative structure-activity relationships (QSAR) descriptors [26], ECFPs, physicochemical as well as ADMET (absorption, distribution, metabolism, excretion, and toxicity) properties from Canvas [27], as well as structural and physicochemical descriptors from ChemoPy [28]. Thus, BitterSweet implements state-of-the-art ML models for bitter-sweet taste prediction, whose performance has been proven on large specialized chemical sets, e. g. FlavorDB [29], FooDB (http://foodb.ca), SuperSweet [19], Super Natural II [20], DSSTox [30], and DrugBank [21]. All datasets for building the BitterSweet models have been made publicly available (https://github.com/cosylabiiit/bittersweet/). In addition, the BitterSweet predictor is implementable in a freely available software for bitter- and sweet-taste prediction.

7.3 Bitter natural products from plants

In general, plants make several bitter principles, including polyphenols (e. g. flavonoids, isoflavonoids, tannins, etc.) and alkaloids. Tannins, for example, are particularly useful in repelling unwanted insect predators, while flavonoids are cytotoxic to the herbivores by interacting with different enzymes through complexation [31]. The defense mechanism of plants against predators and harsh environmental conditions is complex, involving direct defense (e. g. by forming thorns, spines, prickles, hard waxy leaves, etc.), induced resistance (Figure 7.3) and indirect plant defense (Figure 7.4). Bitter principles only play a minor role, since some herbivores are known to tolerate bitter principles [5].

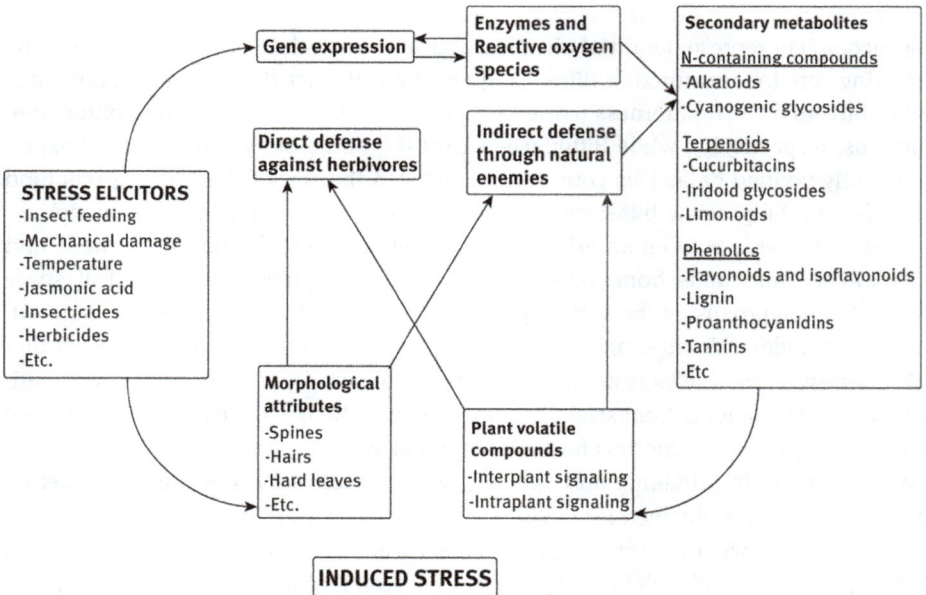

Figure 7.3: Induced resistance in plants. Figure adapted from reference [32].

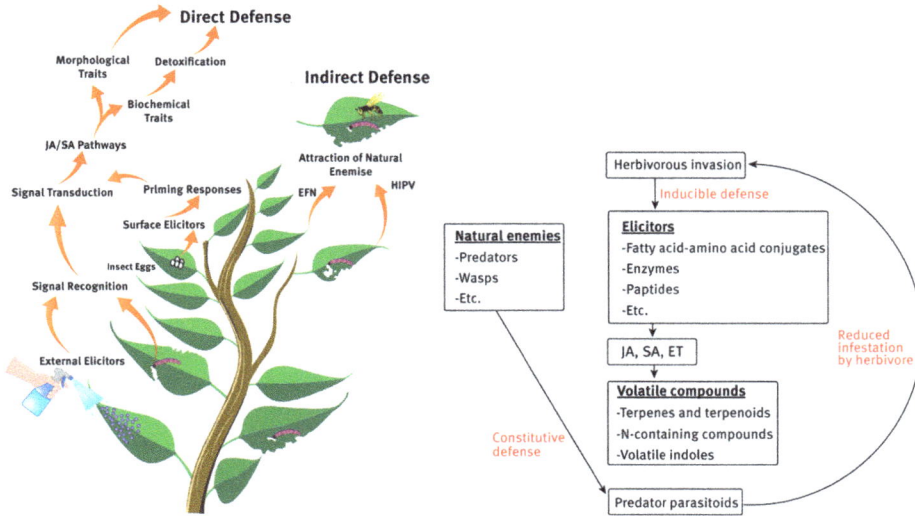

Figure 7.4: (Left) Plant defense against insect pests (EPF = extra floral nectar; HIPV = herbivore induced plant volatiles; JA = jasmonic acid; SA = salicylic acid). (Right) Major components and pathways involved in indirect plant defense (ET = Ethylene). Figure adapted from references [32, 33].

Induced resistance, against insect attack, for example, involves many signal transduction pathways mediated by a network of phytohormones (or plant hormones) [32, 33]. Phytohormones play a critical role in regulating plant growth, development, and defense mechanisms. The signal transduction pathways are mediated by jasmonic acid (JA, an important phytohormone), salicylic acid, and ethylene (Figure 7.4). Specific sets of defense-related genes are activated by these pathways upon wounding the plant or by insect feeding. JA is derived in octadecanoid pathway from linolenic acid. Rising levels of JA in response to herbivore attack often triggers the production of several proteins involved in plant defenses, e. g. proteins that inhibit digestion in the herbivore. Phytohormones may act individually, synergistically, or antagonistically, depending upon the attacker. The phytohormone accumulates upon wounding and or destruction of plant tissue by herbivores. Chewing of plant parts by insects, for example, causes the dioxygenation of linoleic and linolenic acids.

Ethylene is another important phytohormone, which plays an active role in plant defense against many insects, the ethylene signaling pathway playing an important role in induced plant defense against insects and pathogens both directly and indirectly. Ethylene signaling pathway works either synergistically or antagonistically, with JA in expression of plant defense responses against pathogens and herbivorous insects. It has been reported that ethylene and JA work together in tomato in proteinase inhibitors expression.

7.3.1 Why are some bitter plants medicinal?

It has been mentioned from the previous paragraphs that many NPs, including those contained in medicinal plants, are bitter [12]. The bitter taste and hence the therapeutic potential of these plants could be attributed to the presence of the former. The fact that plants do contain bioactive principles does not mean that plant had humans in mind when biosynthesizing the bioactive metabolite. The plants actually biosynthesized the bitter principle to defend itself against predators (including insects, mammal herbivores, etc.) and disease-causing pathogens. Humans only later discovered the curative properties of the plants in the quest to find treatment for their own diseases. The identification of the bioactive principles in the plants responsible for their therapeutic uses only followed later. SMs can either be stored in the inactive form or induced in response to the herbivore or microbe attack.

An example where a plant produces bitter bioactive metabolites as a result of pathogenic attack is seen in the process of cyanogenesis (i. e. the ability of living organisms to liberate a toxic compound hydrogen cyanide from stored nontoxic cyanogenic glycosides) [34–37]. This process involves the conversion of phytoanticipins (e. g. cyanogenic glycosides) to phytoalexins (e. g. antimicrobial compounds synthesized by plants that accumulate rapidly at areas of pathogen infection). The phytoanticipins are mainly activated by β-glucosidase when the plant is consumed by herbivores (Figure 7.5). This in turn mediates the release of various biocidal aglycone metabolites. Phytoalexins include isoflavonoids, terpenoids, alkaloids, etc., which, for example, influence the performance and survival of insects. Apart of the fact that the aforementioned compounds classes have a broad range of known biological activities useful for treatment of several human ailments, such SMs both defend the plants from different stresses and also increase the fitness of the plants [38].

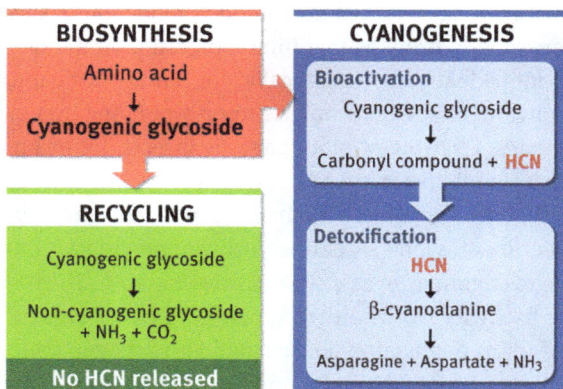

Figure 7.5: Summary of biosynthesis of cyanogenic glucosides and mechanism of cyanogenesis [34]. Figure reproduced by permission.

7.3.2 Understanding the role of natural bitter compounds in plant defense mechanisms

7.3.2.1 Defensive plant-based bitter principles

SMs involved in plant defense (collectively known as antiherbivory compounds) are often classified into three sub-groups: nitrogen-containing compounds (including alkaloids, cyanogenic glycosides, glucosinolates, and benzoxazinoids), terpenoids, and phenolics. The chemical structures of some plant-based bitter principles which are known to play a defensive role are shown in Figure 7.6. Among them, plant phenols are the most common and widespread group of defensive compounds. Their major role is in resistance against insects, microorganisms, and competing plants. Cucurbitacins, for example, are bitter compounds whose hostility to a wide range of herbivores, including lepidopteran larvae, beetles, mites, and vertebrate grazers,

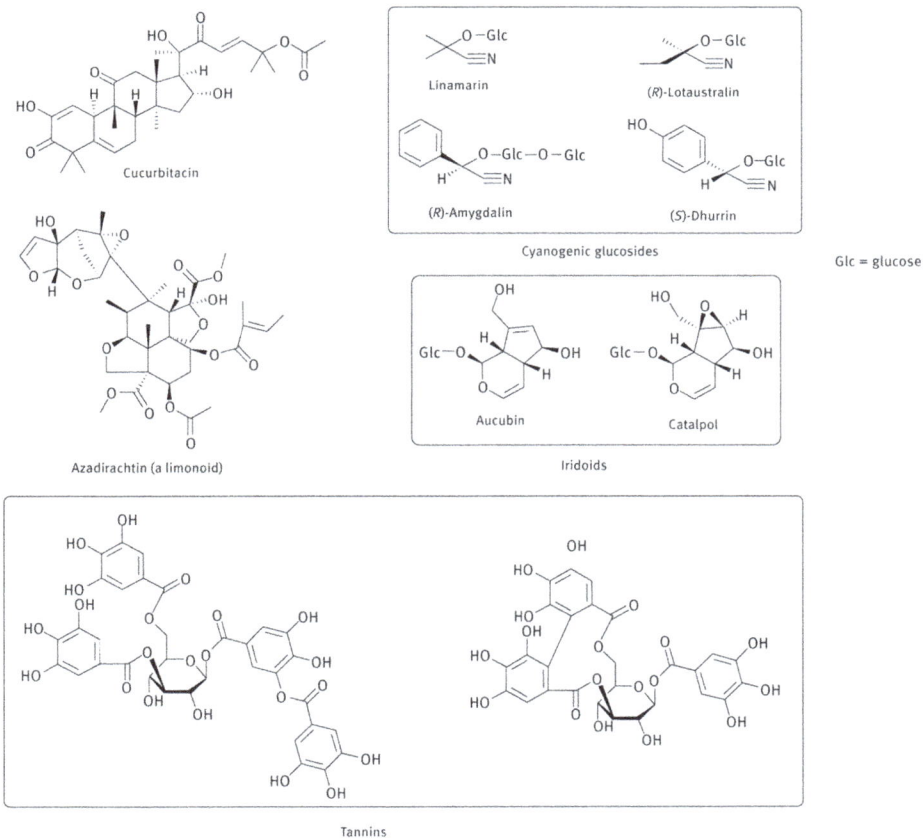

Figure 7.6: Chemical structures of some bitter principles involved in diverse defensive roles in plants.

can be explained by their taste [32]. The known defensive functions of some bitter SMs have been summarized in Table 7.2.

Table 7.2: Defensive role of plant-based bitter compounds.

Compound or compound class	Description	Function/mode of action	References
Alkaloids	NPs containing an N-containing heterocycle and are biosynthesized from various amino acids.	– Can inhibit or activate enzymes. – Alter carbohydrate and fat storage by inhibiting the formation of phosphodiester bonds involved in their breakdown. – Bind to nucleic acids and can inhibit the synthesis of proteins and affect DNA repair mechanisms. – Affect cell membrane and cytoskeletal structure causing the cells to weaken, collapse, or leak, and can affect nerve transmission.	[39, 40]
Cucurbitacins	A bitter class of triterpenes that makes plants hostile to a wide range of herbivores.	– Directly affect insect growth and development. – Act indirectly by acting as oviposition deterrents. – Act as phagostimulants to insect pests. – Cucurbitacins B and D have been reported as major phagostimulants for leaf beetles.	[41–43]
Cyanogenic glycosides	Bitter and toxic principles. Found in more than 2650 different plant species.	– These are bitter compounds that deter herbivores. – They are degraded by enzymes (e. g. β-glucosidases and hydroxynitrile lyases) to liberate (toxic) hydrogen cyanide to deter herbivores. – Formation of their breakdown products is initiated by microbial attack or consumption of plant material by herbivores.	[34–37]

Table 7.2 (continued)

Compound or compound class	Description	Function/mode of action	References
Flavonoids and isoflavonoids	Bitter polyphenols common in several plant families.	– Cytotoxic and interact with different enzymes through complexation. – Both classes of compounds protect the plant against insect pests by influencing the behavior, and growth and development of insects. – They are very important in plant resistance against pathogenic bacteria and fungi. – They may also be directly involved in the inhibition of the pathogen's enzymes, especially those digesting the plant cell wall, by chelating metals required for their activity.	[42–44]
Iridoid glycosides	A class of bitter cyclopentanoid monoterpene-derived compounds.	– In plants, they act primarily as a defense against herbivores or against infection by microorganisms. – Directly toxic to insect pests. – Reduce the nutritional quality of plant tissues, thereby rendering them less digestible to insects. – Denature amino acids, proteins, and nucleic acids by binding covalently to nucleophilic side chains via imine formation. – Inhibit the activity of enzymes involved in the formation of prostaglandins and leukotrienes.	[45]

Table 7.2 (continued)

Compound or compound class	Description	Function/mode of action	References
Lignin	A phenolic heteropolymer whose synthesis has been found to be induced by insect or pathogen attack.	– Limits the entry of pathogens by physically blocking the leaves. – Increases the leaf toughness, hence reducing the number of leaves consumed by insects. – Decreases the nutritional content of the leaf. – Its rapid deposition further reduces the growth of the pathogen or reduces insect fecundity.	[46, 47]
Limonoids	Phytochemicals of the triterpenoid class.	– Azadirachtin, for example, is a naturally occurring insecticide from this class of compounds. – Active as a feeding inhibitor towards the desert locust (Schistocerca gregaria). – Acts mainly as an antifeedant and growth disruptor.	[48, 49]
Proanthocyanidins	Oligomeric or polymeric flavonoids, also known as condensed tannins.	They serve among other chemical and induced defense mechanisms against plant pathogens and predators.	[50–53]
Tannins	Bitter, toxic, and poor tasting polyphenols that deter insects from feeding on plants containing them.	– Known to influence insect growth and development by binding to the predator vital proteins. – Reduce nutrient absorption efficiency and cause midgut lesions. – Precipitate proteins nonspecifically (including the digestive enzymes of insects), by hydrogen bonding or covalent bonding of protein NH_2 groups. – Chelate the metal ions, thereby reducing their bioavailability to insects. – When ingested, they reduce the digestibility of the proteins thereby decreasing the nutritive value of plants and plant parts to insects.	[31, 54]

Popular plant-based alkaloids include nicotine, caffeine, morphine, cocaine, colchicine, ergolines, strychnine, and quinine. Their (mostly) aversively bitter taste is a natural deterrent to herbivores. While alkaloids mostly act on receptors of neurotransmitters, others (such as phenolics and terpenoids) are less specific and attack a multitude of proteins by building hydrogen, hydrophobic, and ionic bonds, thus modulating their 3D structures and in consequence their bioactivities.

Cyanogenic glycosides (often stored in inactive forms in plant vacuoles) become toxic when consumed by herbivores. This is because the rupture of the plant cell membranes releases the glycosides which bring them into contact with HCN-releasing enzymes in the cytoplasm (Figure 7.5). HCN is known to be highly toxic by blocking cellular respiration [55]. Glucosinolates are activated in a much similar manner, but the products of their breakdown rather cause milder effects like gastroenteritis, salivation, diarrhea, and mouth irritation [40]. Benzoxazinoids are also stored as inactive glucosides in plant vacuoles. Upon plant tissue disruption when herbivores feed, these antiherbivory compounds get into contact with β-glucosidases from the chloroplasts and lead to the enzymatic release of toxic aglucones [56]. Some benzoxazinoids are only synthesized when herbivores start feeding. Such SMs are considered to act by an induced plant defense mechanism [57].

Among the terpenoids, diterpenoids are widely distributed in latex and resins and can be quite toxic. The high toxicity of Rhododendron leaves can be attributed to the presence of diterpenoids [58]. Saponins are complex triterpenoids which are known to break down the red blood cells of herbivores [59]. Among the limonoids (a sub-class of terpenoids), azadirachtin is a well-known naturally occurring insecticide, which is active as a feeding inhibitor towards the desert locust (*Schistocerca gregaria*), acting mainly as an antifeedant and growth disruptor [48, 49]. Iridoid glycosides (another sub-class terpenoids), e. g. aucubin and catalpol, prevent the invasion of plants by insects and microorganisms [45].

Phenolics are known to exhibit antiseptic properties, while others disrupt endocrine activity. From simple tannins to the more complex flavonoids which confer on plants much of their colorful pigments, polyphenols are known for several activities, e. g. antioxidant activity. Some polyphenols are involved in plant defense mechanisms, e. g. lignin, silymarin, and cannabinoids [46, 47]. Condensed tannins (e. g. proanthocyanidins), including 2 to >50 flavonoid molecules, inhibit herbivore digestion by binding to the consumed plant proteins, rendering digestion difficult or almost impossible. This is done by the SMs interfering with protein absorption and digestive enzymes [50–53, 60].

7.3.2.2 Case study: investigating human taste receptors of phenolic compounds

Soares et al. undertook an investigation of 6 polyphenols against several human taste receptors, with the view of identifying which receptors are activated by these compounds [61]. The compounds included; the hydrolyzable tannin pentagalloylglucose

(PGG), the precursor of condensed tannins (-)-epicatechin, two procyanidin oligomers or condensed tannins (the dimer B3 and the trimer C2), and the anthocyanins malvidin-3-glucoside and cyanidin-3-glucoside, which are commonly found in plant-based foods and drinks, e. g. red wine, beer, tea, and chocolate. The chemical structures of the investigated compounds have been shown on Figure 7.7.

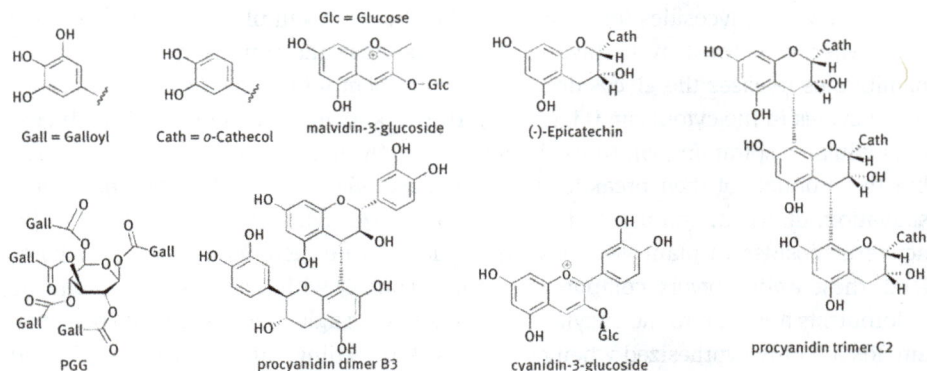

Figure 7.7: Chemical structures of some plant-based bitter principles investigated in the study [61].

The observed EC_{50} values for the different compounds vary 100-fold, the lowest values being for PGG and malvidin-3-glucoside. The compounds were shown to activate different combinations of the 25 hTas2Rs, e. g. (-)-epicatechin activated 3 of the receptors (i. e. hTas2R4, hTas2R5, and hTas2R39), while PGG activated only two receptors (i. e. hTas2R5 and hTas2R39). Meanwhile, malvidin-3-glucoside and procyanidin trimer only stimulated one receptor each (i. e. hTas2R7 and hTas2R5, respectively). The authors remarkably discovered tannins to be the first selective natural agonists for the receptor hTas2R5, with a high potency only toward this receptor. The authors also suggested the catechol and/or galloyl groups to be important structural determinants for mediating the interaction of these polyphenolic compounds with this receptor. This hypothesis could be verified by docking the compounds against the receptor site. The conclusions of this study would lead to the suggestion that the presence of these polyphenols in the food items could explain the bitterness of fruits, vegetables, and derived products even when in low concentrations.

7.4 Conclusions

Predicting bitterness is a very costly and challenging task for both the pharmaceutical and the food industry. In this work, a brief presentation of ML methods and recent ML tools (algorithms, web servers, and software) for taste prediction has been provided, with an emphasis on bitter/sweet compounds. Some of these methods are freely implemented in software packages and in some cases the tools for implementing them, e. g. curated datasets, KNIME workflows are available for download. Such tools/datasets would be quite useful for predicting if a SM is bitter or not and which receptor it is most likely to activate. SMs in general and bitter principles are known to be important in defending plants against predators and harsh environmental conditions. The usefulness of bitter principles from plants in defense against pathogens and herbivores has been highlighted in the second part of this work. Emphasis has been laid on nitrogen compounds (including alkaloids, cyanogenic glycosides, glucosinolates, and benzoxazinoids), terpenoids (iridoids, limonoids, and cucurbitacins), and phenolic compounds (flavonoids, isoflavonoids, lignin, proanthocyanidins, and tannins), which are involved in the defense of plants against herbivores and pathogens.

Acknowledgements: The author acknowledges a return fellowship and an equipment subsidy from the Alexander von Humboldt Foundation, Germany. Financial support for this work is acknowledged from the Ministry of Education, Youth and Sports of the Czech Republic.

List of Abbreviations

AB	adaptive boosting
AN	artificial neurons
ANN	artificial neural networks
AUC	area under the curve
BitterDB	database of bitter compounds and receptors
DNN	deep neuron network
DT	decision tree
ECFP	extended-connectivity fingerprints
FN	false negatives
FP	false positives
hTas2Rs	human bitter taste receptors
JA	jasmonic acid or jasmonate
k-NN	k-nearest neighbors
MCC	Matthews correlation coefficient
ML	machine learning
NPs	natural products
PCA	principal component analysis
PGG	pentagalloylglucose
RF	random forest
ROC	Receiver Operator Characteristics
SVMs	support vector machines
TN	true negatives
TP	true positives
t-SNE	t-distributed stochastic neighbor embedding

References

[1] Drewnowski A, Gomez-Carneros C. Bitter taste, phytonutrients and the consumer: a review. Am J Clin Nutr. 2000;72:1424–35.

[2] Ntie-Kang F. Mechanistic role of plant-based bitter principles and bitterness prediction for natural product studies I: database and methods. *Phys Sci Rev* 2019. DOI: 10.1515/psr-2018-0117.

[3] Shaik FA, Singh N, Arakawa M, Duan K, Bhullar RP, Chelikani P. Bitter taste receptors: extraoral roles in pathophysiology. Int J Biochem Cell Biol. 2016;77:197–204.

[4] Berenbaum MR. Turnabout is fair play: secondary roles for primary compounds. J Chem Ecol. 1995;21:925–40.

[5] Nolte DL, Russell Mason J, Lewis SL. Tolerance of bitter compounds by an herbivore, *Cavia porcellus*. J Chem Ecol. 1994;20:303–8.

[6] Barratt-Fornell A, Drewnowski A. The taste of health: nature's bitter gifts. Nutr Today. 2002;37:144–50.

[7] Bahia MS, Nissim I, Niv MY. Bitterness prediction *in-silico*: a step towards better drugs. Int J Pharm. 2017;536:526–9.

[8] Wiener A, Shudler M, Levit A, Niv MY. BitterDB: a database of bitter compounds. Nucleic Acids Res. 2012;40:D413–9.

[9] Dagan-Wiener A, Di Pizio A, Nissim I, Bahia MS, Dubovski N, Margulis E, et al. Bitter DB: taste ligands and receptors database in 2019. Nucleic Acids Res. 2019;47:D1179–85.

[10] Bitter DB. Institute of Biochemistry, Food Science and Nutrition, Faculty of Agriculture, The Hebrew University of Jerusalem. Available at: http://bitterdb.agri.huji.ac.il/dbbitter.php. Accessed: 15 Jan 2019.

[11] Huang W, Shen Q, Su X, Ji M, Liu X, Chen Y, et al. BitterX: a tool for understanding bitter taste in humans. Sci Rep. 2016;6:23450.

[12] Wiener AD, Nissim I, Abu NB, Borgonovo G, Bassoli A, Niv MY. Bitter or not? BitterPredict, a tool for predicting taste from chemical structure. Sci Rep. 2017;7:12074.

[13] Banerjee P, Preissner R. BitterSweetForest: a random forest based binary classifier to predict bitterness and sweetness of chemical compounds. Front Chem. 2018;6:93.

[14] Zheng S, Jiang M, Zhao C, Zhu R, Hu Z, Xu Y, et al. e-Bitter: bitterant prediction by the consensus voting from the machine-learning methods. Front Chem. 2018;6:82.

[15] Tuwani R, Wadhwa S, Bagler G. BitterSweet: building machine learning models for predicting the bitter and sweet taste of small molecules. bioRxiv. 2018. doi:10.1101/426692.

[16] Rücker C, Rücker G, Meringer M. y-Randomization and its variants in QSPR/QSAR. J Chem Inf Model. 2007;47:2345–57.

[17] Kim S, Thiessen PA, Bolton EE, Chen J, Fu G, Gindulyte A, et al. PubChem substance and compound databases. Nucleic Acids Res. 2016;44:D1202–13.

[18] Berthold MR, Cebron N, Dill F, Gabriel TR, Kötter T, Meinl T, et al. KNIME: the Konstanz information Miner. In: Preisach C, Burkhardt H, Schmidt-Thieme L, Decker R, editors. Data analysis, machine learning and applications SE - 38, Studies in Classification, Data Analysis, and Knowledge Organization. Berlin Heidelberg: Springer, 2008:319–26.

[19] Ahmed J, Preissner S, Dunkel M, Worth CL, Eckert A, Preissner R. SuperSweet – a resource on natural and artificial sweetening agents. Nucleic Acids Res. 2011;39:D377–82.

[20] Banerjee P, Erehman J, Gohlke BO, Wilhelm T, Preissner R, Dunkel M. Super natural II – a database of natural products. Nucleic Acids Res. 2015;43:D935–9.

[21] Wishart DS, Feunang YD, Guo AC, Lo EJ, Marcu A, Grant JR, et al. DrugBank 5.0: a major update to the drugbank database for 2018. Nucleic Acids Res. 2018;46:D1074–82.

[22] Drwal MN, Banerjee P, Dunkel M, Wettig MR, Preissner R. ProTox: a web server for the *in silico* prediction of rodent oral toxicity. Nucleic Acids Res. 2014;42:W53–8.

[23] Chéron JB, Casciuc I, Golebiowski J, Antonczak S, Fiorucci S. Sweetness prediction of natural compounds. Food Chem. 2017;221:1421–5.

[24] Rogers D, Hahn M. Extended-connectivity fingerprints. J Chem Inf Model. 2010;50:742–54.

[25] van der Maaten L, Hinton G. Visualizing data using t-SNE. J Mach Learn Res. 2008;9:2579–605.

[26] Mauri A, Consonni V, Pavan M, Todeschini R. Dragon software: an easy approach to molecular descriptor calculations. Match Commun Math Comput Chem. 2006;56:237–48.

[27] Duan J, Dixon SL, Lowrie JF, Sherman W. Analysis and comparison of 2D fingerprints: insights into database screening performance using eight fingerprint methods. J Mol Graph Model. 2010;29:157–70.

[28] Cao DS, Xu QS, Hu QN, Liang YZ. ChemoPy: freely available python package for computational biology and chemoinformatics. Bioinformatics. 2013;29:1092–4.

[29] Garg N, Garg N, Sethupathy A, Tuwani R, Nk R, Dokania S, et al. FlavorDB: a database of flavor molecules. Nucleic Acids Res. 2018;46:D1210–6.

[30] Richard AM, Williams CR. Distributed structure-searchable toxicity (DSSTox) public database network: a proposal. Mutat Res. 2002;499:27–52.

[31] Belete T. Defense mechanisms of plants to insect pests: from morphological to biochemical approach. Trends Tech Sci Res. 2018;2:555584.

[32] War AR, Taggar GK, Hussain B, Taggar MS, Nair RM, Hari C, et al. Plant defence against herbivory and insect adaptations. AoB Plants. 2018;10:ply037.

[33] War AR, Paulraj MG, Ahmad T, Buhroo AA, Hussain B, Ignacimuthu S, et al. Mechanisms of plant defense against insect herbivores. Plant Signal Behav. 2012;7:1306–20.

[34] Zagrobelny M, Bak S, Møller BL. Cyanogenesis in plants and arthropods. Phytochemistry. 2008;69:1457–68.

[35] Fürstenberg-Hägg J, Zagrobelny M, Bak S. Plant defense against insect herbivores. Int J Mol Sci. 2013;14:10242–97.

[36] Sánchez-Pérez R, Jørgensen K, Olsen CE, Dicenta F, Lindberg Møller B. Bitterness in almonds. Plant Physiol. 2008;146:1040–52.

[37] Malagón J, Garrido A. Relación entre el contenido de glicósidos cianogénicos y la resistencia a Capnodis tenebrionis l. En frutales de hueso. Bol Sanid Veg Plagas. 1990;16:499–503.

[38] Agrawal AA. Induced responses to herbivory in wild radish: effects on several herbivores and plant fitness. Ecology. 1999;80:1713–23.

[39] Roberts MF, Wink M. Alkaloids: biochemistry, ecology, and medicinal applications. New York: Plenum Press, 1998. ISBN 978-0-306-45465-3.

[40] Rhoades DF. Evolution of plant chemical defense against herbivores. In: Rosenthal GA, Janzen DH, editors. Herbivores: their interaction with secondary plant metabolites. New York: Academic Press, 1979:3–54. ISBN 978-0-12-597180-5.

[41] Chen JC, Chiu MH, Nie RL, Cordell GA, Qiu SX. Cucurbitacins and cucurbitane glycosides: structures and biological activities. Nat Prod Rep. 2005;22:386–99.

[42] Treutter D. Significance of flavonoids in plant resistance and enhancement of their biosynthesis. Plant Biol. 2005;7:581–91.

[43] Blount JW, Dixon RA, Paiva NL. Stress responses in alfalfa (*Medicago sativa* L.) XVI. Antifungal activity of medicarpin and its biosynthetic precursors; implications for the genetic manipulation of stress metabolites. Physiol Mol Plant Pathol. 1992;41:333–49.

[44] Dai GH, Nicole M, Andary C, Martinez C, Bresson E, Boher B, et al. Flavonoids accumulate in cell walls, middle lamellae and callose-rich papillae during an incompatible interaction

between *Xanthomonas campestris* pv. *malvacearum* and cotton. Physiol Mol Plant Pathol. 1996;49:285–306.

[45] Stephenson AG. Iridoid glycosides in the nectar of *Catalpa speciosa* are unpalatable to nectar thieves. J Chem Ecol. 1982;8:1025–34.

[46] Bagniewska-Zadworna A, Barakat A, Łakomy P, Smoliński DJ, Zadworny M. Lignin and lignans in plant defence: insight from expression profiling of cinnamyl alcohol dehydrogenase genes during development and following fungal infection in Populus. Plant Sci. 2014;229:111–21.

[47] Liu Q, Luo L, Zheng L. Lignins: biosynthesis and biological functions in plants. Int J Mol Sci. 2018;19:335.

[48] Mordue AJ, Blackwell A. Azadirachtin: an update. J Insect Physiol. 1993;39:903–24.

[49] Lee JW, Jin CL, Jang KC, Choi GH, Lee HD, Kim JH. Investigation on the insecticidal limonoid content of commercial biopesticides and neem extract using solid phase extraction. J Agric Chem Env. 2013;2:81–5.

[50] Laaksonen OA, Salminen JP, Mäkilä L, Kallio HP, Yang B. Proanthocyanidins and their contribution to sensory attributes of black currant juices. J Agric Food Chem. 2015;63:5373–80.

[51] Ferreira D, Marais JP, Coleman CM, Slade D. Comprehensive natural products II 6.18 - proanthocyanidins: chemistry and biology. In: Liu HW, Mander L, editor(s). *Comprehensive natural products II chemistry and biology* Vol. 6. Oxford, UK: Elsevier, 2010:605–61.

[52] Ma X, Yang W, Laaksonen O, Nylander M, Kallio H, Yang B. Role of flavonols and proanthocyanidins in the sensory quality of sea buckthorn (*Hippophaë rhamnoides* L.) berries. J Agric Food Chem. 2017;65:9871–9.

[53] Amil-Ruiz F, Blanco-Portales R, Munoz-Blanco J, Caballero JL. The strawberry plant defense mechanism: a molecular review. Plant Cell Physiol. 2011;52:1873–903.

[54] Wink M. Modes of action of herbal medicines and plant secondary metabolites. Medicines. 2015;2:251–86.

[55] Niemeyer HM. Plant cyanogenic glycosides. Toxicon. 2000;38:11–36.

[56] Niemeyer HM. Hydroxamic acids derived from 2-hydroxy-2*H*-1,4-benzoxazin-3(4H)-one: key defense chemicals of cereals. J Agric Food Chem. 2009;57:1677–96.

[57] Glauser G, Marti G, Villard N, Doyen GA, Wolfender JL, Turlings TC, et al. Induction and detoxification of maize 1,4-benzoxazin-3-ones by insect herbivores. Plant J. 2011;68:901–11.

[58] Wink M, van Wyk BE. Mind-altering and poisonous plants of the world. Portland, OR, USA: Timber Press; 2010.

[59] Herrmann F, Wink M. Synergistic interactions of saponins and monoterpenes in HeLa and Cos7 cells and in erythrocytes. Phytomedicine. 2011;18:1191–6.

[60] van Soest PJ. Nutritional ecology of the ruminant: ruminant metabolism, nutritional strategies, the cellulolytic fermentation, and the chemistry of forages and plant fibers. Corvallis, Oregon: O & B Books, 1982. ISBN 978-0-9601586-0-7.

[61] Soares S, Kohl S, Thalmann S, Mateus N, Meyerhof W, De Freitas V. Different phenolic compounds activate distinct human bitter taste receptors. J Agric Food Chem. 2013;61:1525–33.

Abraham Madariaga-Mazón, Ricardo Bruno Hernández-Alvarado, Karla Olivia Noriega-Colima, Adriana Osnaya-Hernández and Karina Martinez-Mayorga

8 Toxicity of secondary metabolites

Abstract: Secondary metabolites, commonly referred to as natural products, are produced by living organisms and usually have pharmacological or biological activities. Secondary metabolites are the primary source for the discovery of new drugs. Furthermore, secondary metabolites are also used as food preservatives, biopesticides or as research tools. Although secondary metabolites are mainly used by their beneficial biological activity, the toxicity of some of them may limit their use. The toxicity assessment of any compound that is prone to be used in direct contact with human beings is of vital importance. There is a vast spectrum of experimental methods for toxicity evaluation, including *in vitro* and *in vivo* methodologies. In this work, we present an overview of the different sources, bioactivities, toxicities and chemical classification of secondary metabolites, followed by a sketch of the role of toxicity assessment in drug discovery and agrochemistry.

Keywords: secondary metabolites, drugs, agrochemicals, toxicity assessment

8.1 Introduction

Organisms, such as bacteria, fungi, or plants, produce a variety of organic compounds, being some of them essential for the survival of the organism. These compounds include sugars, proteins, and amino acids and are commonly known as primary metabolites. Other types of compounds that are not essential for their normal growth, development, or reproduction, are classified as secondary metabolites or generically also called natural products. Williams et al. [1] defined natural products as substances that have no apparent role in the internal economy of the organism that produces it.

Secondary metabolites have served as drugs or as an inspiration to obtain drugs. Other applications include dyes, polymers, fibers, glues, oils, waxes, flavoring agents, and perfumes. Consequently, awareness of their biological properties has opened the search for new drugs, antibiotics, insecticides, and herbicides, among other uses [2].

This article has previously been published in the journal *Physical Sciences Reviews*. Please cite as: Madariaga-Mazón, A., Hernández-Alvarado, R. B., Noriega-Colima, K. O., Osnaya-Hernández, A., Martinze-Mayorga, K. Toxicity of secondary metabolites. *Physical Sciences Reviews* [Online] **2019** DOI: 10.1515/psr-2018-0116

https://doi.org/10.1515/9783110668896-008

Although natural products have a key role in drug discovery, some of them are highly toxic. In this context, toxicity refers to the harmful effect on a whole organism, such as an animal, bacterium, or plant, as well as the effect on a substructure of the organism, such as a cell or an organ. Nevertheless, the toxic effect of compounds can also be beneficial for the treatment of some illness, for example, cytotoxic compounds are used in cancer treatment. Thus, secondary metabolites may show dual roles [3].

The overview of the different chemical and biological aspects of secondary metabolites presented in this work paves the way to the prediction of toxicity of these relevant compounds.

8.2 Secondary metabolites: Sources, classification, and bioactivities

Secondary metabolites are produced in every single living organism on earth. By their ancestral use, plants are probably the best known and more explored source of bioactive natural products, followed by microorganisms and a smaller fraction from animals. There is evidence on the usage of medicinal plants since the Sumerians (approximately 5000 years ago), referring to a collection of plants including poppy, henbane, and mandrake; or the Chinese (2500 BC) describing the use of camphor, ginseng, cinnamon bark and ephedra [4]. Nevertheless, it was until the nineteenth century that active secondary metabolites were isolated and characterized from medicinal plants. Nowadays, the use of these remedies is a common practice, particularly in countries where alternative medicine is the preferred and available choice to maintain health for the prevention, diagnosis, improvement or treatment of physical and mental illness. This knowledge, accumulated over centuries, represents the basis for modern medicine.

The discovery of penicillin, the first mass-produced antibiotic derived from Penicillium fungi, is the well-known Fleming's serendipity case. Since then, the use of microorganisms for the discovery of secondary metabolites with biological activities has remarkably increased. In particular, bacteria and fungi arise as a promising source of bioactive compounds. It is estimated that only a small fraction of the total fungi species known have been chemically characterized [5]. The diversity in chemical structures of the fungi-derived compounds, and by consequence in biological activities, depends on several factors. One of them is the environment of isolation, i.e. soil, dust, sand, marine sediment, fresh and sea water, tissues of terrestrial and marine plants (endophytes), as well as from marine invertebrates and vertebrates (entomopathogenic) [6].

Many of the biological activities of the newly discovered secondary metabolites remain unknown. With the aid of data mining, new methodologies can be applied to suggest bioactivities of novel structures. For example, based on network pharmacology approaches, a suggestion of possible molecular targets is provided [7].

8.2.1 Chemical classification and bioactivities

Secondary metabolites are structurally and chemically diverse. Traditionally, natural products have been classified according to their structural similarity, and hence similarities in their pathway of biosynthesis: fatty acids and polyketides (from acetate pathway), phenylpropanoids and aromatic amino acids (from shikimate pathway), terpenoids and steroids (from mevalonate pathway) and nitrogen-containing compounds (alkaloids). Attempts to classify secondary metabolites based on their biological effects was not possible since in most cases the activity was (and still is) unknown [1]. The large diversity observed for these compounds derives on a large coverage of the chemical space as recently described [8]. In chemoinformatic terms, this is relevant for the development of predictive models which relies on the chemical space covered by the molecules in training sets.

Polyketides are built by condensation of acetate units deriving in poly-β-keto chains and creating a large class of structurally diverse compounds. This class includes fatty acids, polyacetylenes, prostaglandins, macrolide antibiotics and a large group of aromatic compounds such as anthraquinones and tetracyclines. Examples of fatty acids are alkamides (**1–4**), present in herbal preparations of dried roots of *Echinacea purpurea* are widely used as immunostimulants, commonly used for treating bacterial and viral infections. Among polyketides are the medical important macrolides, a large family of natural products with antibiotic activity, characterized by a macrocyclic lactone.

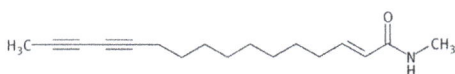

1 Tetradeca-2*E*-ene-10, 12-diynoic acid isobutylamide

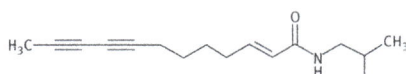

2 Dodeca-2*E*-ene-8, 10-diynoic acid isobutylamide

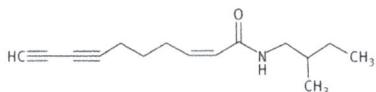

3 Undeca-2*Z*-ene-8, 10-diynoic acid 2-methylbutylamide

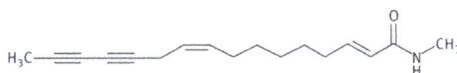

4 Hexadeca-2*E*,9*Z*-diene-12, 14-diynoic acid isobutylamide

Erythromycins are macrolide compounds isolated from *Saccharopolyspora erythraea*. Erythromycin A is the main compound of the commercial product erythromycin, a mixture containing erythromycins A, B, and C (**5–7**). Other examples of macrolides with antibiotic activity are oleandomycin (**8**), produced by fermentation cultures of *Streptomyces antibioticus* and spiramycins (**9**) produced by cultures of *Streptomyces ambofaciens* [9].

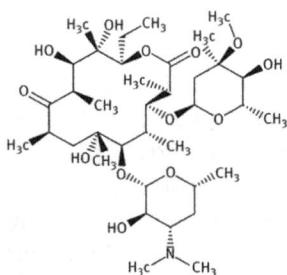

5 Erythromycin A

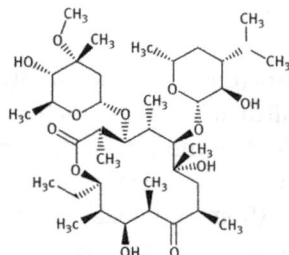

6 Erythromycin B

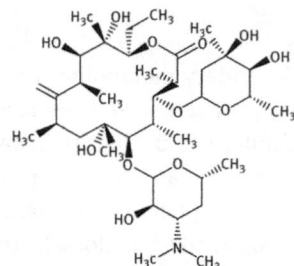

7 Erythromycin C

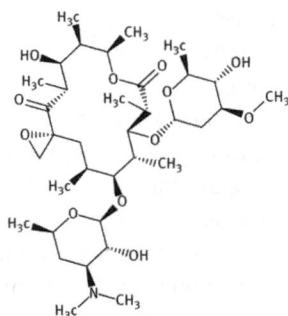

8 Oleandomycin

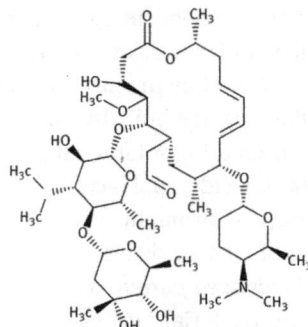

9 Spiramycin

Aflatoxins are mycotoxins produced by the fungi *Aspergillus flavus* and *Aspergillus parasiticus* with an anthraquinone derived biosynthesis (**10–12**). There are four main aflatoxins, but these toxins can be metabolized by animals or other microorganisms to produce other aflatoxins. The most common mycotoxin is aflatoxin B_1 and is also the most acutely toxic. These toxins are associated with peanuts, maize, rice and pistachio nuts [10, 11].

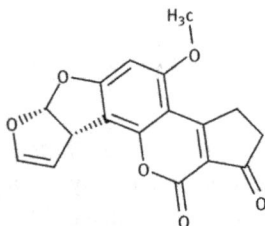

10 Aflatoxin B1

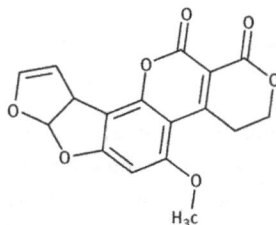

11 Aflatoxin G1

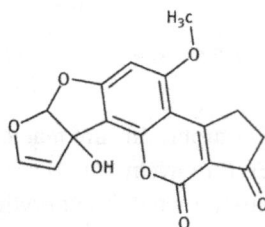

12 Aflatoxin M1

Terpenoids are the largest group of secondary metabolites with over 35,000 known members, and they are classified according to the number of isoprene units (five carbons) in their skeleton. The smaller building block, the isoprene, is thought to

protect the plant from heat. Representative terpenoids are essential oils or higher polymers such as rubber. The monoterpene (ten carbons, two isoprene units) thujone (**13**)
is present in the wormwood oil from *Artemisia absinthium*, one of the main components of the drink absinthe, known by its neurotoxic effect and now banned in several
countries [12]. Essential oils give plants their fragrance, though the good aroma is not
only their main purpose. For example, citronella oil from *Cymbopogon winterianus* and
C. nardus is used as a component of commercial insect repellents; eucalyptus oil from
Eucalyptus globulus (and other species) has antiseptic effect; or the well-known pine
essential oil from *Pinus palustris* or other pines species used in industrial cleaning
chemicals by its disinfectant activity [13]. One of the "molecular stars" in the terpenoids
group is taxol (**14**), a diterpene with anticancer properties isolated in 1971, obtained
from the bark of the Pacific yew, *Taxus brevifolia*. Since the amount needed of taxol for
a course treatment may be as high as 2 g, the harvesting of the bark has been strictly
regulated. Some alternatives have been explored and used, such as semi-synthesis
from more accessible structurally related materials; or obtention by microbial culture from endophytic fungi isolated from the bark of the *T. brevifolia* and *T. wallachi-
ana* species (*Taxomyces andreanae* and *Pestalotiopsis microspora*, respectively).
Nevertheless, the chemical conversion of the derivatives to produce the final product
taxol is still complex, and the yields from fungi culture extractions are so low that is
insignificant in terms of commercial drug production [14, 15]. Cardioactive glycosides
are present in many plants that are grown as ornamentals but must be considered as
toxic and should be treated with caution, including *Digitalis* and *Helleborus* species.
Among these compounds, digitoxin (**15**) is the only compound used with medical
purposes, employed in congestive heart failure and treatment of cardiac arrhythmias
[16]. Rubber, the largest of the terpenoids, contains over 400 isoprene units and it is
obtained from latex (a fluid produced by *Hevea brasiliensis)* [17].

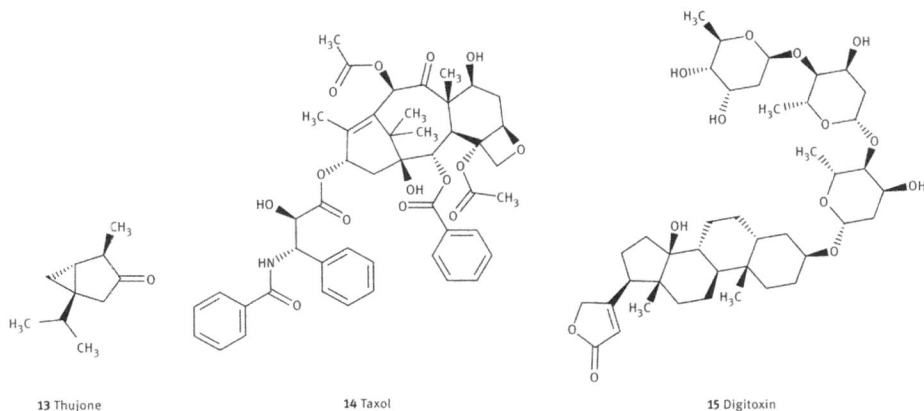

13 Thujone **14** Taxol **15** Digitoxin

From the shikimate pathway, limited to microorganisms and plants, but no animals, a great variety of aromatic compounds can be biosynthesized. These include

aromatic amino acids such as phenylalanine, tyrosine, tryptophan, and benzoic acids. One of the first broad-spectrum antibiotics discovered belongs to this type of compounds: chloramphenicol (**16**), an aminated derivative of L-*p*-aminophenylalanine produced by cultures of *Streptomyces venezuelae*, even though the synthetic compound is now used as the commercial drug [18]. Phenylalanine and tyrosine (in their L- forms) are precursors of important phenylpropanoids such as caffeic and ferulic acids, with significant antioxidant activities [19]. In turn, phenylpropanoids are the building blocks for lignans and lignins, polymers formed by two or more units, respectively. Podophyllotoxin (**17**), one of the main lignans contained in preparations from roots of *Podophyllum* species, demonstrated antimitotic effect by binding to the protein tubulin during mitosis. Due to its significant toxicity, the use in the clinic for anticancer treatment has been banned, increasing the research around derivatives with lesser undesirable effects [20]. Coumarins are also contained in this group, they are biosynthesized from cinnamic acid and later a rearrangement to form a lactone, to form the characteristic two-ring systems of this biologically important metabolites. Dicoumarol (**18**) and warfarin (**19**) are coumarins having toxic effects due to their potent anticoagulant effects. Indeed, warfarin was initially used as a rodenticide, causing internal hemorrhage and death of the rat [21].

Finally, flavonoids, stilbenes, and phenylpropanoids comprise an important and vast group from the shikimate pathway. Although they are known by their noteworthy beneficial properties to human health, several compounds have also attracted attention due to the diverse roles that play in plants, such as antimicrobial agents (phytoalexins), photoreceptors, pigments, insect repellents, regulators of hormones transport, among others [22]. A variety in flavonoids and stilbenes can be found in red wine, rich in resveratrol (**20**) and relevant for its antioxidant activity, and recently described also as an anti-inflammatory, inhibition of platelet aggregation, and protect against cancer and cardiovascular diseases [23].

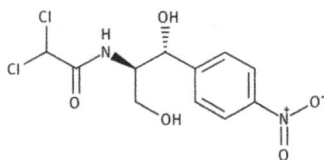

16 Chloramphenicol

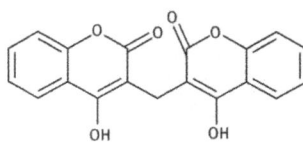

18 Dicoumarol

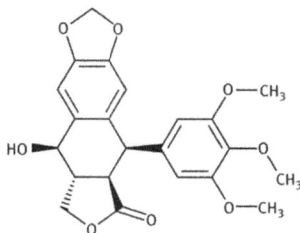

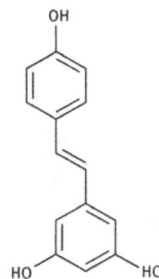

20 Resveratrol

17 Podophyllotoxin

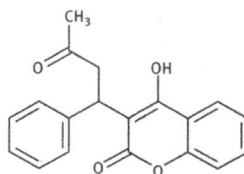

19 Warfarin

The last group of secondary metabolites is alkaloids: low molecular weight nitrogen-containing compounds mainly found in plants and with lesser occurrence in microorganisms and animals. One of the most remarkably known alkaloids is morphine (**21**), extracted from opium poppy *Papaver somniferum*. Morphine is a powerful analgesic and narcotic, and despite its several adverse effects (state of euphoria and mental detachment, together with nausea, vomiting, gastrointestinal constipation, respiratory depression, among others), remains as one of the first options for relief of severe pain. Several investigations have been conducted aiming to design novel structures with the ability to retain the analgesic properties but diminishing adverse effects, resulting in the new paradigm of biased activation of the opioid receptors. Nevertheless, further research is warranted [24]. In our daily used beverages, such as tea, coffee, and cocoa, one can find several xanthine-derivatives alkaloids. The main alkaloids, caffeine (**22**), theobromine (**23**), and theophylline (**24**), inhibit the phosphodiesterase that degrades cyclic AMP, resulting in an increase in cAMP levels and consequent stimulation of the central nervous system, relaxation of bronchial smooth muscle, induction of diuresis, among others [25].

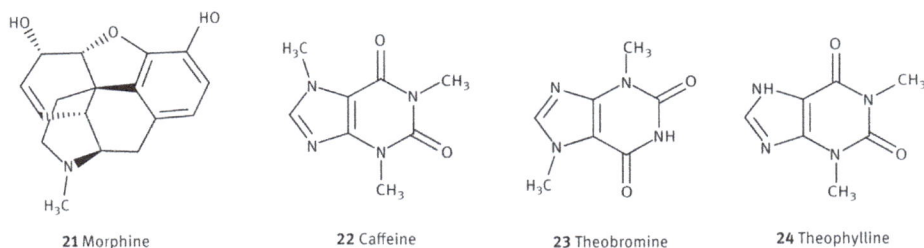

21 Morphine   22 Caffeine   23 Theobromine   24 Theophylline

This overview highlights the importance of natural products and bioactivities as therapeutic agents as well as commonly used foods and beverages.

8.2.2 Toxicity of representative secondary metabolites

As described in the text, natural products are chemical compounds that usually have a beneficial application for human health. Nevertheless, the same organisms also produce toxic molecules. It has been hypothesized that secondary metabolites are biosynthesized as waste or detoxification products. It has also been suggested that secondary metabolites have functional metabolic roles. The most accepted hypothesis is that secondary metabolites are produced as a survival role, e.g. as a defense against other organisms, such as in the case of prey capture [1]. Therefore, high toxicity is an inherent characteristic of this purpose. Interestingly, to whom these compounds are toxic to, needs to be stated.

For centuries, humans have had a fascination (and also a fear) for the toxic effects of organisms: plants, animals, microorganisms. The interest in this area is such that a scientific field is entirely dedicated to the research on toxic compounds (toxins) produced by living organisms, called toxicology [26]. It is worth noting the difference between venom and poison. Venoms referred to animal secretions that contain mixtures of different enzymes, toxins, and other compounds, generated through a specialized organ and that is delivered through a specialized system. In turn, poisons are substances, not only produced by natural sources, that cause noxious effects in organisms. The animal-derived poisons are produced via specialized cells or tissues, and they can be suffused throughout their bodies, so in this case, the toxic effect occurs when the predator organism is exposed to the poisonous organism [27]. Selected examples of natural products with known toxic activity are listed in Table 8.1.

Table 8.1: Selected examples of secondary metabolites with known toxicity expressed as their lethal doses (LD_{50}) values.

Source	Molecule	Toxicological effect	LD_{50}*, mg/kg (animal model, route of administration)	Reference
Clostridium tetani	Botulinum Toxin**	Neurotoxic	0.0000004–0.0000025 (mus, IP)	[28]
Shigelladysenteriae	Shiga toxin	Enterotoxic	0.02 (mus, ip)	[29]
Catharanthusroseus	Vincristine	Cytotoxic	1.9 (rat, ip)	[30]
Colchicum autumnale	Colchicine	Cytotoxic	6.1 (rat, ip)	[30]
Taxusbrevifolia	Paclitaxel	Cytotoxic	32.5 (rat, ip)	[31]
Amanita muscaria	Muscimol	Neurotoxic	45 (rat, oral)	[32]
Nicotiniana spp.	Nicotine	Neurotoxic	50 (rat, oral)	[33]
Psilocybe spp.	Psilocybins	Neurotoxic	280 (rat, iv)	[30]
Gyromitra esculenta	Gyromitrin	Hepatotoxic	320 (rat, oral)	[34]
Cinchona spp.	Quinine	Antimalarial	115 (mus, ip)	[30]
Atropabelladona	Atropine	Neurotoxic	500 (rat, oral)	[29]
Prunus spp.	Amygdalin	Cyanogenic	880 (rat, oral)	[35]

*LD_{50} is the amount of a material, given all at once, which causes the death of one half of a group of test animals.
**Range for different botulinum toxins: A, B, C1, C2, D, E and F.

Humans have found practical applications for toxic secondary metabolites in, at least, two major fields: drug design and agrochemical industry. Analyzing the features that contribute to their toxicity is mandatory in drug discovery since the majority of candidates to be a hit in big pharma are rejected due to either their lack of bioavailability or their inherent toxicity to human organs or cells. In turn, if compounds with exceptional pesticide activity are toxic not only to targeted organisms (insects, fungi, weeds, etc.) but also to humans, they fail in the final agrochemical-development pipeline. Therefore, it is advisable to determine toxicological endpoints early in the development of new molecules. In the following sections, we focus on the relevance of toxicity in drug discovery and agrochemistry.

8.3 Toxicity assessment in agrochemicals and drugs

8.3.1 Toxicity assessment

Typically, the measure of toxicity is based on the LD_{50} value, defined as the lethal dose that kills, on average, 50% of the test animals using a specified administration route [26]. There are, however, modifications to this method, to decrease the number of laboratory animals used. Acute toxicity is, in general, easier and cheaper to evaluate, compared to sub-chronic and chronic exposures. Nonetheless, chronic exposure of toxicants, as well as those with long residence time, should be evaluated properly. In practice, this influences, for example, the decision about dosing. The major difference between repeated dose and subchronic toxicity studies is the duration: repeated dose toxicity studies are conducted over a duration of 28 days, and subchronic toxicity studies are carried out over 90 days [36]. A detailed description of the guidance for the evaluation of chronic effects is provided by the OECD [37]. Relevant definitions used in toxicity assessment are provided below.

Route of administration: The preparation to be used determines if the oral, topical, inhalation or any other via of administration/exposure should be investigated.

Organ or tissue damaged/affected: Commonly damage to DNA is investigated. This section includes mutagenicity, carcinogenicity, teratogenicity, and neurotoxicity, among others.

Pharmacologically based pharmacokinetics (PBPK) models: Moving forward to more complex determinations, identification of PBPK models could allow mitigating tissue-specific effects (e.g. limit exposure to the central nervous system for undesirable on-target neuronal pharmacology). Efforts on this direction have been reported and continue to be developed. Ruiz et al. [38] at the Agency of Toxic Substances & Disease Registry (ATSDR) have reported the on-going work for the development of the ATDSR toolkit, a computational tool to assist with site-specific health assessments. Currently, models of environmental contaminants, including volatile organic compounds (VOCs) and metals, have been developed [39–41].

A review of PBPK models performed by Sager et al. [42] showed that the most commonly reported models were those for drug-drug interaction predictions (28%), followed by inter-individual variability and general clinical pharmacokinetic predictions (23%), formulation or absorption modeling (12%), and predicting age-related changes in pharmacokinetics and disposition (10%). In addition, they identified 106 models of sensitive substrates, inhibitors, and inducers. The lack of consistency in model quality and development, identified by the authors, highlighted a need for the development of best practice guidelines.

Adverse outcome pathway (AOP). The AOP is a way to link existing knowledge to one or more series of causally connected key events (KE) between two points—a molecular initiating event (MIE) and an adverse outcome (AO) that occur at a level of biological organization. AOP is important for risk assessment (evaluation of what can go wrong, how likely it is to happen, what the potential consequences are, and how tolerable the identified risk is) and regulatory applications. AOPs are under development, the definitions, the working workflow, and the details around AOPs are available elsewhere [43].

The testing of acute effects of chemicals "6-pack" include:

Acute Inhalation: Medial lethal concentration (LC50) from short-term exposure to a test substance by the inhalation route.

Acute Oral: Medial lethal dose (LD50) from short-term exposure to a test substance by the oral route.

Acute Dermal: Medial lethal dose (LD50) from short-term exposure to a test substance by the dermal route.

Eye Irritation: Provides information on health hazard likely to arise from exposure to test substance (liquids, solids, and aerosols) by single dose application on the eye.

Skin Irritation: Provides information on health hazard likely to arise from exposure to liquid or solid test substance by dermal application.

Dermal Sensitization: Determine the potential for a test substance to elicit a skin sensitization reaction. There are different methods, for example, the basic principle underlying the Local Lymph Node Assay (LLNA) in mouse is that sensitizers induce a primary proliferation of lymphocytes in the auricular lymph nodes draining the site of chemical application. This proliferation is proportional to the dose applied and provides a measurement of sensitization.

Detailed descriptions of these and other methods are accessible in the OECD Guide documentation.

The extensive use of laboratory animals is a concern. International organizations (OECD, EPA, etc.) spend efforts to the use, standardization, and acceptance of alternative methods. A modernization of the battery of acute toxicity tests "6-pack" has been proposed.

The 6-pack was revised in 2015 to include non-animal (in vitro/ex vivo) test methods for classification and labeling for Eye Irritation, both for commonly used household cleaning products with anti-microbial claims and more conventional pesticides products. In 2018, EPA has continued the modernization of the 6-pack through its Draft Interim Science Policy to allow the Skin Sensitization endpoint to be performed in *in vitro* models as well. Currently, only pure substances are permitted to be used as test materials for submission purposes and one of the Integrated Testing Strategies.

The modernization of the acute toxicity "6-pack" has three main goals:
- Critically evaluate which studies form the basis of Office of Pesticide Programs decisions.
- Expanding acceptance of alternative methods and;
- Reducing barriers such as challenges of data sharing among companies and international harmonization to adopting alternative methods in the U.S. and internationally.

8.3.2 Toxicity of drug candidates

Being toxicity an inherent property of all substances, it needs to be extensively assessed to any drug before it is launched to the market. A battery of *in vitro* and *in vivo* experiments is required. These experiments are costly, time-consuming and demand a large number of laboratory animals and reagents. Toxicological determinations in the discovery process focus on key mechanisms of toxicity. In general, there are a series of assays to be performed depending on the drug discovery stage; for example, studies on earlier stages use high-throughput methods both *in silico* and *in vitro* but lead or preclinical compounds are subjected to advanced procedures focused to off-target selectivity screening or using of animals for dosing experiments [44]. This scenario highlights the convenience of eliminating drug candidates that are likely to fail due to toxicity issues, and for this purpose, *in silico* methods are valuable tools. Nevertheless, the quality and the quantity of the toxicological data or the mechanism of action to be calculated remain to be the workhorse for the development of *in silico* models. Common *in silico* filters include hepatotoxicity, mutagenicity, carcinogenicity, and propensity for developing reactivity reactions. Segall et al. [45] provide predictions for a range of endpoints including hepatotoxicity, hERG-channel inhibition, developmental toxicity, teratogenicity, chromosomal damage (*in vitro* and *in vivo*), mutagenicity (*in vitro*) and carcinogenicity (dataset available upon request). The authors present the analysis of toxicity in the context of balancing multiple factors involved in the discovery of new drugs. Such a scenario is often called multiparameter optimization (MPO), emphasizing on the advantages of avoiding hard filters, by weighting the scoring profiles and considering the uncertainty in the data.

8.3.3 Toxicity of agrochemicals

Pesticides are widely employed around the world. The number and type of pesticide use vary across different crops and countries. Figure 8.1 summarizes the number of different pesticides used for different crops. This information was obtained from the analysis of the Pesticides Properties DataBase (PPDB) [46] developed by the Agriculture & Environment Research Unit (AERU) at the University of Hertfordshire. Notably, there is a large variety of pesticides used to produce cereals, fruits, and vegetables. Clearly, modern agricultural practices require the use of hundreds of pesticides; it is paramount to stick to good practices and ethics.

An evident requirement of the pesticides is their toxicity. Needless to say, the pesticides should be as selective as possible to the pest under control and as harmless as possible to non-target organisms, especially humans and the environment [47]. In a recent study, we analyzed the different ranges of toxicity across different species, modes of exposure and toxicological modes of action [48]. As expected, for

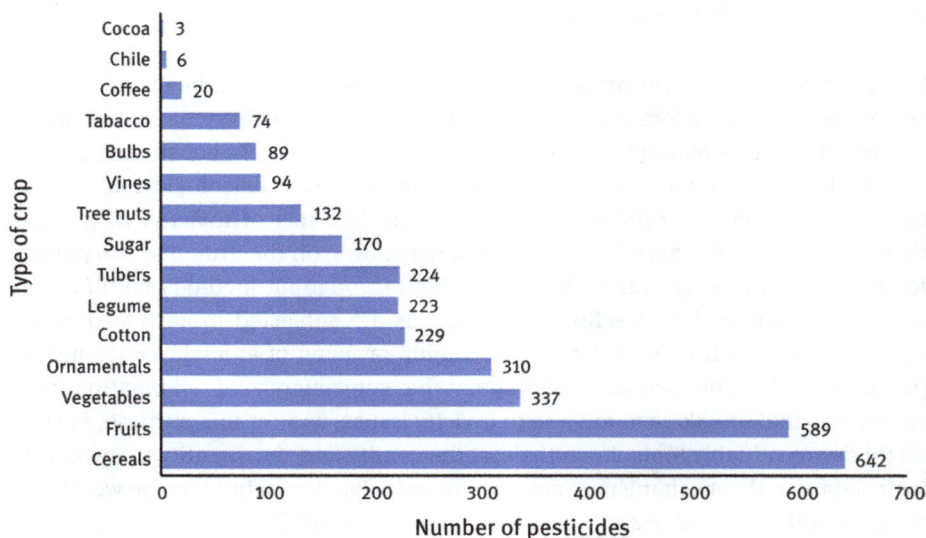

Figure 8.1: The number of different pesticides used for different crops.

sensitive animals, such as birds, the ranges of toxicities obtained are one order of magnitude more toxic, for the same endpoint, than for larger and less sensitive species, such as mammalians.

Most pesticides are synthetic compounds. However, a small group of pesticides obtained from natural sources has shown particularly good toxicological profiles, being more selective to pest and less harmful to humans, animals or the environment. Pesticides with lesser toxicological and more benign-environmental profiles should take high priority in the development of new pesticides. Some natural products fill the characteristics for the development of new pesticides with these needs (for example with molecular target sites different than those of the synthetic pesticides). Table 8.2 show examples of selected secondary metabolites used in pest control.

Table 8.2: Selected examples of toxic compounds used in agrochemical industry.

Classification	Molecule name	Reference
Weed control	Glufosinate	[49]
	AAL-toxin	[50]
	Leptospermone	[51]
Insect control	Spinosad	[52]
	Avermectins	[53]
	Rotenone	[54]

Table 8.2 (continued)

Classification	Molecule name	Reference
Mollusc control	Vulgarone B	[55]
	Phytolacca dodecandra	[56]
Algae control	9,10-anthraquinone	[57]
	Menadione sodium bisulfite	[58]
Plant pathogens control	Azoxystrobin	[59]
	Cinnamaldehyde	[60]
	Flindersine	[61]

Interestingly, a large portion of pesticides, contained in PPDB (one of the most comprehensive pesticide database), are classified as metabolites [46]. In this database, those classified as metabolites correspond to chemical degradation substances and might retain pesticidal activity. This highlights the relevance of metabolic products, in addition to those naturally occurring in nature.

8.4 Conclusions

Secondary metabolites have reached a remarkable place as drugs, agrochemicals, cosmetics, food additives, etc. Knowledge on the structural diversity, natural sources, synthetic feasibility, as well as its inherent toxicities of these compounds will help expand their uses, for the benefit of human and animal health and the environment.

Funding: This work was supported by Instituto de Quimica-UNAM, and DGAPA-UNAM [grant number PAPIIT IN210518].

Acknowledgements: The authors thank ChemAxon for kindly providing academic license of their software.

References

[1]   Williams DH, Stone MJ, Hauck PR, Rahman SK. Why are secondary metabolites (natural products) biosynthesized? J Nat Prod. 1989;52:1189–208.
[2]   Croteau R, Kutchan T, Lewis NG. Natural products (secondary metabolites). In: Buchanan B, Gruissem W, Jones R, editors. Biochemistry & molecular biology of plants. American Society of Plant Physiologists, 2000:1250–319.
[3]   Schoental R. Toxicology of natural products. Food Cosmet Toxicol. 1965;3:609–20.
[4]   Petrovska B. Historical review of medicinal plants usage. Pharmacogn Rev. 2012;6:1–5.

[5]   Kirk JL, Beaudette LA, Hart M, Moutoglis P, Klironomos JN, Lee H, et al. Methods of studying soil microbial diversity. J Microbiol Methods. 2004;58:169–88.

[6]   Debbab A, Aly AH, Proksch P. Bioactive secondary metabolites from endophytes and associated marine derived fungi. Fungal Divers. 2011;49:1–12.

[7]   Zuo H, Zhang Q, Su S, Chen Q, Yang F, Hu Y. A network pharmacology-based approach to analyse potential targets of traditional herbal formulas: an example of Yu Ping Feng decoction. Sci Rep. 2018;8:1–15.

[8]   Saldívar-González FI, Pilón-Jiménez BA, Medina-Franco JL. Chemical space of naturally occurring compounds. Phys Sci Rev. 2018. DOI: 10.1515/psr-2018-0103.

[9]   Pal S. A journey across the sequential development of macrolides and ketolides related to erythromycin. Tetrahedron. 2006;62:3171–200.

[10] Leong YH, Latiff AA, Ahmad NI, Rosma A. Exposure measurement of aflatoxins and aflatoxin metabolites in human body fluids. A short review. Mycotoxin Res. 2012;28:79–87.

[11] Kumar P, Mahato DK, Kamle M, Mohanta TK, Kang SG. Aflatoxins: a global concern for food safety, human health and their management. Front Microbiol. 2017;7:2170.

[12] Lachenmeier DW, Walch SG, Padosch SA, Kröner LU. Absinthe—a review. Crit Rev Food Sci Nutr. 2006;46:365–77.

[13] Edris AE. Pharmaceutical and therapeutic potentials of essential oils and their individual volatile constituents: a review. Phyther Res. 2007;21:308–23.

[14] Altmann KH, Gertsch J. Anticancer drugs from nature – natural products as a unique source of new microtubule-stabilizing agents. Nat Prod Rep. 2007;24:327–57.

[15] Kingston DGI. The shape of things to come: structural and synthetic studies of taxol and related compounds. Phytochemistry. 2007;68:1844–54.

[16] Gobbini M, Cerri A. Digitalis-like compounds: the discovery of the O-aminoalkyloxime group as a very powerful substitute for the unsaturated [gamma]-butyrolactone moiety. Curr Med Chem. 2005;12:2343–55.

[17] Mooibroek H, Cornish K. Alternative sources of natural rubber. Appl Microbiol Biotechnol. 2000;53:355–65.

[18] Piraee M, Magarvey N, He J, Vining LC. The gene cluster for chloramphenicol biosynthesis in Streptomyces venezuelae ISP5230 includes novel shikimate pathway homologues and a monomodular non-ribosomal peptide synthetase gene. Microbiology. 2001;147:2817–29.

[19] Razzaghi-Asl N, Garrido J, Khazraei H, Borges F, Firuzi O. Antioxidant properties of hydroxycinnamic acids: a review of structure- activity relationships. Curr Med Chem. 2013;20:4436–50.

[20] Shareef MA, Duscharla D, Ramasatyaveni G, Dhoke NR, Das A, Ummanni R, et al. Investigation of podophyllotoxin esters as potential anticancer agents: synthesis, biological studies and tubulin inhibition properties. Eur J Med Chem. 2015;89:128–37.

[21] Borges F, Roleira F, Milhazes N, Santana L, Uriarte E. Simple coumarins and analogues in medicinal chemistry: occurrence, synthesis and biological activity. Curr Med Chem. 2005;12:887–916.

[22] Ververidis F, Trantas E, Douglas C, Vollmer G, Kretzschmar G, Panopoulos N. Biotechnology of flavonoids and other phenylpropanoid-derived natural products. Part I: chemical diversity, impacts on plant biology and human health. Biotechnol J. 2007;2:1214–34.

[23] Gambini J, Inglés M, Olaso G, Lopez-Grueso R, Bonet-Costa V, Gimeno-Mallench L, et al. Properties of resveratrol: in Vitro and In Vivo studies about metabolism, bioavailability, and biological effects in animal models and humans. Oxid Med Cell Longev. 2015;2015:837042.

[24] Madariaga-Mazón A, Marmolejo-Valencia AF, Li Y, Toll L, Houghten RA, Martinez-Mayorga K. Mu-Opioid receptor biased ligands: A safer and painless discovery of analgesics? Drug Discov Today. 2017;22:1719–29.

[25] Ashihara H, Sano H, Crozier A. Caffeine and related purine alkaloids: biosynthesis, catabolism, function and genetic engineering. Phytochemistry. 2008;69:841–56.

[26] Wexler P, Fonger GC, White J, Weinstein S. Toxinology: taxonomy, interpretation, and information resources. Sci Technol Libr. 2015;34:67–90.

[27] Fox J, Serrano S. Approaching the golden age of natural product pharmaceuticals from venom libraries: an overview of toxins and toxin-derivatives currently involved in therapeutic or diagnostic applications. Curr Pharm Des. 2007;13:2927–34.

[28] Environmental Health & Safety. University of Florida. http://www.ehs.ufl.edu/.

[29] Lewis RJ. Sax's dangerous properties of industrial materials. J Am Chem Soc. 2005;127:2794–4.

[30] Kim S, Chen J, Cheng T, Gindulyte A, He J, He S, et al. PubChem 2019 update: improved access to chemical data. Nucleic Acids Res. 2019;47:D1102–9.

[31] Wishart DS, Feunang YD, Guo AC, Lo EJ, Marcu A, Grant JR, et al. DrugBank 5.0: a major update to the DrugBank database for 2018. Nucleic Acids Res. 2018;46:D1074–82.

[32] Budavari S. The merck index : an encyclopedia of chemicals, drugs, and biologicals, 11th ed. (Budavari S, ed.). N.J., USA: Rahway, 1989.

[33] Sine C. Nicotine. In: Meister RT, editor. Farm chemicals handbook '93. Willoughby, Ohio: 1993:C245

[34] Patocka J, Pita R, Kuca K. Gyromitrin, mushroom toxin of gyromitra spp. Mil Med Sci Lett. 2012;81:61–7.

[35] Adewusi SR, Oke OL. On the metabolism of amygdalin. 1. The LD50 and biochemical changes in rats. Can J Physiol Pharmacol. 1985;63:1080–3.

[36] Parasuraman S. Toxicological screening. J Pharmacol Pharmacother. 2011;2:74–9.

[37] OECD. Test No. 452: chronic toxicity studies. OECD, 2009. DOI: 10.1787/9789264071209-en.

[38] Ruiz P, Yang X, Lumen A, Fisher J. Quantitative structure-activity relationship (QSAR) models, physiologically based pharmacokinetic (PBPK) models, biologically based dose response (BBDR) and toxicity pathways: computational tools for public health. In: Comput Toxicol. January 2013:5–21.

[39] Ruiz P, Fowler BA, Osterloh JD, Fisher J, Mumtaz M. Physiologically based pharmacokinetic (PBPK) tool kit for environmental pollutants – metals. SAR QSAR Environ Res. 2010;21:603–18.

[40] Ruiz P, Mumtaz M, Osterloh J, Fisher J, Fowler BA. Interpreting NHANES biomonitoring data, cadmium. Toxicol Lett. 2010;198:44–8.

[41] Mumtaz MM, Ray M, Crowell SR, Keys D, Fisher J, Ruiz P. Translational research to develop a human PBPK models tool kit—volatile organic compounds (VOCs). J Toxicol Environ Heal Part A. 2012;75:6–24.

[42] Sager JE, Yu J, Ragueneau-Majlessi I, Isoherranen N. Minireview physiologically based pharmacokinetic (PBPK) modeling and simulation approaches: a systematic review of published models, applications, and model verifications. DRUG Metab Dispos Drug Metab Dispos. 2015;43:1823–37.

[43] Leist M, Ghallab A, Graepel R, Marchan R, Hassan R, Bennekou SH, et al. Adverse outcome pathways: opportunities, limitations and open questions. Arch Toxicol. 2017;91:3477–505.

[44] Kerns EH, Di L, Kerns EH, Di L. Toxicity. Drug-like prop concepts, struct design and methods from ADME to toxicity optimization. Elsevier, 2008:215–23.

[45] Segall MD, Barber C. Addressing toxicity risk when designing and selecting compounds in early drug discovery. Drug Discov Today. 2014;19:688–93.

[46] Lewis KA, Tzilivakis J, Warner DJ, Green A. An international database for pesticide risk assessments and management. Hum Ecol Risk Assess An Int J. 2016;22:1050–64.

[47] Duke SO, Cantrell CL, Meepagala KM, Wedge DE, Tabanca N, Schrader KK. Natural toxins for use in pest management. Toxins (Basel). 2010;2:1943–62.

[48] Madariaga-Mazón A, Osnaya-Hernández A, Chávez-Gómez A, García-Ramos JC, Cortés-Guzmán F, Castillo-Pazos DJ, et al. Distribution of toxicity values across different species and modes of action of pesticides from PESTIMEP and PPDB databases. Toxicol Res. 2019;8:146–56.

[49] Duke SO, Rimando AM, Schrader KK, Cantrell CL, Meepagala KM, Wedge DE, et al. Natural products for pest management. In: Ikan R, editor. Selected topics in the chemistry of natural products. World Scientific, 2008:209–51.

[50] Duke SO, Dayan FE. Modes of action of microbially-produced phytotoxins. Toxins. 2011;3:1038–64.

[51] Owens DK, Nanayakkara NPD, Dayan FE. In planta mechanism of action of leptospermone: impact of its physico-chemical properties on uptake, translocation, and metabolism. J Chem Ecol. 2013;39:262–70.

[52] Williams T, Valle J, Viñuela E. Is the naturally derived insecticide Spinosad® compatible with insect natural enemies? Biocontrol Sci Technol. 2003;13:459–75.

[53] Putter I, Mac Connell JG, Preiser FA, Haidri AA, Ristich SS, Dybas RA. Avermectins: novel insecticides, acaricides and nematicides from a soil microorganism. Experientia. 1981;37:963–4.

[54] Gupta RC. Rotenone. In: Wexler P, editor. Encyclopedia of toxicology. Elsevier, 2014:185–7.

[55] Prabhakaran G, Bhore S, Ravichandran M. Development and evaluation of poly herbal molluscicidal extracts for control of apple snail (Pomacea maculata). Agriculture. 2017;7:22.

[56] Kariuki ST, Kariuki JM, Mailu BM, Muchiri DR. Phytolacca octandra (L.), Phytolacca dodecandra (L'Herit) and Balanites aegypiaca (L.) extracts as potential molluscicides of schistosomiasis transmitting snails. J Med Plants Res. 2017;10:823–8.

[57] Schrader KK, De Regt MQ, Tidwell PR, Tucker CS, Duke SO. Selective growth inhibition of the musty-odor producing Cyanobacterium Oscillatoria cf. chalybea by natural compounds. Bull Environ Contam Toxicol. 1998;60:651–8.

[58] O'Neil J, Scarrott B, Svalheim RA, Elliott J, Hodges SJ. Vitamin K2 in animal health: an overview. In: Gordeladze J, editor. Vitamin K2 – vital for health and wellbeing. InTech, 2017:215–36.

[59] Liang S, Xu X, Lu Z. Effect of azoxystrobin fungicide on the physiological and biochemical indices and ginsenoside contents of ginseng leaves. J Ginseng Res. 2018;42:175–82.

[60] Xie Y, Huang Q, Wang Z, Cao H, Zhang D. Structure-activity relationships of cinnamaldehyde and eugenol derivatives against plant pathogenic fungi. Ind Crops Prod. 2017;97:388–94.

[61] Cantrell CL, Case BP, Mena EE, Kniffin TM, Duke SO, Wedge DE. Isolation and identification of antifungal fatty acids from the basidiomycete Gomphus floccosus. J Agric Food Chem. 2008;56:5062–8.

Part III: **Chemoinformatics Tools for Natural Products Discovery in the Modern Age and Case Studies**

Ramsay Soup Teoua Kamdem*, Omonike Ogbole, Pascal Wafo,
Philip F. Uzor, Zulfiqar Ali, Fidele Ntie-Kang, Ikhlas A. Khan
and Peter Spiteller

9 Rational engineering of specialized metabolites in bacteria and fungi

Abstract: Bacteria and fungi have a high potential to produce compounds that display large structural change and diversity, thus displaying an extensive range of biological activities. Secondary metabolism or specialized metabolism is a term for pathways and small molecule products of metabolism that are not mandatory for the subsistence of the organism but improve and control their phenotype. Their interesting biological activities have occasioned their application in the fields of agriculture, food, and pharmaceuticals. Metabolic engineering is a powerful approach to improve access to these treasured molecules or to rationally engineer new ones. A thorough overview of engineering methods in secondary metabolism is presented, both in heterologous and epigenetic modification. Engineering methods to modify the structure of some secondary metabolite classes in their host are also intensively assessed.

Keywords: natural products, secondary metabolites, synthetic biology, heterologous expression, combinative biosynthesis

9.1 Introduction

Specialized metabolites present in plants and microorganism cultures are described as compounds that are not important for the host life, even though they likely confer an evolutionary benefit to the producer organism. Though glycolysis and amino-acid (AA) biosynthesis, known as primary metabolic processes, are present in most species of life, the biosynthesis process that leads to secondary metabolites are specific and the principal process is found in a taxonomically closed group of organisms. Compounds derived from different natural sources are different in terms of their structure and therefore exhibit an extensive range of fascinating biological activities. Due to noteworthy successful outcome of microbial secondary metabolites as an early first step for the development of such bioactive metabolites, there has been upsurge curiosity in rain forest medicinal plants and microorganisms for new

This article has previously been published in the journal Physical Sciences Reviews. Please cite as: Teoua Kamdem, R. S., Ogbole, O., Wafo, P., Uzor, P. R., Ali, Z., Ntie-Kang, F., Khan, I. A., Spiteller, P. Rational engineering of specialized metabolites in bacteria and fungi. *Physical Sciences Reviews* [Online] 2021 DOI: 10.1515/psr-2018-0170

https://doi.org/10.1515/9783110668896-009

drugs discovery [1]. In conjunction with the award of the 2015 Nobel Prize to two scientists in the field of Physiology and Medicine, respectively, for their works on avermectins and artemisinin, two therapeutic principles that have positively changed the treatment of overwhelming microbial ailments, and with the current research in the field of biotechnology and bioinformatics, an important step of natural products drug discovery has evolved [2]. Though the usual bioactivity based screening of medicinal plant and microbial extracts are recognized as prolific makers of bioactive compounds these technics have so far demonstrated drawback owing to the tendency of these methods to re-isolate known metabolites [3]. Furthermore, recent genetic investigations of such microorganisms show that their genomes often encode 10–20 times more gene clusters encoding the biosynthesis of these useful compounds than are found in laboratory fermentations. This issue has encouraged the researchers to investigate other approaches to detect novel lead compounds, including high-throughput screening of synthetic compound libraries and fragment-based design. Despite efforts make, these technics have had many disappointments [3], raising the important research interrogation: From which sources will novel leads compounds for antibacterial, antitumor or agrochemical agents be isolated? Some interesting approaches to answer such question have come from microbial genomics, transcriptomics and heterologous methods which have demonstrated that many filamented fungi and microorganisms have faraway high probability to yield new metabolites than have been revealed by classic screening methods and that systematic methods can reveal these new compounds. It is important to note that other approaches, such as One Strain Many Compounds (OSMAC) and co-cultivation, to activate biosynthetic gene clusters, are empirical and need to be improved since there has been a development of genomic mining and bioinformatics which include an innovative approach for rational engineering of new secondary metabolites.

9.2 Roles and definition of secondary metabolites

Secondary metabolites traditionally look like they are not indispensable for the growing or subsistence of plants or microorganisms. Although primary metabolism is common for most organisms, the general aspect of natural products is that they have a slanted phylogenetic repartition owing to numerous influences that had led to the destruction and diversification of their cluster and encoding genes across slender evolutionary lines.

On the bench experiment, genes responsible for high metabolite biosynthesis can be sometimes silenced, or underexpressed with apparently less effect on the growth of the host (plants, microorganisms) [4, 5]. Despite that, most secondary metabolites that fall in this class meet higher difficulty to survive in nature, for instance, the case of most pigments, iron-chelating compounds (siderophores), and plant repelling compounds (pheromones). Their importance to the host and producer

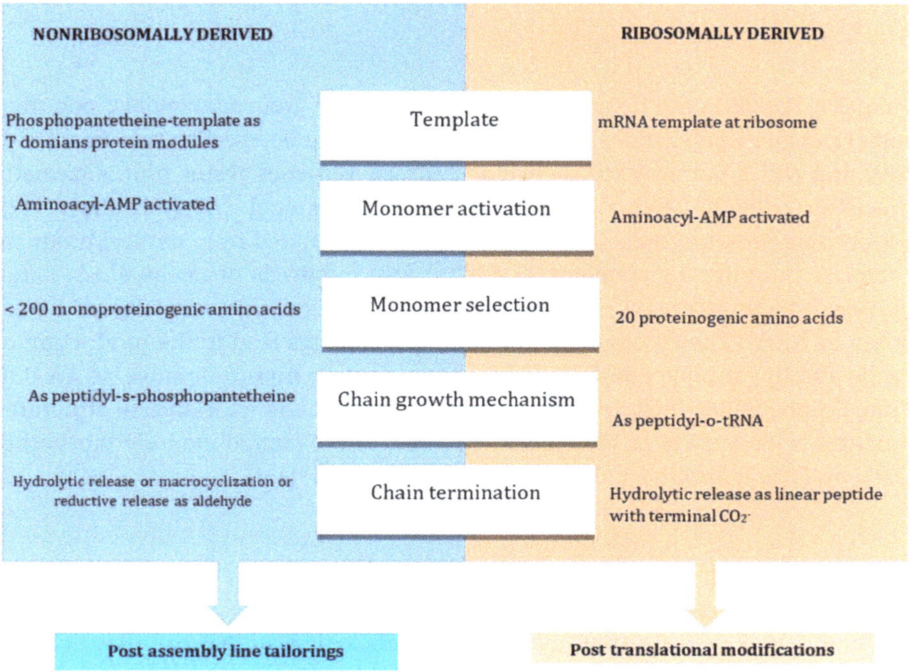

NONRIBOSOMALLY DERIVED		RIBOSOMALLY DERIVED
Phosphopantetheine-template as T domians protein modules	Template	mRNA template at ribosome
Aminoacyl-AMP activated	Monomer activation	Aminoacyl-AMP activated
< 200 monoproteinogenic amino acids	Monomer selection	20 proteinogenic amino acids
As peptidyl-s-phosphopantetheine	Chain growth mechanism	As peptidyl-o-tRNA
Hydrolytic release or macrocyclization or reductive release as aldehyde	Chain termination	Hydrolytic release as linear peptide with terminal CO_2^-
Post assembly line tailorings		Post translational modifications

Figure 9.1: Characteristics of ribosomal and nonribosomal pathways.

is undervalued due to their partial repertoire or misapprehended role. It has been reported that the host producer of secondary metabolites and molecule of small size receive from them a highly vital and protective benefit [6]. Sometimes, the frontier between primary metabolites and secondary metabolites is blurry because of the compatibility of their behaviour, for example, the case of ethanedioic acid (oxalic acid) [7].

The fungi secondary metabolites are mostly released from their cells in volatile form. Their metabolites are spread into the environment or used in the fungi cells. Such compounds are described as "extrolites" based on the fact that their biological role is supposed to be ostensively directed [8]. But the classical term "secondary metabolite" is preferred because the role and sites of cellular biosynthesis remain unknown. According to the structural point of view, secondary metabolites are biosynthesized according to the pathways that engaged primary metabolites as starting point to shape complex molecular edifice, such as NRKS/PKS compounds. It can be strongly regulated because their genes expression can be triggered by a small molecules (elicitors) or co-cultivation [9, 10].

9.3 Main classes of secondary metabolites and their biosynthetic pathway

Fungal metabolites are frequently classified into four groups: alkaloids, non-ribosomal peptides, polyketides, and terpenes [11]. Alkaloids always use a mixture of NPK and NRKS pathway for their biosynthesis. Terpenes result from enzymatic precursor dimethylallylpyrophosphate and isopentenyl pyrophosphate [12]. Concerning proteins, most of the small peptide originated from natural products comprise not only the common 20 AA but also hundreds of unusual AA. Large classes of enzymes are responsible for the assembling of this natural peptide, denoted as NRPS. These complex enzymatic mechanisms lead to the production of cyclic and linear polypeptides which usually contain non-proteinogenic AA [13]. Fungal polyketides include a large group of compounds with diverse structures having a wide variety of biological activities. These compounds are biosynthesized by Type II iterative polyketide synthases (NPKS) [14].

9.3.1 Metabolites that originated from the shikimate pathway

Shikimic acid is the trihydroxy cyclohexene carboxylic acid that contributes to the carbon skeleton formation of the aromatic AA using multistage process foremost across the shikimate pathway [15]. Consequently, shikimate-derivatives are any metabolites that own aromatic AA moiety in their skeleton. The shikimate pathway originated in fungi, plants and bacteria. And it is an important process in which primary AA are involved. This pathway has led to the production of a considerable number of secondary metabolites, largely quinones and aromatic compounds [16].

9.3.2 Altered peptides obtained through the ribosomal and post-translational pathway

Fungi are the rich source of NRPs among all-natural source types. This high potentiality to produce natural products reaches its higher point in the filamentous Ascomycota [17, 18]. Ribosomes are effectively RNA machines, polymerizing AA monomers at rates of from 10 to 30 peptide bonds per second. The order of AA selected and incorporated into elongating chains of peptides bonds is dictated by a particular messenger RNA (mRNA) serving as a template. Twenty (20) canonical AA are proteinogenic, although in some organisms, selenocysteine and pyrrolysine can constitute 21st and 22nd building blocks. The AA are selected by cognate aminoacyl transfer (t)RNA synthetase enzymes and activated as mixed carboxylic-phosphoric anhydrides in the form of aminoacyl-AMPs. The thermodynamically activated aminoacyl moieties are transferred to cognate tRNAs in the active site of the aminoacyl-tRNA synthetases. These loaded aminoacyl-tRNAs (oxoesters) are kinetically stable enough to form chain-elongating peptide bond in the peptidyl transferase centres of the 50S ribosomal subunit.

In contrast, the non-ribosomal peptide synthetase assembly lines are independent of any mRNA or any other RNA template. The unique scent of the adenylation domains in each NRPS module dictates what AA will be selected and activated (Figure 9.1). The same logic of cleaving an ATP molecule into pyrophosphate (PPi) as each AA is activated as an aminoacyl-AMP is utilized with the adenylation domains, which are thought to represent convergent evolution with respect to aminoacyl-tRNA synthetases. The AA units are then transferred to the thiol groups at the end of phosphopantetheinyl arms that have been installed on each carrier protein domain in each NRPS module. Thus, two distinctions apply; the aminoacyl groups to be incorporated into growing peptidyl chains are thioesters not oxoesters. Also, they are covalently tethered, one to each peptidyl carrier protein (PCP).

As the chain grows the elongating peptidyl chain is transferred to the next downstream PCP domain where the incoming aminoacyl unit acts as a nucleophile. The analogy to macroscopic assembly lines comes as from the concept of each PCP domain filled by an elongating peptidyl thioester chain. The further downstream the PCP domain is in the NRPS assembly line, the more finished the peptidyl chain. When the length of peptidyl chain reaches the most downstream module, it undergoes a catalyzed disconnection from its thioester linkage. We will note a variety of intramolecular (such as macrocyclization) and intermolecular (such as hydrolysis) routes that break the covalent thioester attachment and release either cyclic or linear peptide products.

Because the NRPS assembly lines do not use aminoacyl-tRNA synthetases and their aminoacyl-specifying adenylation domains have evolved convergently, NRPS enzymes can select up to 200 different, nonproteinogenic AAs as well as the 20 proteinogenic AAs [19].

9.3.3 Polyketides

Polyketides represent the largest and structurally diverse group of secondary metabolites belonging mostly to fungal reign. Their carbon scaffolds (backbone) derive from the polymerization of short-chain carboxylic acid units (e.g., acetate, malonate) and also share a common biosynthetic pathway. Biosynthesis of polyketide is performed by large multifunctional enzymes called polyketide synthases (PKSs) which rearrange into four main groups according to the mechanisms involved in their formation (catalytic and enzymatic) [20]. Filamentous fungal polyketides enzymes are similar to enzymes encoded by FASN gene. Considering higher fungi, the widespread categories of polyketide enzyme synthetase belong to the modified class of polyketides. In which the domain act repetitively during the polymer extension cycle. The carbon chain is built through Claisen condensation accelerated by the enzyme β-ketoacylsynthase and use malonyl thioesters as extender units. The acyl-carrier-protein (ACP) domain is used as an anchor where the budding polyketide chain is covalently attached. The polyketide skeleton is increased by one *ketide* unit for every chain-extension step. The newly formed polyketide can be selectively modified by additional domains consisting of dehydratases, β-ketoreductases, and enoyl reductases [21, 22].

Fungal-related polyketide synthetase can be grouped into three major subtypes: non-reducing (aromatic) (NRPKS), partial reducing (PR-PKS), and highly reducing (HR-PKS) PKSs.

9.4 Secondary metabolic engineering

For a number of years, secondary metabolites from natural source have been used in different industrial field including pharmaceutical, cosmetic, and agrochemical. Recent literature showed that between 1981 and 2014, 174 new compounds used for treatment of cancer were approved for commercialization, among them, 56 were natural unmodified products or direct derivatives, 37 inspired by chemical structures of natural products, 43 were synthetic chemical entities, 33 of non-biological origin and 5, vaccines. Thus, a total of 93 (53%) are natural products or its derivates [23]. Moreover, natural specialized metabolites and their derivatives are the most effective source in the field of drug discovery. Although many pharmaceutical companies have turned to relatively cheap synthetic routes as the origin of new medicine leads, in the past decade, there has been renewed interest in secondary metabolites [24].

This is likely due to the increasing recognition that secondary metabolites are evolutionarily preselected for biological activity; they are dubbed privileged scaffolds. Regrettably, many compounds are formed in fewer quantities by the host organism, on the other side, the genes accountable for their biogenesis can occasionally be deleted, under-expressed or cryptic, the host organisms can be also be delayed to growing or sometimes impossible to culture under laboratory environment. Furthermore, regular bioactivity-based screening experimentation of plant or microorganism extract that is recognized as productive sources of interesting producers of bioactive natural products, for example, during most previous researches, actinobacteria and filamentous fungi, has demonstrated unsatisfactory due to the propensity of these methods for characterization and re-isolation of previously known metabolites [3]. This has occasioned the researcher for the investigation of new and alternative technics to identify novel lead compounds, such as high-throughput screening of synthetic compound libraries and fragment-based design. These approaches had limited success so far, upbringing the paramount interrogation: *Where will new drugs sources against bacterial/ microbes and, cancer and agrochemicals come from?* [3]

Many attempts have been made to find an answer to the above question, such as OSMAC approach, co-cultivation of fungi-fungi, fungi-bacteria and by addition of chemical elicitors in the fermentation medium (epigenetic modifiers). These methods that are previously viewed as a great solution are now thought to be too empirical, and in many cases, the results are not duplicative.

Nowadays, with the advent of the post-genomic era and bioinformatics, it has been demonstrated that metabolic engineering and genome mining of microorganism strain systems has the potential to be a change, selective, and low-cost way to access these bioactive molecules with good profits and higher purity. Additionally, the chemical or biotransformation of natural products structure can frequently increase or change their biological activity.

The secondary metabolites structure can be modified or altered through biological pathway engineering, in other to build wholly novel molecules with novel or improved biological activities.

9.5 Bioinformatic and computer predictions of biosynthetic gene clusters

One of the obstinate surprises from the early genome sequences of fungi and bacteria with large genomes, such as actinomycetes and myxobacteria, but not small ones, such as *Escherichia coli*, was the prediction of PKs and NRPs and hybrid NRPS clusters that surpassed the known biosynthetic capacity of the producer strains. For instance, the industrial strain of *Streptomyces avermitilis*, used to produce avermectins, was known to make three to five natural products. Genome analysis, however, indicated the presence of 22 nonribosomal peptide NPSs clusters and 30 PKSs clusters [25]. The situation was comparable when A*spergillus fumigatus* and *Aspergillus nidulans* genomes were sequenced, with a predicted PKS, NRPS, NRPS-PKS capacity of 30–50 natural products [11, 26–28].

The current opinion is that, under standard axenic culture conditions, the microorganism (bacteria and fungi) usually express only 10% of their biosynthetic capacity. This is the genetic underpinning of the OSMAC (one strain many compounds) protocols. The silent pathway can be turned on by tweaking growth conditions, signal input, chromatin modifications, *etc.* Figure 1.2 shows several inputs to fungal cells and their intracellular signalling protein partners that can affect transcription of cryptic PKS and NRPS genes clusters. The OSMAC approach and related approaches have been implemented to trigger cryptic biosynthetic microorganism genes, as well as overexpression of transcription factors around the gene clusters, promoter exchange for specific induction, exchange or modulation of global regulators, chromatin modifiers that affect histone methylation and acetylation states, and the array of physiologic conditions (Figure 9.2 and Figure 9.3) [29].

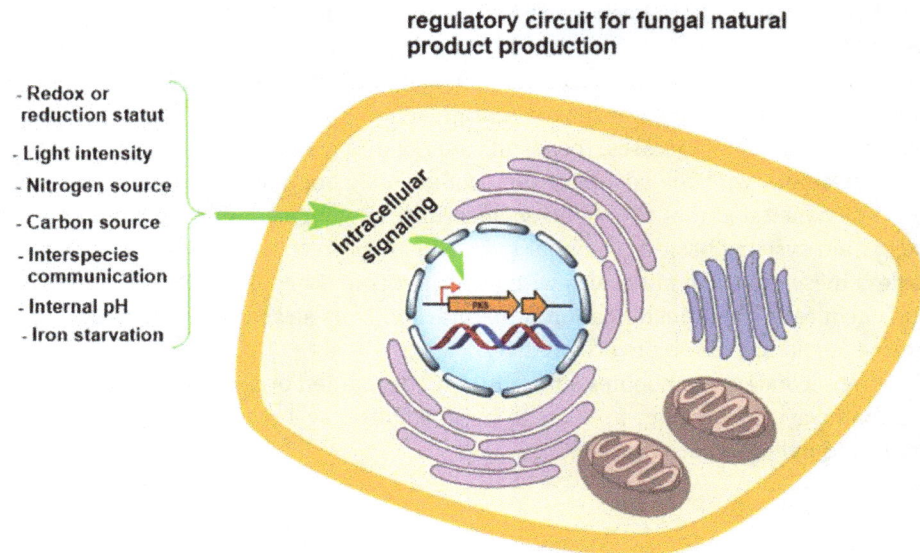

regulatory circuit for fungal natural product production

- Redox or reduction statut
- Light intensity
- Nitrogen source
- Carbon source
- Interspecies communication
- Internal pH
- Iron starvation

Intracellular signaling

Figure 9.2: Inputs to fungal cells and subsequent intracellular signalling can affect transcription of natural product gene clusters.

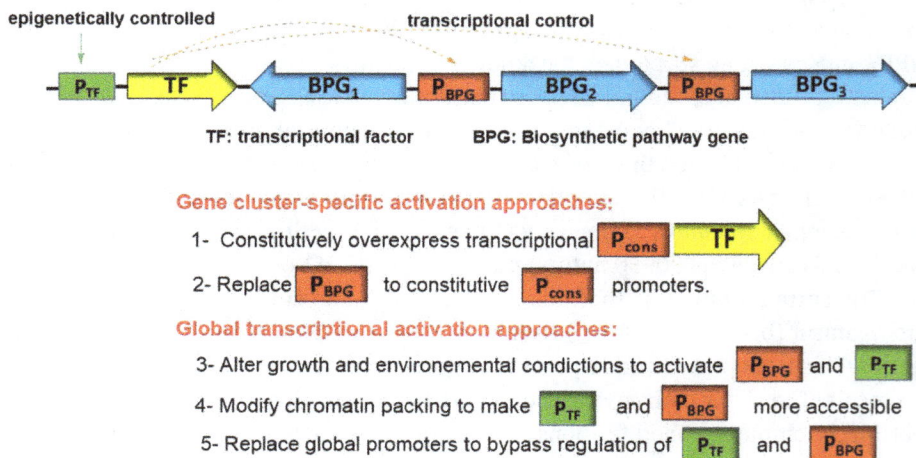

epigenetically controlled   transcriptional control

P_{TF} TF BPG_1 P_{BPG} BPG_2 P_{BPG} BPG_3

TF: transcriptional factor   BPG: Biosynthetic pathway gene

Gene cluster-specific activation approaches:

1- Constitutively overexpress transcriptional P_{cons} TF

2- Replace P_{BPG} to constitutive P_{cons} promoters.

Global transcriptional activation approaches:

3- Alter growth and environemental condictions to activate P_{BPG} and P_{TF}

4- Modify chromatin packing to make P_{TF} and P_{BPG} more accessible

5- Replace global promoters to bypass regulation of P_{TF} and P_{BPG}

Figure 9.3: Proposed bioinformatic methods used to activate cryptic genes in microorganisms.

Recently, computational analysis of 2700 microbial genomes by the means of algorithms as cluster finder and antiSMASH [30, 31] which look for NRPS, PKS, and terpene biosynthetic genes, among others, revealed 3300 predicted gene family for PKS, NRPS, and NRPS-PKS metabolites, essentially all (>90%) unknown [32]. The latest compilation done by Li et al. shows 581 fungal genomes computed, 4984 PKS clusters, 2983 NRPS, 550 dimethyltransferase, and 336 geranylgeranyl-PP synthases [33]. Provisionally, for the near future point of 20 000 sequenced genomes of fungi and potential bacterial secondary metabolites producers, there may be up to 1 million predicted biosynthetic gene cluster that encodes more than 99% unknown secondary metabolites.

9.6 Genomics-driven discovery of secondary metabolites

In the postgenomic era, it is obvious that the metabolites production potential of microorganisms has been greatly underestimated. The majority of the genes accountable for the biosynthesis of bioactive compounds are simply kept inactive under lab environments. New technics are usually used to specifically trigger such enigmatic biosynthesis gene clusters, which allow the discovery of compounds with structural novelty including those with potent antibiotic activities. The complete genome sequences of actinobacteria and filamentous fungi in the years 2000s has begun to demonstrate that they possess huge potential to yield complex metabolites compared with those obtained by traditional bioactivity-based screening methods [25, 34, 35]. For example, *Streptomyces coelicolor* has been studied as a prototypical antimicrobial-producing actinobacterium for more than five decades. Recently, the conclusion about the study of genome sequencing of *S. coelicolor* in 2002 reveals six classes of compounds with a distinct structure that had been identified using molecular genetics methods [36]. Complete genome studies of other actinobacteria demonstrated that they bear many enigmatic biosynthetic gene clusters that possibly encode novel compounds [37]. The same remarks were described for *Aspergilus sp* which seems to contain even more numbers of silent biosynthetic gene clusters than the case of actinobacteria. The observation that most bacteria and microbes strain harbour cryptic genes and can probably yield novel secondary metabolites has stimulated the expansion of many technics for detecting metabolic products of these clusters [38, 39]. This has prompted researcher to investigate the state-of-art genomic-guide natural product discovery (Figure 9.4).

9.7 Engineering approaches to produce novel natural products

9.7.1 Pathway rebuilding in heterologous hosts

A state of art approach in natural product design is to rebuild a whole metabolic pathway in a novel host. Indeed, despite that many hosts (producers) are hard to ferment or transform, the appropriate host is permitted by heterologous expression and can be genetically transformed. These technics are usually cheap and effective. Figure 9.5 explains different steps included in heterologous expression technics [41, 42].

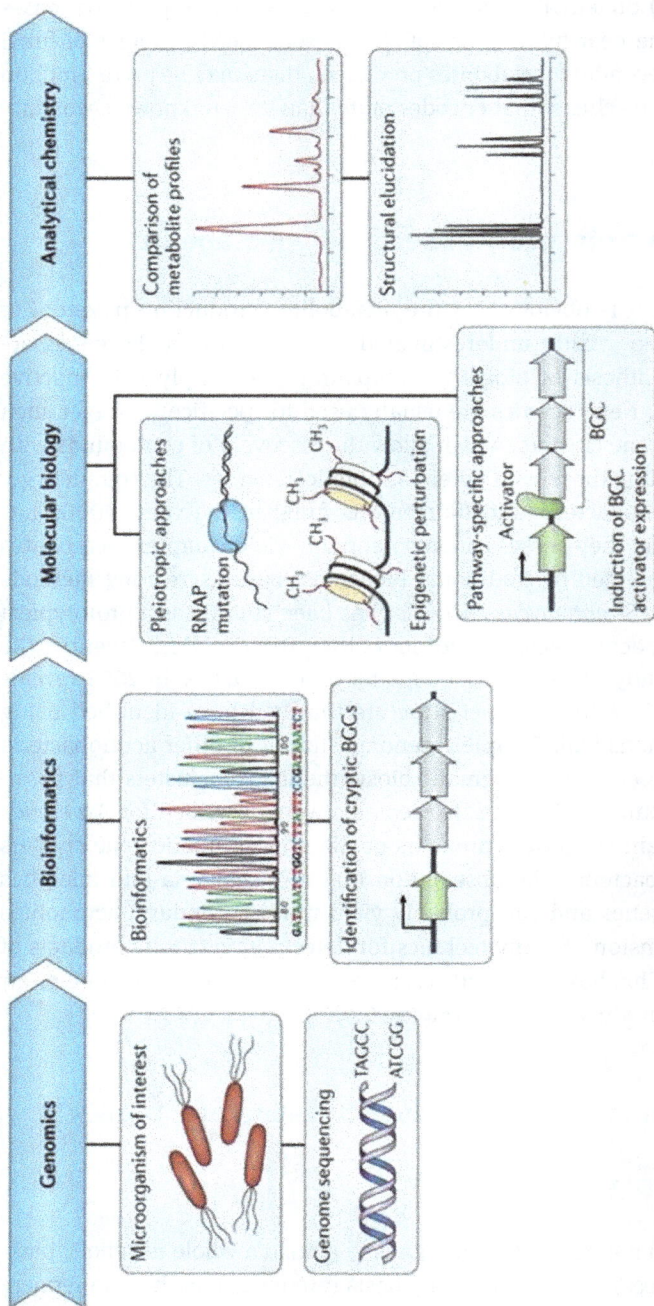

Figure 9.4: Genome drawing methods for new natural product discovery in microbes (Rutledge, P. J. and G. L. Chalis. "Discovery of microbial natural products by activation of silent biosynthetic gene clusters" [40].

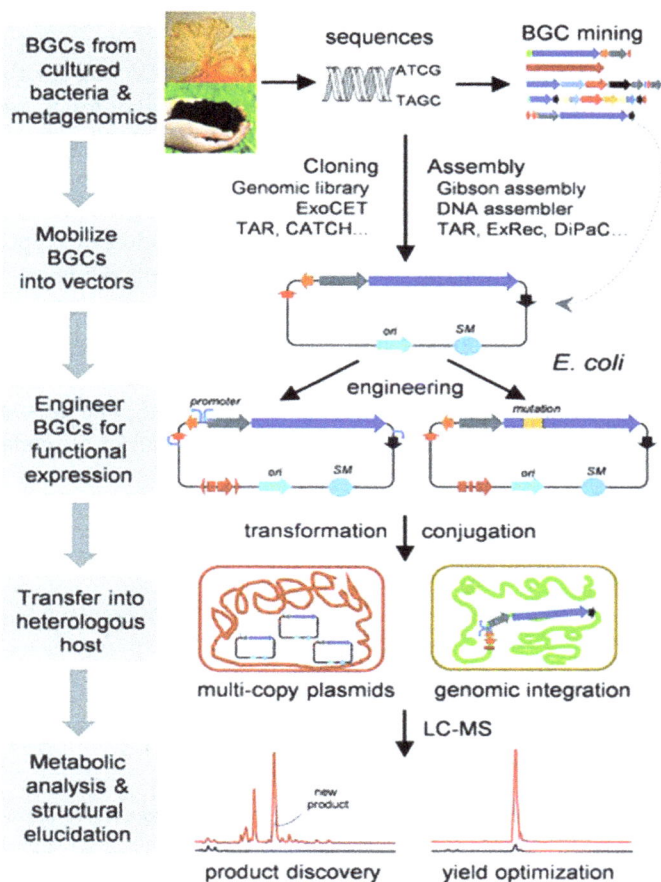

Figure 9.5: Road map for the heterologous expression approach. Reprinted by the permission from the Royal Society of Chemistry: Nature product report [43].

9.7.2 Genetic manipulation in the natural hosts

Natural products from fungi have been improved by accidental change or discriminating reproduction of the natural host, nevertheless genetic manipulation of the natural producer if the host is exhibiting the desired reaction to genetic operation – some time result in an interesting transformation to the metabolites chromatography profile (upregulation, down-regulation, new compounds signal, etc.). Due to their tightly and redundantly regulation, they require innovative strategies or state of the art technics to modify or change titer levels in the native producer host.

9.7.3 Activation of cryptic genes clusters in a native microbial host

A huge number of secondary metabolites are predictable to await finding for the reason that the synthesizing microorganism host has not been isolated or fermented yet, or even if it has been isolated or fermented, it did not show its real biosynthetic potential in the lab because many of its cryptic biosynthetic gene clusters remain silent or under expressed [44]. Currently, many methods founded on gene manipulation have been elaborated and described according to the activation of cryptic biosynthetic gene groups in fungi. Bergmann and collaborators described a respectable advancing method throughout which the identification of silent gene clusters of the fungus *Aspergillus nidulans* by genome manipulation leads to a novel cytotoxic compounds named aspyridones were identified through the higher expression of the transcription factor gene *apdR* that was previously silent *apd* gene cluster. The experimentation was done by inducible promoter and activation of all genes through transcription. (Figure 9.6).

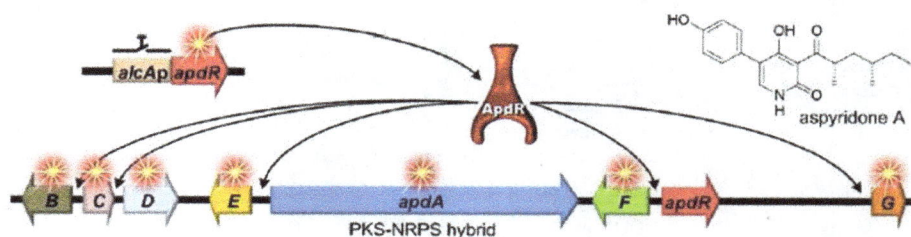

Figure 9.6: Overexpression of apdR transcription in *Aspergillus nidulans leading to novel cytotoxic agent* aspyridone. Reprinted by the permission from the American Society of Microbiology: Applied and Environmental Microbiology [45].

9.7.4 Domain exchange in multi-modular enzymes

Two principal classes of enzyme modulators are involved in the biosynthetic pathway of secondary metabolites, cited as non-ribosomal peptide synthetases (NRPSs) responsible for the synthesis of compounds with nitrogen content and polyketide synthases (PKSs) [46] which are arranged in a module similar to building block kit. This inherent feature is considered as a possible initial point for enzyme manipulation and combinative recombination of such modules. The existing certain numbers of examples using these techniques for engineering novel metabolites: the Type I PKS genes are built-in module, each translating to specific enzymatic functions [47]. An unusual natural product that can be considered as "Un-natural" can be biologically synthesized by two possible ways: – rearrangement or – exchange of certain polyketide synthetase modules. The Domain communicators, so called

"Domain switches" were described for the PKS/NRPS, programming the biosynthesis of the closely related metabolites tenellin and desmethylbassianin [48]. In the filamentous fungus *Aspergelus oryzae,* the conduction of genes expression analysis with heterologous host results to the detection of unprecedented isolated compounds building according to PKS/NRPS hybrid gene., which showed different characteristics features as methylation pattern or chain length that could be related with the corresponding area communicators (Figure 7(a)). According to other examples of reasonable domain exchange, the exchange of the filamentous fungus *Aspergelus nidulans* was used to generate a hybrid PKS asperfuranone acyl group protein (ACP) transacylase domain starter unit acyl transferase (SAT) with the sterigmatocystin SAT domain of the same organism [48]. Novel metabolites which have the same chain length as asperfuranone (Figure 7(b)) were biosynthesized by this method. Deepl lore of reducing PKS accountable for hypothemycin biosynthesis and development of different chimaera enzymes guided to the formation of an unusual diastereomer [49].

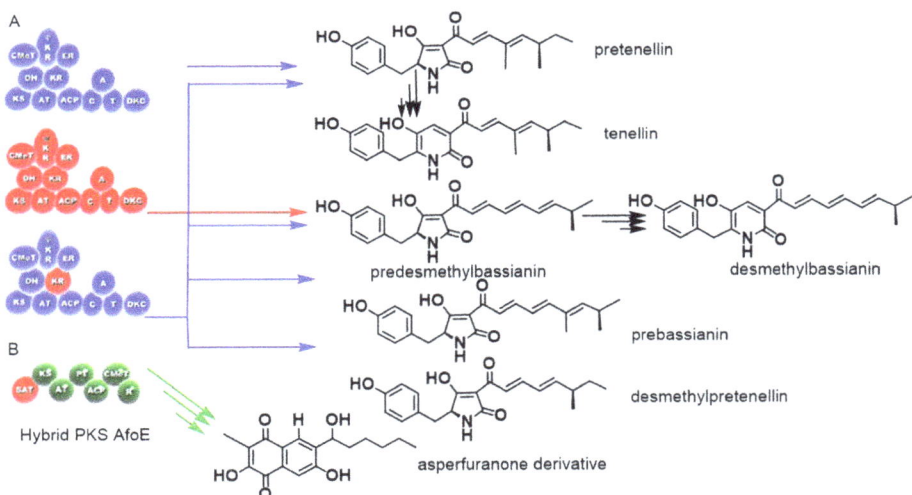

Figure 9.7: Domain exchange in multi-modular enzymes leading to the formation of novel compouonds [48, 51].

The use of the concept of domain reorganizing is to appropriate to both PKSs and NRPSs. The novel surfactin related compounds production was noticed when the transformation of the NRPS domains was engaged in the surfactin biogenesis in *Bacillus subtilis* and with domains of fungal δ-(l-α-aminoadipyl)-l-cysteinyl-d-valine synthetase engaged in penicillin/cephalosporin biosynthesis [50].

9.8 Plug and play approach to cloning transcriptionally silent gene clusters of unknown function from bacteria

The intersection between synthetic biology methods and microbial genomic sequencing has resulted in the assembly of full molecular toolboxes to go from DNA to protein regulation to facilitate the identification of unknown natural products predicted to be encoded in the transcriptionally silent gene cluster, at one end from fully synthetic DNA to the other end where natural DNA sequences are revamped for promoter optimization, transcriptional regulation, installation of improved ribosome binding sites, and even codon usage changes to fit the preference of the selected heterologous host [52]. One example that gives insights into lowering the barriers for capturing and moving an unknown gene cluster into a *Streptomycete* strain, *S. coelicolor* M1146 was performed with a silent 67 kb gene cluster homologous to the highly successful lipopeptide antibiotic daptomycin from a marine actinomycete, *Saccharomonospora sp* [53]. The cloning approach has been termed transformation-associated recombination (TAR) and in this case involved a vector with the element that allowed homologous recombination in the yeast, constituting the foreign DNA capture step with 73 kb of entrained CNQ-490 DNA. Potential repressor genes among the 20 in the captured cluster could be replaced *in vivo* in *S. cerevisiae* by marker replacement. Then, the circular construct could be shuttled into *E. coli* for any addition recombination plans. Finally, the tailored DNA construct could be conjugally transferred from *E. coli* to the *S. coelicolor* host and examined for expression (Figure 9.8). A lipopeptide homologous to daptomycin, termed taromycin, was produced. It is chlorinated at two residues, Trp1 and kyneurinine14, and has calcium-dependent antibiotic activity against Gram-positive bacteria by the action of daptomycin.

The ability to seize foreign DNA with a size of 70 kb suitable to encode full biosynthetic pathways for the great majority of polyketides, non-ribosomal peptides, and terpenes, and to change from yeast to *E. coli* to a selected bacterial host strain decreases many technical barriers for investigation of the silent gene cluster. This demonstrates how genome mining and synthetic biology have converged to enable researchers to take on almost any silent, unknown microbial gene cluster and give it a high probability of heterologous expression.

9.9 Conclusions

Many have referred to our generation as "the Golden Age of Natural Products Drug Discovery". This could be explained by the increasing use of advanced genomic, transcriptomic and metabolomic methods for analyzing the metabolic and biosynthetic processes leading to the secondary metabolites found within the source organisms. Thus, several software and algorithms have been lately published, which

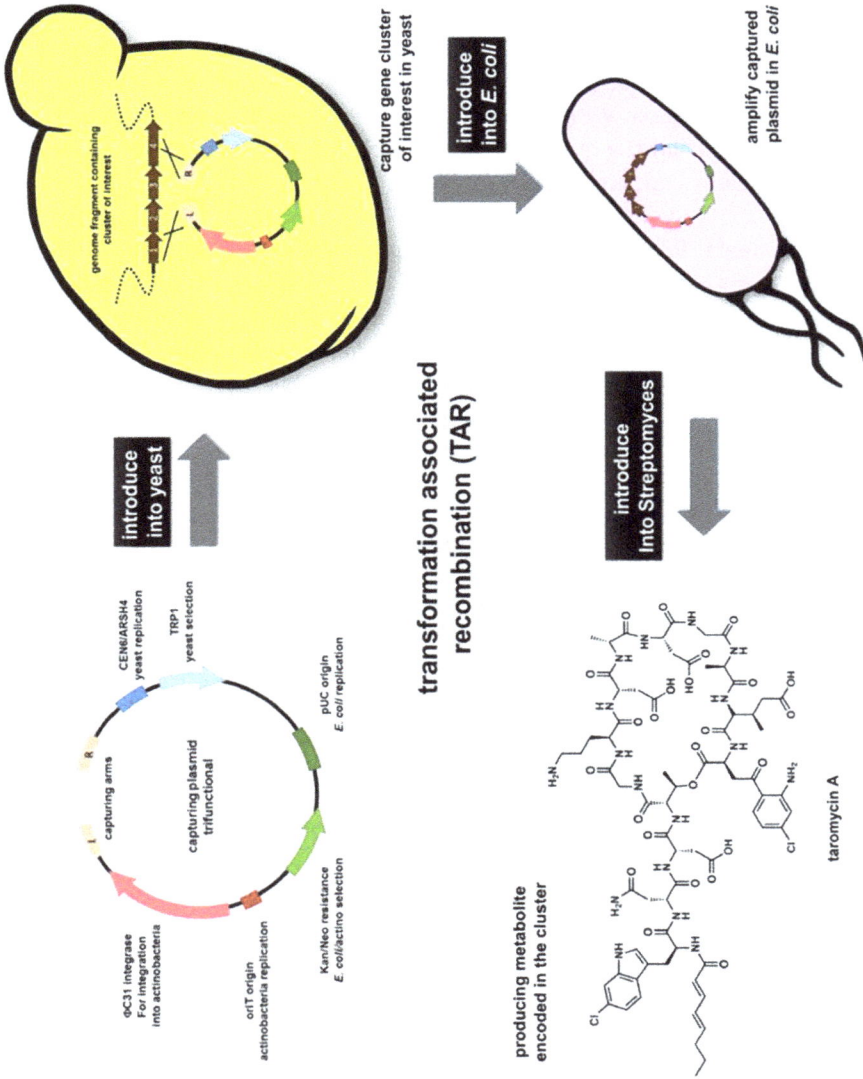

Figure 9.8: Transformation-associated recombination (TAR) can be used to rapidly capture large gene clusters [53].

accelerate the analysis of such data. This review aims to vividly demonstrate that metabolic engineering is a powerful strategy for the enhancement of the discovery of valuable molecules in a rational manner. An overview of the processes of engineering secondary metabolism is provided, which focuses on heterologous and transcriptomic expression. With a focus on selected bacterial and fungal species, engineering methods for the modification of the chemical structures of key secondary metabolite classes are provided.

Acknowledgements: RSTK and PS acknowledge financial support from the Alexander von Humboldt (AvH) through return fellowship. FNK acknowledges a return fellowship and an equipment subsidy from the Alexander von Humboldt Foundation, Germany. Financial support for this work is acknowledged from a ChemJets fellowship from the Ministry of Education, Youth and Sports of the Czech Republic awarded to FNK.

References

[1] Demain AL, Sanchez S. Microbial drug discovery: 80 years of progress. J Antibiot (Tokyo). 2009;62:5–16.
[2] Shen B. A new golden age of natural products drug discovery. Cell. 2015;163:1297–300.
[3] Tulp M, Bohlin L. Rediscovery of known natural compounds: nuisance or goldmine? Bioorg Med Chem. 2005;13:5274–82.
[4] Bills GF, Gloer JB. Biologically active secondary metabolites from the fungi. In: The fungal kingdom. 2017:1087–119.
[5] Chiang YM, Ahuja M, Oakley CE, Entwistle R, Asokan A, Zutz C, et al. Development of genetic dereplication strains in Aspergillus nidulans results in the discovery of aspercryptin. Angewandte Chemie. 2016;128:1694–7.
[6] Schreiber SL. Small molecules: the missing link in the central dogma. Nat Chem Biol. 2005;1:64–6.
[7] Dutton MV, Evans CS. Oxalate production by fungi: its role in pathogenicity and ecology in the soil environment. Can J Microbiol. 1996;42:881–95.
[8] Frisvad JC, Larsen TO. Chemodiversity in the genus Aspergillus. Appl Microbiol Biotechnol. 2015;99:7859–77.
[9] Lu M-Y, Fan W-L, Wang W-F, Chen T, Tang Y-C, Chu F-H, et al. Genomic and transcriptomic analyses of the medicinal fungus Antrodia cinnamomea for its metabolite biosynthesis and sexual development. Proc National Acad Sci. 2014;111:E4743–E4752.
[10] Brakhage AA. Regulation of fungal secondary metabolism. Nat Rev Microbiol. 2013;11:21–32.
[11] Hoffmeister D, Keller NP. Natural products of filamentous fungi: enzymes, genes, and their regulation. Nat Prod Rep. 2007;24:393–416.
[12] McDaniel R, Welch M, Hutchinson CR. Genetic approaches to polyketide antibiotics. 1. Chem Rev. 2005;105:543–58.
[13] Sieber SA, Marahiel MA. Learning from nature's drug factories: nonribosomal synthesis of macrocyclic peptides. J Bacteriol. 2003;185:7036–43.
[14] Kishimoto S, Hirayama Y, Watanabe K. Polyketide synthase–nonribosomal peptide synthetase hybrid enzymes of fungi. In: Physiology and genetics, Springer, 2018:367–383.

[15] Knaggs AR. The biosynthesis of shikimate metabolites. Nat Prod Rep. 2003;20:119–36.

[16] Herrmann KM, Weaver LM. The shikimate pathway. Annu Rev Plant Biol. 1999;50:473–503.

[17] Wang H, Sivonen K, Fewer DP. Genomic insights into the distribution, genetic diversity and evolution of polyketide synthases and nonribosomal peptide synthetases. Curr Opin Genet Dev. 2015;35:79–85.

[18] Bushley KE, Turgeon BG. Phylogenomics reveals subfamilies of fungal nonribosomal peptide synthetases and their evolutionary relationships. BMC Evol Biol. 2010;10:26.

[19] Walsh CT, O'Brien RV, Khosla C. Nonproteinogenic amino acid building blocks for nonribosomal peptide and hybrid polyketide scaffolds. Angewandte Chemie Int Ed. 2013;52:7098–124.

[20] Shen B. Polyketide biosynthesis beyond the type I, II and III polyketide synthase paradigms. Curr Opin Chem Biol. 2003;7:285–95.

[21] Hertweck C. The biosynthetic logic of polyketide diversity. Angewandte Chemie Int Ed. 2009;48:4688–716.

[22] Cox RJ, Simpson TJ. Fungal type I polyketide synthases. Methods Enzymol. 2009;459:49–78.

[23] Newman DJ, Cragg GM. Natural products as sources of new drugs from 1981 to 2014. J Nat Prod. 2016;79:629–61.

[24] Li JW, Vederas JC. Drug discovery and natural products: end of an era or an endless frontier? Science. 2009;325:161–5.

[25] Ōmura S, Ikeda H, Ishikawa J, Hanamoto A, Takahashi C, Shinose M, et al. Genome sequence of an industrial microorganism Streptomyces avermitilis: deducing the ability of producing secondary metabolites. Proc National Acad Sci. 2001;98:12215–20.

[26] Galagan JE, Calvo SE, Cuomo C, Ma L-J, Wortman JR, Batzoglou S, et al. Sequencing of Aspergillus nidulans and comparative analysis with A. fumigatus and A. oryzae. Nature. 2005;438:1105–15.

[27] Nierman WC, Pain A, Anderson MJ, Wortman JR, Kim HS, Arroyo J, et al. Genomic sequence of the pathogenic and allergenic filamentous fungus Aspergillus fumigatus. Nature. 2005;438:1151–6.

[28] Inglis DO, Binkley J, Skrzypek MS, Arnaud MB, Cerqueira GC, Shah P, et al. Comprehensive annotation of secondary metabolite biosynthetic genes and gene clusters of Aspergillus nidulans, A. fumigatus, A. niger and A. oryzae. BMC Microbiol. 2013;13:91.

[29] Bode HB, Bethe B, Höfs R, Zeeck A. Big effects from small changes: possible ways to explore nature's chemical diversity. ChemBioChem. 2002;3:619–27.

[30] Blin K, Medema MH, Kazempour D, Fischbach MA, Breitling R, Takano E, et al. antiSMASH 2.0 —a versatile platform for genome mining of secondary metabolite producers. Nucleic Acids Res. 2013;41:W204–W212.

[31] Weber T, Blin K, Duddela S, Krug D, Kim HU, Bruccoleri R, et al. antiSMASH 3.0—a comprehensive resource for the genome mining of biosynthetic gene clusters. Nucleic Acids Res. 2015;43:W237–W243.

[32] Wang H, Fewer DP, Holm L, Rouhiainen L, Sivonen K. Atlas of nonribosomal peptide and polyketide biosynthetic pathways reveals common occurrence of nonmodular enzymes. Proc National Acad Sci. 2014;111:9259–64.

[33] Li YF, Tsai KJ, Harvey CJ, Li JJ, Ary BE, Berlew EE, et al. Comprehensive curation and analysis of fungal biosynthetic gene clusters of published natural products. Fungal Genet Biol. 2016;89:18–28.

[34] Bentley SD, Chater KF, Cerdeño-Tárraga A-M, Challis GL, Thomson N, James KD, et al. Complete genome sequence of the model actinomycete Streptomyces coelicolor A3 (2). Nature. 2002;417:141–7.

[35] Keller NP, Turner G, Bennett JW. Fungal secondary metabolism—from biochemistry to genomics. Nat Rev Microbiol. 2005;3:937–47.

[36] Challis GL. Exploitation of the Streptomyces coelicolor A3 (2) genome sequence for discovery of new natural products and biosynthetic pathways. J Ind Microbiol Biotechnol. 2014;41:219–32.

[37] Nett M, Ikeda H, Moore BS. Genomic basis for natural product biosynthetic diversity in the actinomycetes. Nat Prod Rep. 2009;26:1362–84.

[38] Zerikly M, Challis GL. Strategies for the discovery of new natural products by genome mining. ChemBioChem. 2009;10:625–33.

[39] Weber T, Charusanti P, Musiol-Kroll EM, Jiang X, Tong Y, Kim HU, et al. Metabolic engineering of antibiotic factories: new tools for antibiotic production in actinomycetes. Trends Biotechnol. 2015;33:15–26.

[40] Rutledge PJ, Challis GL. Discovery of microbial natural products by activation of silent biosynthetic gene clusters. Nat Rev Microbiol. 2015;13:509–23.

[41] Phelan RM, Sachs D, Petkiewicz SJ, Barajas JF, Blake-Hedges JM, Thompson MG, et al. Development of next generation synthetic biology tools for use in Streptomyces venezuelae. ACS Synth Biol. 2017;6:159–66.

[42] Ross AC, Gulland LE, Dorrestein PC, Moore BS. Targeted capture and heterologous expression of the Pseudoalteromonas alterochromide gene cluster in Escherichia coli represents a promising natural product exploratory platform. ACS Synth Biol. 2015;4:414–20.

[43] Ongley SE, Bian X, Neilan BA, Müller R. Recent advances in the heterologous expression of microbial natural product biosynthetic pathways. Nat Prod Rep. 2013;30:1121–38.

[44] Hertweck C. Hidden biosynthetic treasures brought to light. Nat Chem Biol. 2009;5:450–2.

[45] Bergmann S, Funk AN, Scherlach K, Schroeckh V, Shelest E, Horn U, et al. Activation of a silent fungal polyketide biosynthesis pathway through regulatory cross talk with a cryptic nonribosomal peptide synthetase gene cluster. Appl Environ Microbiol. 2010;76:8143–9.

[46] Walsh CT, Fischbach MA. Natural products version 2.0: connecting genes to molecules. J Am Chem Soc. 2010;132:2469–93.

[47] Moyer A, Coghill R. Production of penicillin in surface culture. press.

[48] Liu T, Chiang Y-M, Somoza AD, Oakley BR, Wang CC. Engineering of an "unnatural" natural product by swapping polyketide synthase domains in Aspergillus nidulans. J Am Chem Soc. 2011;133:13314–16.

[49] Zhou H, Gao Z, Qiao K, Wang J, Vederas JC, Tang Y. A fungal ketoreductase domain that displays substrate-dependent stereospecificity. Nat Chem Biol. 2012;8:331–3.

[50] Stachelhaus T, Schneider A, Marahiel MA. Rational design of peptide antibiotics by targeted replacement of bacterial and fungal domains. Science. 1995;269:69–72.

[51] Fisch KM, Bakeer W, Yakasai AA, Song Z, Pedrick J, Wasil Z, et al. Rational domain swaps decipher programming in fungal highly reducing polyketide synthases and resurrect an extinct metabolite. J Am Chem Soc. 2011;133:16635–41.

[52] Walsh CT, Garneau-Tsodikova S, Howard-Jones AR. Biological formation of pyrroles: nature's logic and enzymatic machinery. Nat Prod Rep. 2006;23:517–31.

[53] Yamanaka K, Reynolds KA, Kersten RD, Ryan KS, Gonzalez DJ, Nizet V, et al. Direct cloning and refactoring of a silent lipopeptide biosynthetic gene cluster yields the antibiotic taromycin A. Proc National Acad Sci. 2014;111:1957–62.

Sergi Herve Akone, Cong-Dat Pham, Huiqin Chen,
Antonius R. B. Ola, Fidele Ntie-Kang and Peter Proksch

10 Epigenetic modification, co-culture and genomic methods for natural product discovery

Abstract: Fungi and bacteria are encountered in many habitats where they live in complex communities interacting with one another mainly by producing secondary metabolites, which are organic compounds that are not directly involved in the normal growth, development, or reproduction of the organism. These organisms appear as a promising source for the discovery of novel bioactive natural products that may find their application in medicine. However, the production of secondary metabolites by those organisms when cultured axenically is limited as only a subset of biosynthetic genes is expressed under standard laboratory conditions leading to the search of new methods for the activation of the silent genes including epigenetic modification and co-cultivation. Biosynthetic gene clusters which produce secondary metabolites are known to be present in a heterochromatin state in which the transcription of constitutive genes is usually regulated by epigenetic modification including DNA methylation and histone deacetylation. Therefore, small-molecule epigenetic modifiers which promote changes in the structure of chromatin could control the expression of silent genes and may be rationally employed for the discovery of novel bioactive compounds. Co-cultivation, which is also known as mixed-fermentation, usually implies two or more microorganisms in the same medium in which the resulting competition is known to enhance the production of constitutively present compounds and/or to lead to the induction of cryptic metabolites that were not detected in axenic cultures of the considered axenic microorganism. Genomic strategies could help to identify biosynthetic gene clusters in fungal genomes and link them to their products by the means of novel algorithms as well as integrative pan-genomic approaches. Despite that all these techniques are still in their infancy, they appear as promising sources for the discovery of new bioactive compounds. This chapter presents recent ecological techniques for the discovery of new secondary metabolites that might find application in medicine.

Keywords: fungi, bacteria, epigenetic modification, co-culture, genomics, silent genes

This article has previously been published in the journal *Physical Sciences Reviews*. Please cite as: Akone, S. H., Pham, C., Chen, H., Ola, A. R. B., Ntie-Kang, F., Proksch, P. Epigenetic modification, co-culture and genomic methods for natural product discovery. *Physical Sciences Reviews* [Online] 2019 DOI: 10.1515/psr-2018-0118

https://doi.org/10.1515/9783110668896-010

10.1 Introduction

Natural products have been playing a vital role in the pharmaceutical industry. In fact, in drug development, they are used either directly as drugs, either as templates for structural modification or as a key scaffold for structure activity relation studies [1, 2]. The revolutionary discovery of penicillin from the fungus *Penicillium notatum* by Alexander Fleming in 1929 sparked immense interest in microorganism derived natural products. Penicillin was widely used as an antibiotic agent since the 1940s [3]. Since then, microbial secondary metabolites have gained wide attention and contributed greatly for the development of therapeutic agents for the treatment of cancer, infectious diseases, cardiovascular diseases etc. as exemplified by the discovery of griseofulvin, lovastatin, and echinocandin B [4]. Interestingly, the biosynthesis of those natural products more often occurs within discrete localized sections of the fungal genome known as biosynthetic gene clusters [5]. These gene clusters encode for enzymes such as polyketide synthases (PKS) or non-ribosomal peptide synthases (NRPS) [6, 7]. However, the discovery rate of new potent natural products has plateaued in recent years indicated by an increase in the rate of re-isolated known compounds using standard laboratory conditions [5, 8]. Whole-genome sequencing of various fungi suggested that the fungi's biosynthetic capacity is far from being exhausted [6, 9]. One reason for this observation might be that only a subset of biosynthetic pathway genes is expressed under standard laboratory conditions while the rest remains silent. Therefore, only a miniscule number of secondary metabolites are produced. Such silent loci are referred to as cryptic pathways [7]. It is worth to mention that these secondary metabolites served as chemical signals to communicate with other species, to defend the habitat of the producers or to inhibit the growth of competitors [5, 10]. Several effective approaches have been reported to expand the natural biogenetic capability of microorganisms including epigenetic modification, co-cultivation, and the genomic approach [10, 11].

One approach that has been proven highly successful in triggering the activation of silent biosynthetic gene clusters for the production of unknown secondary metabolites is the treatment of the microorganism, usually fungi, with small molecule epigenetic modifiers. "Epigenetics" is a term that describes heritable changes in gene expression which occur without changes in DNA sequence. Posttranslational modifications which include DNA methylation, histone acetylation, and phosphorylation determine time and location of gene expression [12]. Hypoacetylation of histones is usually associated with gene silencing while hyperacetylation is more commonly associated with gene activation [13, 14]. To date, several epigenetic modifying enzymes have been identified, such as DNA methyltransferases (DNMT), histone acetyltransferases and histone deacetyltransferases (HDAC). Small molecule inhibitors and activators thereof have found application not only as probes for biological studies and as potential therapeutic agents, but also as powerful tools for inducing cryptic biosynthetic gene clusters for the production of novel natural products [15, 16]. Examples of

DNMT inhibitors include 5-azacytidine, 5-aza-2'-deoxycytidine, methylthioadenosine, and S-adenosylhomocysteine, while inhibitors of HDAC are trichostatin A, suberoyla-nilide hydroxamic acid (SAHA), trapoxin B, apicidin, HC-toxin, sodium butyrate and valproic acid [17]. HDACs and DNMTs have been shown to impact the biosynthesis of fungal natural products as highlighted in studies by Williams et al. [15].

In their natural habitat, microorganisms rarely occur in axenic form, but rather co-exist with other micro- or macro- organisms in dynamic and constantly changing environments [5, 18]. Thus the production of bioactive secondary metabolites is considered as a natural response to competition for limited territory and nutrients [8, 19, 20]. Co-cultivation also was known as mixed-fermentation is a widely used approach for the activation of silent genes for the discovery of new natural products. Cultivation of different microbial strains within the same culture vessel allows the microorganisms to directly interact with each other and is an approach to mimic their natural habitat in hopes to induce cryptic secondary metabolites or to enhance the production of known compounds [11, 21, 22]. In fact, the reason why certain novel compounds have been reported only in co-culture might be that in axenic culture, they are produced only in trace amounts thus making their detection difficult [5]. The mechanism behind the activation of silent biosynthetic pathways during co-cultivation can be explained by either unilateral stimulation through physical interaction or chemical signals; or by unilateral induction of the biosynthesis of signaling molecules which then trigger the production of cryptic metabolites [8].

Genomic methods have been shown to be powerful tools for the discovery of new natural products. This approach relies on the development of bioinformatic tools and genome sequencing techniques especially for microorganisms for the discovery of new drugs [23]. While all before mentioned strategies follow the "compound-to-gene" route which is often time consuming and limited to the use of one strain only, this approach is based on a "gene-to-compound" route in which a large library of strains can be screened for potential novel bioactive compounds [23]. Bioinformatic tools can help to identify key players in biosynthetic pathways involving PKS or NRPS and thus novel polyketides and peptides based metabolites can be predicted [24]. The combination of isotope-guided fractionation and bioinformatic analysis can be used for the identification of potential end products of orphan gene clusters [25]. This approach allows an understanding of the target substrate and physicochemical properties of the end products [10].

The above approaches have been widely accepted and frequently employed in the field of natural products in recent years. This chapter presents examples from 2000–2017 of cultured microorganisms in which either the production of cryptic secondary metabolites has been successfully induced or the accumulation of known compounds has been enhanced using small molecules epigenetic modifiers, co-cultivation, and genomic mining. The corresponding structures are shown in Figure 1.1–Figure 1.9.

10.2 Production of natural compounds through epigenetic modification

10.2.1 Fungal cryptic metabolites induced by small molecules epigenetic modifiers

Fungi have been the primary target of epigenetic modification to gain access to cryptic natural product genes which in turn can lead to the accumulation of new compounds [6, 26–28]. Modification of gene transcription can be achieved by different strategies including the addition of epigenetic modifying small molecules such as SAHA or 5-azacytidine. The following paragraphs describe the discovery of novel cryptic secondary metabolites and provide mechanistic insight into epigenetic modification.

A novel pentacyclic polyketide, daldinone E (**1**), was discovered from a *Daldinia sp.* fungal isolate treated with SAHA, along with a biosynthetically related epoxide-containing daldinone analogue (**2**) (Figure 10.1) [27]. Both compounds showed DPPH radical scavenging activities with potency comparable to the positive control ascorbic acid [27].

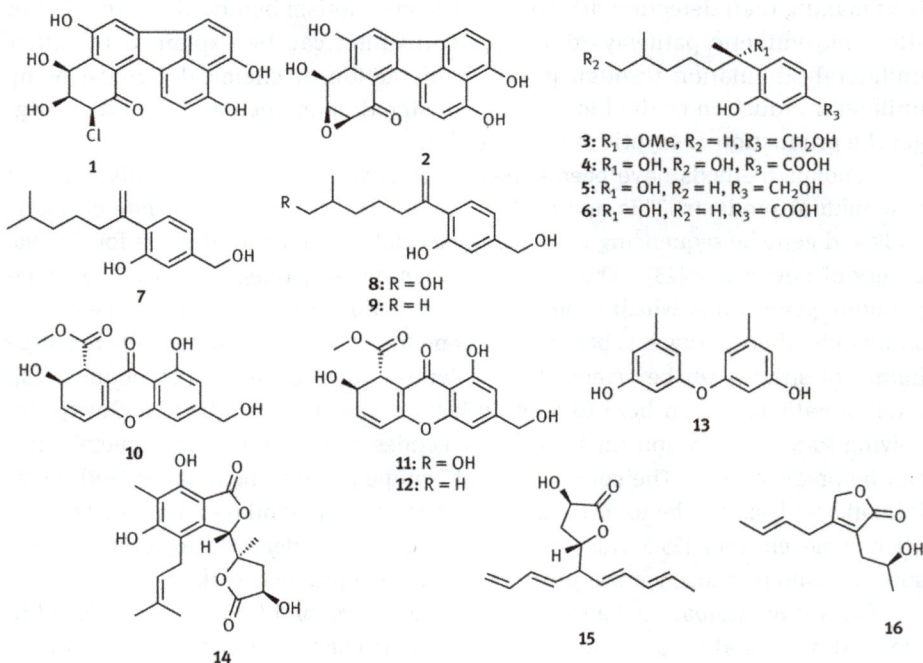

Figure 10.1: Compounds **1–16**.

An *Aspergillus sydowii* culture was treated with a DNA methyltransferase inhibitor, 5-azacytidine, resulting in the isolation of three new bisabolane-type sesquiterpenoids:

(7S)-(+)-7-O-methylsydonol (**3**), (7S,11S)-(+)-12-hydroxysydonic acid (**4**) and 7-deoxy-7,14-didehydrosydonol (**5**), along with eight known compounds (**6–13**) (Figure 10.1) [29]. The authors evaluated the anti-diabetic and anti-inflammatory activities of the isolated compounds and found that among those (S)-(+)-sydonol not only increased insulin-stimulated glucose consumption but also prevented the accumulation of lipids in 3T3-L1 adipocytes. Furthermore, (S)-(+)-sydonol showed significant anti-inflammatory activity via inhibition of superoxide anion generation and elastase release by *N*-formyl-*L*-methionyl-*L*-leucyl-*L*-phenylalanine/Cytochalasin B (FMLP/CB)-induced human neutrophils [29].

Exposure of *Chaetomium cancroideum* to nicotinamide, an NAD^+-dependent HDAC inhibitor, enhanced the production of aromatic and branched aliphatic polyketides, thus yielding the novel compounds chaetophenol G (**14**) and cancrolides A (**15**) and B (**16**) (Figure 10.1) [30]. Feeding experiments with ^{13}C labeled precursors provided evidence of the biosynthesis of the branched aliphatic polyketide skeletons in cancrolides A and B. None of the isolated compounds, however, showed cytotoxic, antibacterial or anti-viral activity [30].

10.2.2 Bacterial cryptic secondary metabolites induced by epigenetic modifications

As prokaryotes lack histone, many of the targets and enzymes that could be altered in fungi to induce epigenetic change are nonexistent in bacteria. However, that is not to say that bacteria do not possess their own epigenetic regulatory system. Currently, the best understood bacterial epigenetic modifying system is DNA methylation [31]. Different DNMT target unique DNA sequences at which they attach methyl groups on cytosine bases. Methylation near gene promoter regions causes variations in gene expression [32]. It may, therefore, be possible to access silent natural product gene clusters in bacteria through treatment with DNMT inhibitors such as 5-azacytidine and procaine.

To date, only one paper has been published with regard to epigenetic modification in bacteria to elicit metabolite production [33]. In that work, the authors described the treatment of *Pseudomonas aeruginosa* with epigenetic modifiers including kinase inhibitors manzamine A, kahalalide F, sceptrin, and the cell transport regulator ilimaquinone. In addition, growth factors such as carbon sources and temperature were altered and co-culturing with four other antibiotic producing microbes was performed. From this organism, two derivatives of phenazine (**17, 18**) and six 2-alkyl-4-quinolones (**19–24**), and seven rhamnolipids (**25–31**) were isolated among others (Figure 10.2) [33]. This study was helpful for understanding the metabolic profile of biologically active secondary metabolites from *P. aeruginosa* and provided the basis for comparison to identify the differences in metabolite production in epigenetically modified *P. aeruginosa*.

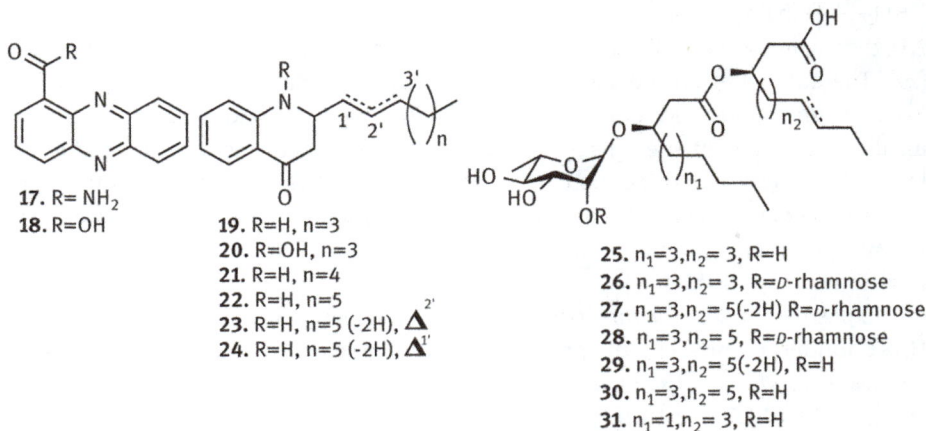

17. R= NH$_2$
18. R=OH

19. R=H, n=3
20. R=OH, n=3
21. R=H, n=4
22. R=H, n=5
23. R=H, n=5 (-2H), Δ
24. R=H, n=5 (-2H), $\Delta^{1'}$

25. n$_1$=3,n$_2$= 3, R=H
26. n$_1$=3,n$_2$= 3, R=D-rhamnose
27. n$_1$=3,n$_2$= 5(-2H) R=D-rhamnose
28. n$_1$=3,n$_2$= 5, R=D-rhamnose
29. n$_1$=3,n$_2$= 5(-2H), R=H
30. n$_1$=3,n$_2$= 5, R=H
31. n$_1$=1,n$_2$= 3, R=H

Figure 10.2: Compounds **17–31**.

10.3 Induction of natural compounds by co-cultivation

10.3.1 Fungal-bacterial co-cultivation: A powerful tool for cryptic metabolites induction

The first fungal-bacterial co-cultivation experiment which resulted in the production of cryptic metabolites is the one reported for the production of pestalone (**32**) (Figure 10.3) [34]. Indeed, pestalone (**32**) a new chlorinated benzophenone antibiotic was only induced in the mixed fermentation of an unicellular marine α-proteobacterium strain CNJ-328 (later identified as *Thalassospira* sp. CNJ-328) with a marine-derived fungus *Pestalotia* sp. strain CNL-365 isolated from the surface of the brown alga *Rosenvingea* sp [34]. In fact, when the fungus and the bacterium were cultured individually, the production of pestalone was not observed. Moreover, the treatment of the fungal axenic culture with the organic extract or the cell-free broth of the bacterium did not induce pestalone (**32**) production suggesting that the induction is not due to chemical signals produced by the bacterium. Pestalone (**32**) showed potent antibacterial activity against methicillin-resistant *Staphylococccus aureus* (MRSA) and vancomycin-resistant *Enterococcus faecium* with MIC values of 84 and 178 nM respectively [34]. Furthermore, pestalone showed moderate cytotoxic activities against 60 human tumor cell lines with a mean IC$_{50}$ value of 6 μM. However, in a recent study, pestalone was tested against diverse MRSA strain but failed to display the expected activity [35].

Schroeckh et al. studied the bacterial-fungal interplay by individually co-culturing the model fungus *Aspergillus nidulans* with a series of soil-dwelling bacteria of the actinomycetes family sharing the same habitat [22]. During this study, only one strain

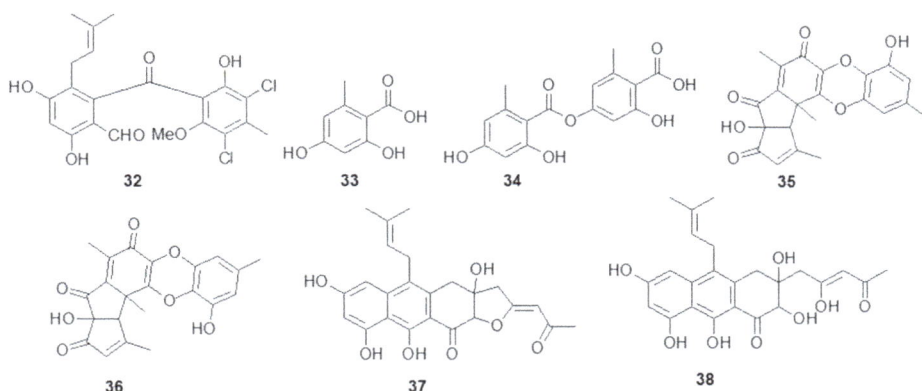

Figure 10.3: Compounds **32–38**.

which was identified as *Streptomyces hygroscopicus* (later renamed as *Streptomyces rapamycinicus*) was found to specifically induce fungal biosynthesis genes. The co-cultivation of *S. rapamycinicus* with *A. nidulans* resulted in the *de novo* production of four known polyketides including orsellinic acid (**33**), lecanoric acid (**34**), and the cathepsin K inhibitors F-9775 (**35**) and F-9775B (**36**) by the co-cultured fungus [22] (Figure 10.3). The monitoring of the effect of the expression of cryptic fungal biogenetic gene cluster was undertaken through transcriptome analysis [22]. Accordingly, full-genome microarrays revealed gene expression in a specific region including the up-regulation of two particular biosynthetic gene clusters. Northern blot analysis and quantitative (q) RT-PCR confirmed the specific induction of that region [22]. Moreover, the expression of the PKS gene (orsA) was found to be required for the production **2–5** as shown during Northern blot analysis. This result was then confirmed through targeted gene inactivation experiments of the respective PKSs gene in the genome of *A. nidulans*, in which the production of the compounds **33–36** was not observed during co-cultivation of the resulting knockout mutant of *S. rapamycinicus*. In order to determine whether the biosynthesis of **32–35** by the fungus was due to bacterial signaling molecules, three sets of experiments were performed. In the first set, the fungal axenic culture was treated with the bacterial culture supernatant, with co-culture extracts, and with heat-inactivated bacteria. In the second set, the co-cultivation of the fungus and the bacteria was carried out in the way that the fungus and the bacterium were separated by a dialysis tube. The effect of the supernatant of the co-culture of *S. rapamycinicus* with *A. nidulans* lacking orsA was studied to eliminate the involvement of some signaling molecules that cannot diffuse through the membrane or are produced in co-culture [22]. Surprisingly, in all these experiments no fungal PKS gene expression was observed thus concluding that fungal PKS gene expression was triggered only by direct physical interaction [22]. In fact, *S. rapamycinicus* induces histone modification by means of histone acetyltranferase complex Saga/Ada in *A. nidulans* upon direct physical contact [36]. Consistently, the Saga/ada complex was found to be required for specific induction of biosynthesis gene including orselinic acid gene clusters as shown by posttranslational histone modifications experiments. The addition of the histone deacetylase (HDAC) inhibitor, SAHA in the fungal cultures, activated *orsA* in the absence of co-cultivation of *A. nidulans* with *S. rapamycinicus* as evidenced by HPLCs analysis of the fungal axenic culture. These two studies provide evidence on posttranslational histone modification as a potential cause of genes expression during mixed-fermentation [22, 36].

The human pathogen *Aspergillus fumigatus* was challenged with *S. rapamycinicus* in a mixed-fermentation for the induction of new natural products [37]. During this co-cultivation, an intimate physical contact between the fungus and the bacterium was observed through electron microscopy analyses as bacterial filament was attached to mycelia previously; this observation was previously reported for *A. nidulans* with the same bacterium [22, 37]. This co-cultivation led to the induction of the prenylated polyphenol fumicycline A (**37**) and its hydrated congener fumicycline B (**38**) only detected in the co-culture extract (Figure 10.3). The full genome microarray revealed an upregulation of many genes during the co-culture [37]. In order to gain further insight into the mechanism of the upregulation of the isolated compounds, the axenic fungal culture was treated with pure bacterial culture medium or supernatant of bacterial. Furthermore, co-cultivation of the pure fungus with the bacterial in which both were separated by a dialysis tube was undertaken. All experiments failed to induce **37** and **38** suggesting that an intimate physical contact is responsible for the activation of the PKS gene cluster [37]. Moreover, the isolated compounds fumicycline A (**37**) and B (**38**) displayed moderate activity against *S. rapamycinicus*, thus suggesting that both compounds were produced as fungal defense response during the co-cultivation of the fungus with the bacterium. These findings shed new light to microbial crosstalk which can elicit the expression of dormant genes [37].

The α-proteobacterium strain (*Thalassospira* sp. CNJ-328) that triggered pestalone (**32**) production was added to a 3-day-old culture of the marine-derived fungus *Libertella* sp. leading to the induction of a series of new pimaran diterpenes, libertellenones A-D which were suggested to derive from the fungus (**39–42**) (Figure 10.2) [38]. Interestingly, the new pimarane diterpenes **39–42** were not detected when the fungus or the bacterium were axenically cultured. Moreover, the treatment of the fungal axenic culture with heat sterilized or cell-free supernatant or the organic extract of the bacterium, failed to induce the biosynthesis of the pimarane diterpenes **39–42** suggesting that their induction was not caused by signaling molecules from the bacteria, rather through cell-cell interactions as previously reported from *A. nidulans* and *A. fumigatus* [22, 37]. However, no significant antibacterial activity was observed even at a higher concentration when the libertellenones were assayed against the bacterium CNJ-328. Interestingly, the cytotoxic evaluation of the libertellenones against HCT-116 human colon carcinoma cancer cell line revealed that libertellenone D was the most potent cytotoxic compounds with an IC_{50} value of 0.76 µM indicating that the propane ring plays an important role for the observed activity; thus this compound could be considered as a potent cytotoxic lead compound [38].

Mixed fermentation of the *Sphingomonas* bacterial strain KMK-001 and *Aspergillus fumigatus* strain KMC-901 isolated from an acidic coal mine drainage resulted in the isolation of a new diketopiperazine disulfide, glionitrin A (**43**) (Figure 10.4) [39]. These compounds were not produced when either microorganism was cultured axenically [39]. It is worth to mention that the axenic fungal culture yielded another closely related diketopiperazine, gliotoxin (**44**). In order to determine the origin of glionitrin A a series of experiments were performed. Treatment of the bacterial culture with the fungal organic extract or with pure gliotoxin failed to induce glionitrin A indicating that its production was neither due to bacterial enzymatic biotransformation nor due to postmodification of gliotoxin [39]. Furthermore, time course production experiments of **43** in which each microbe was isolated from the mixture then cultured individually after detecting glionitrin A in the co-culture, also failed to show the presence of glionitrin A (**43**). Moreover, the treatment of the fungal axenic culture with diverse stress factors including sulfuric acid, phenol, and nystatin, bacterial supernatant, culture extract, and cell lysate failed to trigger glionitrin A(**43**) production [39]. Therefore **43** was likely the result of long-term fungal-bacterial interactions and was suggested to originate from the fungus [39].

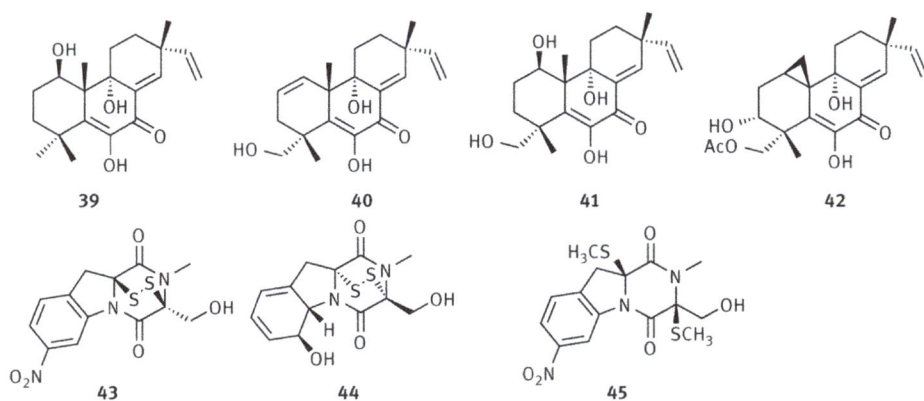

Figure 10.4: Compounds **39–45**.

Two years later, in a subsequent study, the same co-cultivation involving the above mentioned microorganisms was undertaken and led to the induction of not only glionitrin A but also of another new diketopiperazine glionitrin B (**45**) (Figure 10.4) [40]. The only difference was that glionitrin A (**43**) was detected after 8-days of incubation while glionitrin B (**45**) was produced as a minor compound after more than 18 days of incubation [40]. Glionitrin A (**43**) exhibited potent cytotoxic activity against several cancer cells line including HTC-116 (colorectal carcinoma), A549 (lung carcinoma), AGS (gastric adenocarcinoma), DU 145 (prostate carcinoma) MCF-7 (breast adenocarcinoma) and HepG2 (hepatocellular carcinoma) with IC_{50} values in the range of 0.2 to 2.3 µM [40]. Moreover, glionitrin A showed significant antibacterial activity when tested against the gram-positive methicillin-resistant *Staphylococcus aureus* with MIC values of 2.2 µM. On the other hand, glionitrin B (**45**) did not show any activity toward the DU145 cell line thus underlining the key role of the disulfide bridge for the cytotoxicity of these diketopiperazine derivatives [40].

A mixed fermentation of a *Streptomyces bullii* strain isolated from hyper-arid Atacama desert soil with the fungal strain *Aspergillus fumigatus* MBC-F1-F10 afforded the production of seven diketopiperazine derivatives including brevianamide F (**46**) spirotryprostatin A (**47**), 6-methoxy spirotryprostatin B (**48**), fumitremorgin B (**49**), fumitremorgin C (**50**), 12,13-dihydroxyfumitremorgin C (**51**) and verruculogen (**52**), in addition to two pseurotins derivatives including 11-O-methylpseurotin A (**53**) and its new derivative 11-O-methylpseurotin A_2 (**54**) (Figure 10.5) as well as ergosterol [41]. All isolated compounds were suggested to originate from the fungus and were likely induced through direct physical contact and/or long-term fungal-bacterial interactions. No antibacterial activity was found for any of the isolated compounds when tested against *Staphylococcus aureus* or *Escherichia coli* [41]. On the other hand, antitrypanocidal and antileishmanicidal assays of the induced compounds against *Trypanosoma brucei* and *Leishmania donovani* revealed that fumitremorgin B and verruculogen had the strongest activity with EC_{50} values of 0.2 and 3–4 µM, respectively [41]. Furthermore, all fungal metabolites were not selective and exhibited strong cytotoxic activity towards MRC5 cells (normal human fibroblasts), thus preventing their use as potential drugs candidates.

Co-cultivation of the soil-dwelling fungus *Aspergillus terreus* which was isolated from the sediment of the lake of Wadi El Natrun in Egypt with the bacterium *Bacillus subtilis* or *Bacillus cereus* on solid rice medium resulted in the induction of two new butyrolactone derivatives isobutyrolactone II (**55**) and 4-O-demethylisobutyrolactone II (**56**) together with the known compound N-(carboxymethyl)anthranilic acid (**57**) (Figure 10.6) [42]. All isolated compounds were only detected in the co-culture extract of *A. terreus* with *B. subtilis* or *B. cereus*, and not in the fungal or in the bacterial axenic cultures. Moreover, an up to 34-fold increase in the accumulation of constitutively present fungal metabolites was observed when the fungus was co-cultured either with *B. subtilis* or *B. cereus* [42]. A similar increase in the accumulation of constitutive fungal metabolites was also previously reported by Ola et al. [43], when the endophytic fungus *Fusarium tricintum* was co-cultured with the same strain of *B. subtilis*. During the study with *A. terrus*, the effects of the co-culture in fungal metabolite accumulation were found to be specific. **55** and **56** were found to originate from the fungus *A. terreus* as it is well-known to produce butyrolactone [44]. In addition, N-(carboxymethyl)anthranilic acid (**57**) which was induced only during the co-culture of *A. terreus* with live *B. subtilis* but not with *B.*

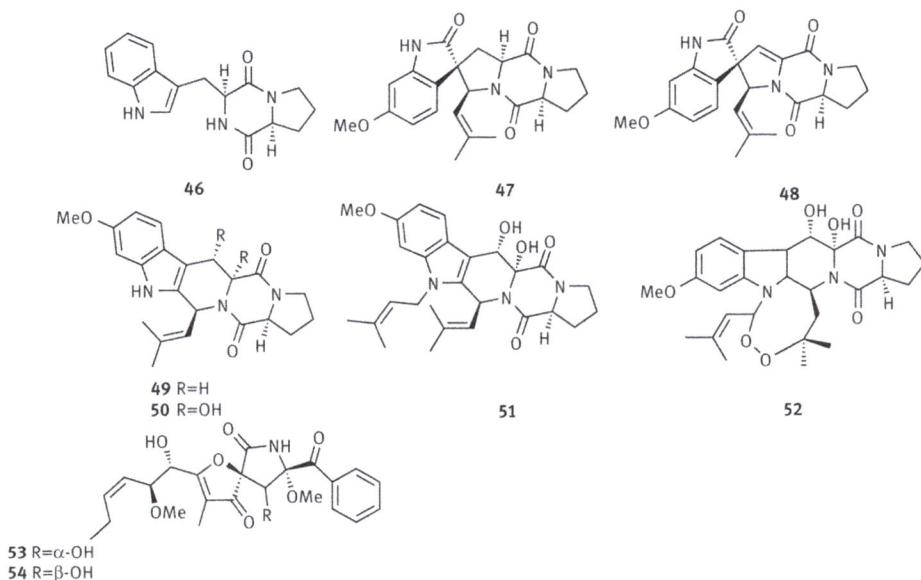

Figure 10.5: Compounds **46–54**.

cereus, nor with autoclaved *B. subtilis* and previously reported as a new metabolite only detected when the fungus *Fusarium tricintum* was co-cultured with the same strain, *B. subtilis* [43] was found to originate from the bacterium as it is a derivative of anthranilic acid, which is a well-known bacterial natural product [45, 46]. Subsequent, co-cultivation of *A. terreus* with other bacteria such as *Streptomyces lividans* or *Streptomyces coelicolor* did not lead to the induction of the fungal metabolites indicating that the response of the fungus toward different bacteria is specific [42]. All compounds were evaluated for their antibacterial activity against the strains used for co-cultivation namely *B. subtilis* or *B. cereus*, but no significant activity was observed. Furthermore, no cytotoxic activity was observed when the isolated compounds were tested toward mouse lymphoma cell line L5178Y in MTT assays [42].

A mixed-fermentation of *Fusarium pallidoroseum* with *Saccharopolyspora erythraea* led to the isolation of four decalin-type tetramic acid analogues related to equisetin (**58**) including N-demethyl-ophiosetin (**59**), pallidorosetin A (**60**) and pallidorosetin B (**61**) as well as the known ophiosetin (**62**) (Figure 10.6) [47]. These compounds were not detected in the axenic fungal or bacterial culture. All isolated compounds were found to originate from the fungus as they are structurally related to equisetin produced by *Fusarium pallidoroseum* in both the axenic culture as well as in the co-culture. However, none of these compounds displayed antibacterial or cytotoxic activity except for the known compound equisetin (**58**) when tested against gram-positive bacteria *Staphylococcus erythraea* and *S. aureus*. Indeed, **58** inhibited both bacteria at

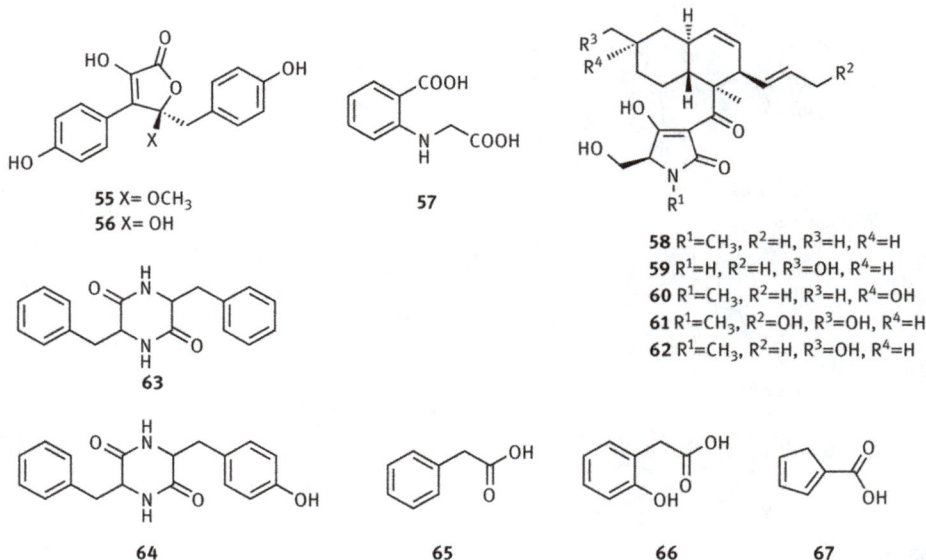

Figure 10.6: Compounds **55–67**.

≤ 2.5 ug/disk in the Kirby-Bauer disk diffusion assay suggesting that the increase of its production was a fungal defense response in the presence of *S. erythraea* [47]. Furthermore, **58** showed strong activity against the leukemia cell line CCRF-CEM with an IC_{50} value of 144 nM [47].

A co-cultivation experiment undertaken between *Aspergillus niger* N402 and *Streptomyces coelicolor* A3(2) M145 resulted in the *de novo* production of five secondary metabolites including *cyclo*-(Phe-Phe) (**63**), *cyclo*-(Phe-Tyr) (**64**), phenylacetic acid (**65**), 2-hydroxyphenylacetic acid (**66**), and furan-2-carboxylic acid (**67**) only detected in the co-cultivation (Figure 10.6) [48]. Compounds **63** and **64** were assumed to originate from the fungus as *Aspergillus* spp. are well known as producers of 2,5-diketopiperazines derivatives. The treatment of *A. niger* with the bacterial culture filtrate resulted in an increased production of **63** and **65** by the fungus. This observation is indicative of the secreted bacterial compounds acting as elicitors of cryptic genes or biotransformation substrates [48]. Interestingly, this result contradicts findings of the previous co-cultivation experiments of *A. nidulans* with *S. rapamycinicus* in which intimate physical contact was found to be the only trigger for the production of the aromatic compounds orsellinic acid, lecanoric acid, F-9775A and F-977B [22].

Mixed-fermentation of the endophytic fungus *Chaetomium* sp. isolated from the Cameroonian medicinal plant *Sapium ellipticum* with viable or autoclaved *Bacillus subtilis* on solid rice medium resulted in the *de novo* production of seven compounds including isosulochrin (**68**), protocathechuic acid methyl ester (**69**) as well as five new natural products among which shikimeran A (**70**), bipherin A (**71**), chorismeron (**72**), quinomeran (**73**), and serkydayn (**74**) (Figure 10.7) [49]. Compounds **70, 71, 72**, and **74** were suggested to originate from the fungus as they are chemically related to other known fungal compounds including the 1:1 mixture of 3- and 4-hydroxybenzoic acid methyl ester, acremonisol A, SB236050, and SB238569 isolated from the axenic culture of the fungus. The presence of **73** as the only quinoline derivative in this study highlights co-cultivation as a powerful tool for the activation of silent genes [49]. All isolated compounds were assayed against the gram-positive bacteria *Staphylococcus aureus* ATCC 25,923, *Enterococcus faecalis* ATCC 29,212, and *B. subtilis* 168 trpC2; the latter being the strain used for co-cultivation. Only **74** displayed a moderate activity against *B. subtilis* with an MIC of 53 μM whereas the remaining compounds were inactive. All isolated compounds were further evaluated for their cytotoxic activity against the mouse lymphoma L517Y cell line. Compound **74** likewise showed the strongest activity with an IC_{50} value of 1 μM whereas **73** and **69** displayed weak activity with an IC_{50} value of 38.6 and 20.8 μM respectively. As **74** was only induced during the co-cultivation of the *Chaetomium* sp. with *B. subtilis*, it can be assumed that the fungus undertook its biosynthesis as a stress response to suppress its competitor [49].

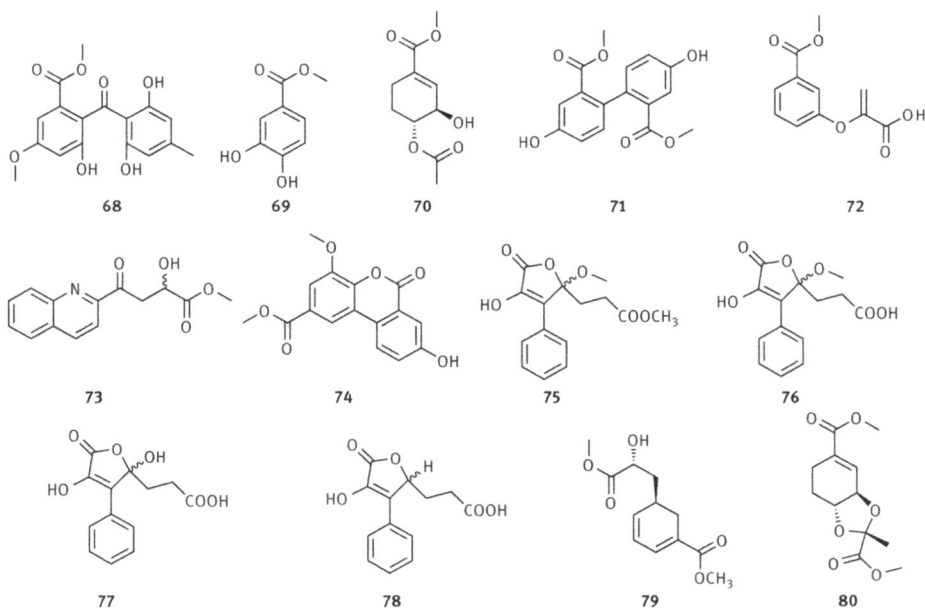

Figure 10.7: Compounds **68–80**.

In a subsequent study, the same strain *Chaetomium* sp. was co-cultured with autoclaved *Pseudomonas aeruginosa* on solid rice medium [50]. This co-cultivation led to an enhancement of the accumulation of constitutively present fungal metabolites which included the new butenolide derivatives chaetobutenolide A (**75**), B (**76**), and C (**77**), and WF-3681 methyl ester (**78**) (Figure 10.7) [50]. The co-cultivation resulted in an up to 33-fold increase of the fungal axenic metabolites. New shikimic acid derivatives chaetoisochorismin (**79**) and shikimeran B (**80**) were exclusively obtained from the co-cultivation of *Chaetomium* sp. with autoclaved *P. aeruginosa*. The new butenolide derivatives were not detected when the same strain was co-cultured with autoclaved *B. subtilis* [49] but only when co-cultured with autoclaved *Pseudomonas aeruginosa* thus highlighting the specificity of the fungal response toward different bacteria [50]. All isolated compounds were inactive when evaluated for their antibacterial activity against *Acinetobacter baumanii* (ATCC BAA1605) or *Staphylococcus aureus* (ATCC 29,213) and for their cytotoxic activity against the mouse lymphoma L517Y cell line [50]. This study highlights co-cultivation as a powerful tool for expanding the metabolic profile of microorganisms [50].

10.3.2 Fungal-fungal co-cultivation

Two mangrove-derived endophytic fungi (strain nos. 1924 and 3893) were isolated from the South China Sea and co-cultured to afford two novel 4-quinolone analogs including marinamide (**81**) and its methyl ester (**82**) (Figure 10.6) which were not produced when the fungi where cultured individually [51, 52]. The evaluation of their antibacterial activities against *Escherichia coli, Pseudomonas pyocyanea*, and *Staphylococcus aureus* revealed a moderate activity with MIC values of 3.9 and 3.7 mM, respectively. Furthermore, both compounds were evaluated for their cytotoxicity toward HepG2 (hepatocellular carcimona), 95-D (lung cancer), MGC832 (gastric cancer), and Hela (cervical carcinoma) cells and proved to be potent cytotoxic agents with IC_{50} values ranging from 0.4 nM to 2.5 µM [52]. The secondary metabolite profile of these microorganisms was expanded in a subsequent co-cultivation study, which resulted in the induction of two new metabolites only produced in the co-culture, including 6-methylsalicylic acid (**83**) and cyclo-(Phe-Phe) dipeptide Compound (62). **Sixty-two** was also reported as being *de novo* produced during a mixed fermentation of the fungus *Aspergillus niger* N402 and *Streptomyces coelicolor* [48]. The ethyl acetate extract showed strong activity against the pest insects *Heliothis armigera* (Huehner) and *Sinergasilus* sp. which was found to be caused by **83** [51].

Mixed-fermentation of *Fusarium tricintum* with *Fusarium begoniae* resulted in the induction of two new linear depsipeptides, subenniatins A (**84**) and B (**85**) (Figure 10.8) which were not detected in either of the axenic fungal cultures but only in the mixed-culture [53]. Compounds 84 and **85** were found to originate from *F. trincintum* as they were shown to be biogenetic building blocks of the cyclic depsipeptides enniatins which are constitutive metabolites of *F. trincintum*. Moreover, *Fusarium equiseti* being another *Fusarium* species was co-cultured with *F. trincintum* or *F. begoniae* but no changes in the HPLC profile were observed. Surprisingly, **84** and **85** did not inhibit the growth of the L517Y mouse lymphoma cell line [53]. Co-cultivation of the same strain *F. trincintum* with *B. subtilis* has previously been reported to cause *de novo* induction of new secondary metabolites [43], thus co-cultivation of *F. trincintum* with different microorganisms might expand its secondary metabolites profile.

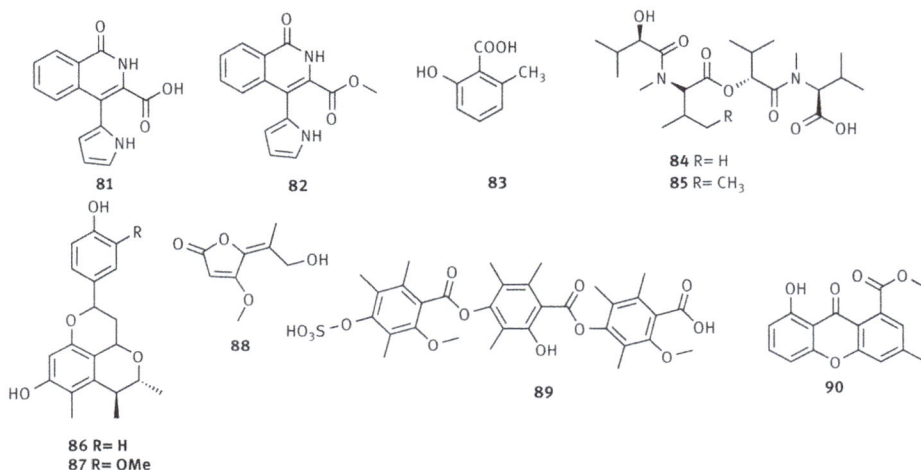

Figure 10.8: Compounds **81–90**.

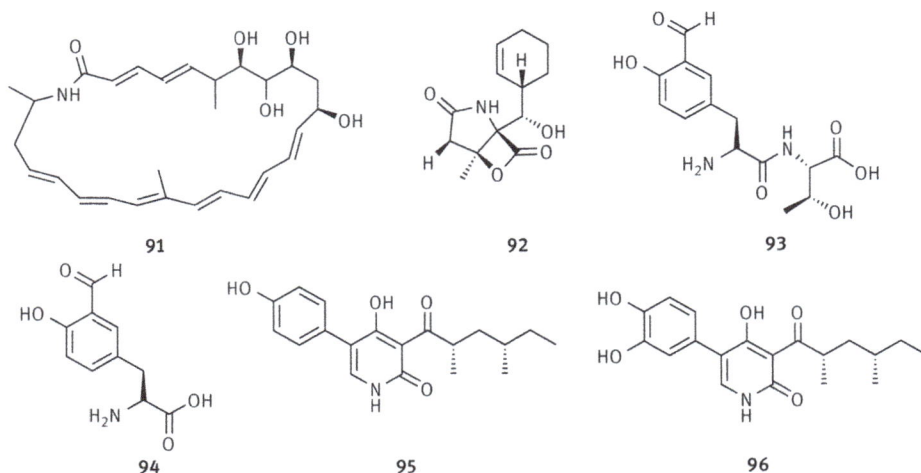

Figure 10.9: Compounds **91–96**.

Two citrinin adducts namely citrifelins A (**86**) and B (**87**) (Figure 10.8) harboring a unique tetracyclic framework, were induced during the co-cultivation of the fungi *Penicillium citrinum* MA-197 and *Beauveria felina* EN-135 [54]. None of the new citrinin adducts were detected in either of the axenic fungi. It can be assumed that these citrinin derivatives originate most likely from *P. citrinum* MA-197 as it is known to produce structurally similar compounds [54]. When evaluated for their antibacterial activity against *Escherichia coli* and *Staphylococcus aureus*, both compounds proved to be active with MICs ranging from 5.6 to 24.5 μM [54].

Co-cultivation of two soil-derived fungi *Fusarium solani* FKI-6853 and *Talaromyces* sp. FKA-65 led to the induction of a new penicillic acid derivative, coculnol (**88**) not produced when the fungi were grown axenically [55].

A mixed-fermentation of *Trichophyton rubrum* and *Bionectra ochroleuca* resulted to the isolation of a new orselinic acid derivative, 4″-hydroxysulfoxy-2,2″-dimethylthievalin P (**89**) only in the co-culture which was suggested to originate from *Bionectra ochroleuca* as its non-sulfonated form was found in the axenic culture of *Bionectra ochroleuca* [56].

Co-cultivation of two mangrove fungi, strain No. K38 and E33 isolated from the South China Sea Coast yielded a novel xanthone derivative **90** which displayed antifungal activities against *Gloeasporium musae*, *Peronophtthora cichoralearum*, *Blumeria graminearum*, and *Fusarium oxyporium*, and *Colletotrichum glocosporioides* [57].

10.4 Genomic approaches

Basically, a genome is the collection of genes containing the genetic information (DNA base pair sequences) that an organism requires to build up proteins [58]. Proteins are then involved in several pathways, which lead to the formation of natural products. The study of genomics is just one of the many recent sub-disciplines within the "omics" world, including transcriptomics, proteomics, metabolomics, etc. Transcriptomics is the study of the total pool of messenger RNA (mRNA) in an organism. It must be noted that during protein biosynthesis, the stored genetic information in the DNA is transcribed unto the mRNA molecule. Meanwhile, proteomics is the study of the entire pool of proteins in an organism and metabolomics is the study of all the molecules present in a given organism. It must be noted that the general genome structure of microbes like prokaryotes is different from those of higher organisms, e. g. plants and other marine species that biosynthesize secondary metabolites. In general, prokaryotic genetic information is encoded in a continuous stretch of DNA molecule while those or higher organisms (eukaryotes) are described by coding sections called exons which are interrupted at intervals by non-coding sections known as introns [59]. Moreover, the genes of certain organisms have been observed to be arranged into groups known as "gene clusters", which are involved in "specialized metabolism" for the biosynthesis of secondary metabolites [60]. Metabolite gene clusters were first observed in bacteria and filamentous fungi, and only recently in plants [61, 62]. A summary of database resources for retrieving genomic data for biosynthesis prediction has been summarized in a recent chapter [63]. Computational tools for biochemical pathway prediction have been summarized in excellent reviews [64]. Besides, several bioinformatic tools for the analysis of organism "omic" data have recently been developed. A full study of this is beyond the scope of this chapter. We shall rather briefly describe some useful tools for the analysis of plant and microbial genomes in relation to metabolic data. These include tools for analysis of specialized metabolism in microbes (e. g. antiSMASH, CompGen, GNP, PRISM and WebAUGUSTUS), for plant metabolism prediction (e. g. AraNet, MADIBA, miP3v2, PlantClusterFinder, SAVI and WikiPathways for plants and plantiSMASH). A third category of tools are more useful for both microbial and plant metabolism prediction and "gene cluster" analysis (e. g. E-zyme, KEGG, PathPred and PathComp). Among these tools, the most cited and user friendly are those developed within the Medema group in Groningen; antiSMASH and plantiSMASH for gene cluster analysis of microbial and plant genomes, respectively [65–67]. The current version of antiSMASH includes 1172 clusters with known end products (compounds) and at the enzyme level, active sites of key biosynthetic enzymes have been pinpointed through a curated pattern-matching procedure and Enzyme Commission numbers, which are assigned to functionally classify all enzyme-coding genes. As for plantiSMASH, 48 high-quality plant genomes have been included, with a rich diversity of candidate plant gene clusters. Besides, the tools allows for the automated identification of plant gene clusters, thus facilitating their comparison across genomes. It also helps users to predict the functional interactions of pairs of genes within and between the gene clusters based on co-expression analysis. A well-documented guide on the use of the plantiSMASH tool is also available in the literature [67].

The genomic approach encompasses the isolation of the DNA of the microorganism, whole genome sequencing for the identification of biosynthetic gene clusters, and genome mining using bioinformatics tools. Genome-guided study of the marine bacterium *Salinispora tropica* CNB-440 resulted in the discovery of the polyene macrolactam salinilactam A (**91**) [68]. Interestingly, genomic mining of *Salinispora pacifica* led to the discovery of salinosporamide K (**92**) by comparative genomic analysis with *S. tropica* [69]. One strategy of genome mining is the comparative metabolic profiling of mutants with the addition or inactivation of the biosynthetic genes [23]. This strategy was also applied to *Pseudoalteromonas tunicata* which led to the identification of two new metabolites including **93** and **94** [70]. Another strategy involves the activation of the biosynthetic genes by manipulation of regulatory genes which was applied to *A. nidulans* and led to the discovery of aspyridones A and B [71].

10.5 Conclusions

The requirement for new drugs is urgent due to an increase in drug resistance pathogens. Therefore, new sources for drug discovery, such as microorganisms from unusual and less investigated habitats should be explored. Fungi and bacteria are well-known producers of a multitude of small compounds with diverse bioactivities, many of which have found application in medicine [72]. However, the discovery of new bioactive compounds has stalled due to the re-isolation of known compounds under standard laboratory conditions. Moreover, genome sequencing studies have shown that the potential of fungi to produce new bioactive compounds has been far underestimated. In fact, under standard laboratory conditions, only a subset of biosynthetic gene clusters is expressed thus limiting the number of compounds [71]. Therefore, new strategies are needed aiming at the activation of the silent *loci*. In this regard, methods such as the use of small molecule epigenetic modifiers, co-culture of two or more microorganisms, and genomic methods have been explored during the last two decades.

Epigenetic modification for the activation of silent gene clusters usually involves the treatment of the microorganism with small compounds which inhibit enzymes during DNA post-translational modifications such as DNA methyltransferase or histone acetyltransferase leading to the production of new compounds [27, 29]. On the other hand, co-cultivation (also known as mixed-fermentation) which is still in its infancy appears to be a promising approach with regard to the discovery of new bioactive compounds despite its reproducibility issues. Manipulation of the target microorganism's DNA (also known as genome mining) is also a promising method for the discovery of new natural products. However, the underlying molecular mechanisms of activation of fungal gene clusters via epigenetic modification, natural mutualistic and/or antagonistic microbial relationships during co-culture or genomic approach are still underexplored.

The recent publications described in this chapter, present powerful strategies to harness the true potential of the microbes of interest in our quest to find novel drugs leads.

Acknowledgements: The authors also gratefully acknowledge the helpful comments and suggestions of the reviewers, which have improved the presentation.

References

[1]   Cragg GM, Newman DJ, Snader KM. Natural products in drug discovery and development. J Nat Prod. 1997;60:52–60.
[2]   Newman DJ, Cragg GM, Snader KM. Natural products as sources of new drugs over the period 1981–2002. J Nat Prod. 2003;66:1022–37.

[3]   Schaefer B. Natural products in the chemical industry. Berlin Heidelberg:Springer-Verlag 2015.

[4]   Balashov SV, Park S, Perlin DS. Assessing resistance to the echinocandin antifungal drug caspofungin in Candida albicans by profiling mutations in FKS1. Antimicrob Agents Chemother. 2006;50:2058–63.

[5]   Reen FJ, Romano S, Dobson AD, O'Gara F. The sound of silence: activating silent biosynthetic gene clusters in marine microorganisms. Mar Drugs. 2015;13:4754–83.

[6]   Brakhage AA, Schuemann J, Bergmann S, Scherlach K, Schroeckh V, Hertweck C. Activation of fungal silent gene clusters: a new avenue to drug discovery. In: Natural compounds as drugs. Switzerland:Birkhäuser Basel, 2008: 1–12.

[7]   Hertweck C. The biosynthetic logic of polyketide diversity. Angew Chem Int Ed. 2009;48:4688–716.

[8]   Scherlach K, Hertweck C. Triggering criptic natural product biosynthesis in microorganism. Org Biomol Chem. 2009;7:1753–60.

[9]   Netzker T, Fischer J, Weber J, Mattern D, König C, Valiante V, et al. Microbial communication leading to the activation of silent fungal secondary metabolite gene clusters. Front Microbiol. 2015;6:299.

[10] Brakhage AA, Schroeckh V. Fungal secondary metabolites–strategies to activate silent gene clusters. Fungal Genet Biol. 2011;48:15–22.

[11] Marmann A, Aly AH, Lin W, Wang B, Proksch P. Co-cultivation—a powerful emerging tool for enhancing the chemical diversity of microorganisms. Mar Drugs. 2014;12:1043–65.

[12] Byrum SD, Raman A, Taverna SD, Tackett AJ, et al. "ChAP-MS: a method for identification of proteins and histone posttranslational modifications at a single genomic locus". Cell Rep. 2012;2:198–205.

[13] Trojer P, Brandtner EM, Brosch G, Loidl P, Galehr J, Linzmaier R, et al. Histone deacetylases in fungi: novel members, new facts. Nucleic Acids Res. 2003;31:3971–81.

[14] Robyr D, Suka Y, Xenarios I, Kurdistani SK, Wang A, Suka N, et al. Microarray deacetylation maps determine genome-wide functions for yeast histone deacetylases. Cell. 2002;109: 437–46.

[15] Williams RB, Henrikson JC, Hoover AR, Lee AE, Cichewicz RH. Epigenetic remodeling of the fungal secondary metabolome. Org Biomol Chem. 2008;6:1895–7.

[16] Li X, Tang J, Wu J, Tong J, Li M, Ju J, et al. Identification and biological evaluation of secondary metabolites from marine derived Fungi-Aspergillus sp. SCSIOW3, Cultivated in the presence of epigenetic modifying agents. Molecules. 2017;22:1302.

[17] Cichewics RH. Epigenome manipulation as a pathway to new natural product scaffolds and their congeners. Nat Prod Rep. 2010;27:11–22.

[18] Blackwell M. The Fungi: 1, 2, 3. . . 5.1 million species? Am J Bot. 2011;98:426–38.

[19] Ebrahim W, El-Neketi M, Lewald LI, Orfali RS, Lin W, Rehberg N, et al. Metabolites from the fungal endophyte Aspergillus austroafricanus in axenic culture and in fungal-bacterial mixed cultures. J Nat Prod. 2016;79:914–22.

[20] Ka AK, Keerthi TR. Co-culture as the novel approach for drug discovery from marine environment. Nov Appro Drug Des Dev. 2017;2:1–4.

[21] Pettit RK. Mixed fermentation for natural product drug discovery. Appl Microbiol Biotechnol. 2009;83:19–25.

[22] Schroeckh V, Scherlach K, Nützmann HW, Shelest E, Schmidt-Heck W, Schuemann J, et al. Intimate bacterial - fungal interaction triggers biosynthesis of archetypal polyketides in *Aspergillus nidulans*. Proc Natl Acad Sci U S A. 2009;106:14558–63.

[23] Zhao XQ. Genome-based studies of marine microorganisms to maximize the diversity of natural products discovery for medical treatments. Evid Based Complement Alternat Med. 2011;2011:1–11.

[24] Deepika VB, Murali TS, Satyamoorthy K. Modulation of genetic clusters for synthesis of bioactive molecules in fungal endophytes: A review. Microbiol Res. 2016;182:125–40.

[25] Gross H, Stockwell VO, Henkels MD, Nowak-Thompson B, Loper JE, Gerwick WH. The genomisotopic approach: a systematic method to isolate products of orphan biosynthetic gene clusters. Chem Biol. 2007;14:53–63.

[26] Asai T, Chung YM, Sakurai H, Ozeki T, Chang FR, Yamashita K, et al. Tenuipyrone, a novel skeletal polyketide from the entomopathogenic fungus, Isaria tenuipes, cultivated in the presence of epigenetic modifiers. Org Lett. 2012;14:513–5.

[27] Du L, King JB, Cichewicz RH. Chlorinated polyketide obtained from a Daldinia sp. treated with the epigenetic modifier suberoylanilide hydroxamic acid. J Nat Prod. 2014;77:2454–8.

[28] Wang X, Sena Filho JG, Hoover AR, King JB, Ellis TK, Powell DR, et al. Chemical epigenetics alters the secondary metabolite composition of guttate excreted by an atlantic-forest-soil-derived *Penicillium citreonigrum*. J Nat Prod. 2010;73:942–8.

[29] Chung YM, Wei CK, Chuang DW, El-Shazly M, Hsieh CT, Asai T, et al. An epigenetic modifier enhances the production of anti-diabetic and anti-inflammatory sesquiterpenoids from *Aspergillus sydowii*. Bioorganic Med Chem. 2013;21:3866–72.

[30] Asai T, Morita S, Taniguchi T, Monde K, Oshima Y. Epigenetic stimulation of polyketide production in Chaetomium cancroideum by an NAD+-dependent HDAC inhibitor. Org Biomol Chem. 2016;14:646–51.

[31] Casadesús J, Low D. Epigenetic gene regulation in the bacterial world. Microbiol Mol Biol Rev. 2006;70:830–56.

[32] Suzuki MM, Bird A. DNA methylation landscapes: provocative insights from epigenomics. Nat Rev Genet. 2008;9:465.

[33] Wang B, Waters AL, Sims JW, Fullmer A, Ellison S, Hamann MT. Complex marine natural products as potential epigenetic and production regulators of antibiotics from a marine *Pseudomonas aeruginosa*. Microb Ecol. 2013;65:1068–75.

[34] Cueto M, Jensen PR, Kauffman C, Fenical W, Lobkovsky E, Clardy J. Pestalone, a new antibiotic produced by a marine fungus in response to bacterial challenge. J Nat Prod. 2001;64:1444–6.

[35] Augner D, Krut O, Slavov N, Gerbino DC, Sahl HG, Benting J, et al. On the antibiotic and antifungal activity of pestalone, pestalachloride A, and structurally related compounds. J Nat Prod. 2013;76:1519–22.

[36] Nützmann HW, Reyes-Dominguez Y, Scherlach K, Schroeckh V, Horn F, Gacek A, et al. Bacteria induced natural product formation in the fungus *Aspergillus nidulans* requires Saga/Ada-mediated histone acetylation. Proc Natl Acad Sci U S A. 2011;108:14282–7.

[37] König CC, Scherlach K, Schroeckh V, Horn F, Nietzsche S, Brakhage AA, et al. Bacterium induces cryptic meroterpenoid pathway in the pathogenic fungus *Aspergillus fumigatus*. Chem Bio Chem. 2013;14:938–42.

[38] Oh DC, Jensen PR, Kauffman CA, Fenical W. Libertellenones A–D: induction of cytotoxic diterpenoid biosynthesis by marine microbial competition. Bioorg Med Chem. 2005;13:5267–73.

[39] Park HB, Kwon HC, Lee CH, Yang HO. Glionitrin A, an antibiotic – antitumor metabolite derived from competitive interaction between abandoned mine microbes. J Nat Prod. 2009;72:248–52.

[40] Park HB, Kim YJ, Park JS, Yang HO, Lee KR, Kwon HC. Glionitrin B, a cancer invasion inhibitory diketopiperazine produced by microbial coculture. J Nat Prod. 2011;74:2309–12.

[41] Rateb ME, Hallyburton I, Houssen WE, Bull AT, Goodfellow M, Santhanam R, et al. Induction of diverse secondary metabolites in *Aspergillus fumigatus* by microbial co-culture. RSC Adv. 2013;3:14444–50.

[42] Chen H, Daletos G, Abdel-Aziz MS, Thomy D, Dai H, Brötz-Oesterhelt H, et al. Inducing secondary metabolite production by the soil-dwelling fungus *Aspergillus terreus* through bacterial co-culture. Phytochem Lett. 2015;12:35–41.

[43] Ola AR, Thomy D, Lai D, Brötz-Oesterhelt H, Proksch P. Inducing secondary metabolite production by the endophytic fungus Fusarium tricinctum through coculture with Bacillus subtilis. J Nat Prod. 2013;76:2094–9.

[44] Sommit D, Boonyuen N, Pudhom K. Butenolide and furandione from an endophytic Aspergillus terreus. Chem Pharm Bull. 2010;58:1221–3.

[45] Abdelfattah MS, Toume K, Arai MA, Masu H, Ishibashi M. Katorazone, a new yellow pigment with a 2-azaquinone-phenylhydrazone structure produced by *Streptomyces* sp. IFM 11299. Tetrahedron Lett. 2012;53:3346–8.

[46] Yang SW, Cordell GA. Metabolism studies of indole derivatives using a staurosporine producer, *Streptomyces staurosporeus*. J Nat Prod. 1997;60:44–8.

[47] Whitt J, Shipley SM, Newman DJ, Zuck KM. Tetramic acid analogues produced by coculture of *Saccharopolyspora erythraea* with *Fusarium pallidoroseum*. J Nat Prod. 2014;77:173–7.

[48] Wu C, Zacchetti B, Ram AFJ, van Wezel GP, Claessen D, Hae Choi YH. Expanding the chemical space for natural products by *Aspergillus-Streptomyces* co-cultivation and biotransformation. Sci Rep. 2015;5:10868.

[49] Akone SH, Daletos G, Kurtán T, Hartmann R, Lin W, Proksch P. Inducing secondary metabolites production by the endophytic fungus *Chaetonium* sp. through bacterial co-culture and epigenetic modification. Tetrahedron. 2016;72:6340–7.

[50] Ancheeva E, Küppers L, Akone SH, Ebrahim W, Liu Z, Mándi A, et al. Expanding the Metabolic Profile of the Fungus *Chaetomium* sp. through Co-culture with Autoclaved *Pseudomonas aeruginosa*. Eur J Org Chem. 2017;2017:3256–64.

[51] Zhu F, Yc L, Jh D, Xp W, Ls H. Secondary metabolites of two marine-derived mangrove endophytic fungi (strain nos. 1924 and 3893) by mixed fermentation. Chem Ind Forest Prod. 2007;27:8–10.

[52] Zhu F, Chen G, Wu J, Pan J. Structure revision and cytotoxic activity of marinamide and its methyl ester, novel alkaloids produced by co-cultures of two marine-derived mangrove endophytic fungi. Nat Prod Res. 2013;27:1960–4.

[53] Wang JP, Lin W, Wray V, Lai D, Proksch P. Induced production of depsipeptides by co-culturing *Fusarium tricinctum* and *Fusarium begoniae*. Tetrahedron Lett. 2013;54:2492–6.

[54] Meng LH, Liu Y, Li XM, Xu GM, Ji NY, Wang BG. Citrifelins A and B, citrinin adducts with a tetracyclic framework from cocultures of marinederived isolates of *Penicillium citrinum* and *Beauveria felina*. J Nat Prod. 2015;78:2301–5.

[55] Nonaka K, Chiba T, Suga T, Asami Y, Iwatsuki M, Masuma R, et al. Coculnol, a new penicillic acid produced by a coculture of *Fusarium solani* FKI-6853 and *Talaromyces* sp. FKA-65. J Antibiot. 2015;68:530–2.

[56] Bertrand S, Schumpp O, Bohni N, Monod M, Gindro K, Wolfender JL. *De novo* production of metabolites by fungal co-culture of *Trichophyton rubrum* and *Bionectria ochroleuca*. J Nat Prod. 2013;76:1157–65.

[57] Li C, Zhang J, Shao C, Ding W, She Z, Lin Y. A new xanthone derivative from the co-culture broth of two marine fungi (strain No. E33 and K38). Chem Nat Compd. 2011;47:382.

[58] Selzer PM, Marhöfer RJ, Rohwer A. Applied bioinformatics: an introduction. Heidelberg: Springer, 2008: 35.

[59] Lewin B. Genes IX. U.S: Jones & Bartlett Publishers, 2007.

[60] Medema MH, Kottmann R, Yilmaz P, Cummings M, Biggins JB, Blin K, et al. Minimum information about a biosynthetic gene cluster. Nat Chem Biol. 2015;11:625–31.

[61] Nützmann HW, Osbourn A. Gene clustering in plant specialized metabolism. Curr Opin Biotechnol. 2014;26:91–9.

[62] Osbourn A. Secondary metabolic gene clusters: evolutionary toolkits for chemical innovation. Trends Genet. 2010;26:449–57.

[63] Moumbock AFA, Simoben CV, Wessjohann L, Sippl W, Günther S, Ntie-Kang F. Computational studies and biosynthesis of natural products with promising anticancer properties. In: Badria FA, editor(s). *Phytochemistry – natural products and cancer*. Rijeka: InTech, 2017:257–85. ISBN 978-953-51-5277-4. DOI: 10.5772/67650.

[64] Medema MH, van Raaphorst R, Takano E, Breitling R. Computational tools for the synthetic design of biochemical pathways. Nat Rev Microbiol. 2012;10:191–202.

[65] Weber T, Blin K, Duddela S, Krug D, Kim HU, Bruccoleri R, et al. antiSMASH 3.0 – a comprehensive resource for the genome mining of biosynthetic gene clusters. Nucleic Acids Res. 2015;43:W237–43.

[66] Kautsar SA, Suarez Duran HG, Blin K, Osbourn A, Medema MH. plantiSMASH: automated identification, annotation and expression analysis of plant biosynthetic gene clusters. Nucleic Acids Res. 2017;45:W55–63.

[67] Kautsar SA, Suarez Duran HG, Medema MH. Genomic identification and analysis of specialized metabolite biosynthetic gene clusters in plants using PlantiSMASH. Methods Mol Biol. 2018;1795:173–88.

[68] Udwary DW, Zeigler L, Asolkar RN, Singan V, Lapidus A, Fenical W, et al. Genome sequencing reveals complex secondary metabolome in the marine actinomycete Salinispora tropica. Proc Natl Acad Sci USA. 2007;104:10376–81.

[69] Eustáquio AS, Nam SJ, Penn K, Lechner A, Wilson MC, Fenical W, et al. The discovery of salinosporamide K from the marine bacterium *Salinispora pacifica* by genome mining gives insight into pathway evolution. Chem Bio Chem. 2011;12:61–4.

[70] Blasiak LC, Clardy J. Discovery of 3-formyl-tyrosine metabolites from Pseudoalteromonas tunicata through heterologous expression. J Am Chem Soc. 2009;132:926–7.

[71] Bergmann S, Schümann J, Scherlach K, Lange C, Brakhage AA, Hertweck C. Genomics-driven discovery of PKS-NRPS hybrid metabolites from *Aspergillus nidulans*. Nat Chem Biol. 2007;3:869.

[72] Chin YW, Balunas MJ, Chai HB, Kinghorn AD. Drug discovery from natural sources. The AAPS J. 2006;8:239–53.

Part IV: **Case Studies**

Conrad V. Simoben, Fidele Ntie-Kang, Dina Robaa and
Wolfgang Sippl

11 Case studies on computer-based identification of natural products as lead molecules

Abstract: The development and application of computer-aided drug design/discovery (CADD) techniques (such as structured-base virtual screening, ligand-based virtual screening and neural networks approaches) are on the point of disintermediation in the pharmaceutical drug discovery processes. The application of these CADD methods are standing out positively as compared to other experimental approaches in the identification of hits. In order to venture into new chemical spaces, research groups are exploring natural products (NPs) for the search and identification of new hits and more efficient leads as well as the repurposing of approved NPs. The chemical space of NPs is continuously increasing as a result of millions of years of evolution of species and these data are mainly stored in the form of databases providing access to scientists around the world to conduct studies using them. Investigation of these NP databases with the help of CADD methodologies in combination with experimental validation techniques is essential to identify and propose new drug molecules. In this chapter, we highlight the importance of the chemical diversity of NPs as a source for potential drugs as well as some of the success stories of NP-derived candidates against important therapeutic targets. The focus is on studies that applied a healthy dose of the emerging CADD methodologies (structure-based, ligand-based and machine learning).

Keywords: CADD, drugs, ligand-based, machine learning, natural products, structure-based

11.1 Introduction

A typical drug discovery and development process from concept to market takes about 13–15 years requiring approximately \$2–3 billion on average [1, 2]. However, the increasing cost of the drug discovery and development process nowadays has not produced an exponential increase in the success rate of drugs approved annually as it has remained relatively flat or decreased over the past decade [1, 3]. Nevertheless, the development and application of new methodologies are on the point of disintermediation

This article has previously been published in the journal *Physical Sciences Reviews*. Please cite as: Simoben, C. V., Ntie-Kang, F., Robaa, D., Sippl, W. Case studies on computer-based identification of natural products as lead molecules. *Physical Sciences Reviews* [Online] **2020** DOI: 10.1515/psr-2018-0119

https://doi.org/10.1515/9783110668896-011

in the pharmaceutical drug discovery processes. Thereby, reducing billions of dollars of the industry's cost for the search of new drugs and a cut down on the time taken to get new medicines approved to just a few processing cycles (few years). The application of computational methodologies has been of tremendous importance at various stages of drug discovery especially in the identification of hits as compared to experimental approaches alone [4–15]. Thus, CADD methods have proven to be successful approaches for finding ligand hits, as well as in assisting in the lead optimization steps in discovery projects.

Exploration of natural products (NPs) for the discovery of new and more efficient leads, repurposing of known NPs, targeting of new targets on the basis of genome analysis, the revelation of mechanisms of action, and optimization of lead compounds are being achieved via the application of some of the developed CADD methodologies [16–24]. Humans had been using crude and/or pure medicinal plant extracts as well as isolated NPs for millennia in the treatment of several ailments before the advent of synthetic medicinal chemistry [17, 21, 24–28]. With the development of novel methodologies for the identification, isolation, and characterization of new and/or unique NPs [16], the chemical space has continuously increased as a result of millions of years of evolution of species. Thereby, opening new avenues for the discovery of new molecules that can target several ailments. The data for most of the identified NPs are made available as databases for usage in research and development (R&D) projects including drug discovery procedures [18, 29–33]. Numerous research groups have provided detailed reports on the growth of NP databases as well as insights into understanding some of the complex molecular scaffolds unparalleled in function, chemical diversity, and sample availability for some of these NPs [17, 19, 22, 25, 31, 34–38]. It is also interesting to note that the sampling of NPs over the past decades has yielded many significant discoveries for modern medicine in the form of drugs or supplements [17, 27, 36, 39]. Nowadays, investigation of the chemical diversity of NPs with the help of CADD methodologies in combination with experimental validation techniques is essential to identify and propose new drug molecules [23, 40]. In this chapter, we highlight the importance of the chemical diversity of NPs as a source for potential drugs as well as some of the success stories of NP-based drugs against important therapeutic targets. We are going to focus on success stories, where emerging CADD methodologies were significantly applied in their discovery.

11.2 Drugs approved by the FDA and the proportion of NPs

It has often been assumed that NPs do not represent a promising source of drugs in the twenty-first century. However, in a chapter of the previous volume of this book [41], Newman showed that even today, NP structures have played a key role in the search for novel agents against many diseases [42]. The discussion was focused on recently published data (within the last 5–10 years) on approved NPs currently in clinical trials against many human disease forms, e. g. cancer and infectious diseases. Newman focused his discussion on agents approved from 2010 to the end of 2017 by the U.S. FDA (and its equivalent agencies or organizations in other countries), along with some agents that have entered clinical trials, which are based on "natural product structures", even if the final product represents a synthetic compound in nature.

11.3 CADD of some NPs identified as lead compounds

The diverse therapeutic activity of bioactive molecule(s) present in the numerous organisms applied in traditional medicine justifies their use in the treatment of ailments for centuries. Thus, NPs remain a vital source for the discovery of new molecules to treat diseases in the modern system of medicine [43, 44]. Several publications have reported the use and application of CADD protocols for the search, design, and development of new drug molecules [13, 45–52]. In the following section, we will highlight a few case studies for NPs identified as hit/lead molecules with the applications of CADD methods.

11.3.1 Structure-based CADD methods

In the application of structured-based CADD methods, knowledge of the target binding site (mainly obtained from X-ray crystallographic or NMR studies) is vital for the construction of a three-dimensional (3D) model that can be used e. g. for molecular docking or pharmacophore generation and screening or molecular dynamics simulations [5, 8, 11, 12, 24, 46, 47]. An example of the use of structure-based CADD methods to identify NP or NP-like protein inhibitors of the human epidermal growth factor receptor 2 (HER2) is presented below.

The work of Li et al. [53] led to the identification of novel effective NP inhibitors against HER2 as their contribution in the search of new molecules to treat breast cancer. Breast cancer has been reported as one of the most common malignancies among women worldwide with substantially higher incidence (43.3 per 100,000) than any other form of cancer [54]. Early-stage breast cancer patients (about 30% of patients) have been linked with the overexpression of HER2, thus, confirming the potential of HER2 as a target to treat breast cancer. Although the efforts in the search of NP inhibitors led to the U.S. FDA approval of one molecule (lapatinib); lapatinib has already received some backlashes due to observed side effects within one year of treatment initiation in a majority of the patients [55–59]. Hence, there is a need to look for alternative therapies to inhibit HER2 urgently.

In the study, Li et al. performed a docking-based virtual screening (Figure 11.1) of 11,247 NPs (obtained from AnalytiCon Discovery NP) against the kinase domain of HER2 crystal structure (PDB ID: 3PP0, with a resolution of 2.25 Å) [60]. From the screening results, 12 compounds together with lapatinib and ATP (Figure 11.2) were selected based on their screening scores (Amber score from the DOCK6.5 program and Vina score from AutoDockVina v1 program) for further evaluation [61, 62]. Prediction of the ADMET properties of these 14 compounds was done using ADMET predictor (an effective tool to predict physicochemical and biological attributes of potential drug-like compounds). Subsequently, they evaluated the stability of the docked conformers of the NP hits in complex with the HER2 crystal structure by running MD simulations using Gromacs 5.0 (the program for Chemical Simulation) [63]. Analysis of the 50 ns MD simulation revealed that most of the selected molecules (Table 11.1) were relatively stable throughout the MD simulation run time, with significantly low root-mean-square deviation (rmsd) value as early as from the 5th ns (Figure 11.3).

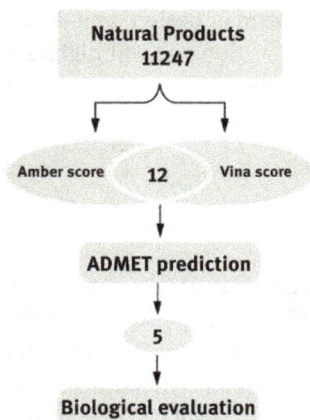

Figure 11.1: The summarized workflow for the identification of novel potential inhibitors targeting HER2 [53]. Figure reproduced with permission.

Among the tested compounds, ZINC15122021 showed submicromolar inhibition of HER2 (Table 1.1). In order to explain the observed difference in activities, Li et al. [53] analyzed the binding pose adopted by the selected molecules within the binding pocket of the HER2 enzyme. In the case of ZINC15122021, it could interact with the residues Ser728 and Asp863 within the HER2 ATP-binding pocket whereas lapatinib could interact with Leu726 and Met801 residues (Figure 11.4). According to the docking study, van der Waals interactions played a key role in the molecular recognition between ZINC15122021 and HER2. In summary, the results presented by Li et al. show that the proposed NPs have favorable ADMET properties. They additionally possessed low cytotoxicity to normal breast cells and interesting activities against HER2. This indicates that these

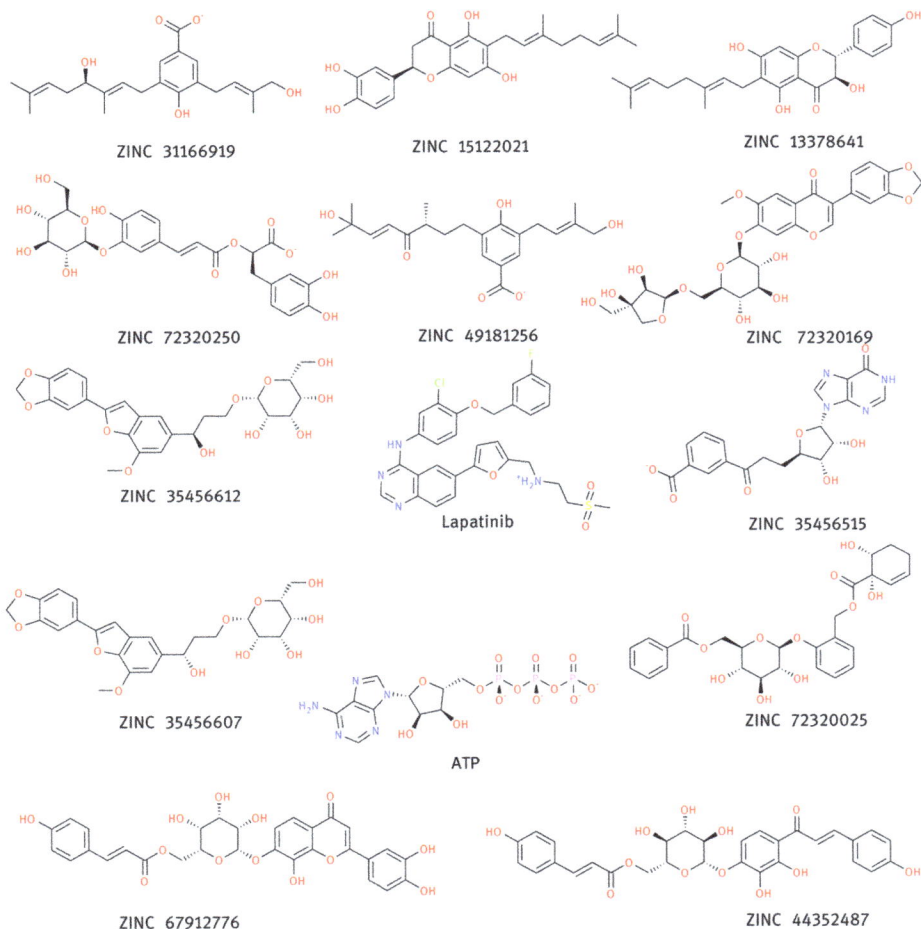

Figure 11.2: Selected compounds from virtual screening step [53]. Figure adapted with permission.

proposed natural compounds can act as a good starting point for optimization and modifications in the drug discovery process to search for novel HER2 inhibitors.

Musyoka et al. [64] highlighted the potential of non-peptidic natural compounds from South Africa (SA) using structure-based CADD methods. Recent fatality (global mortality and morbidity levels) records of malaria have decreased substantially over the last decade due to the relentless efforts put in place to control and prevent *Plasmodium* spp. as well as the treatment of infected individuals [65, 66]. However, the five major artemisinin-based combination therapies (ACTs) representing the first-line treatment of malaria could become ineffective in the near future. This is in accordance with reported cases of resistance in several places in Asia [67–69]. Despite the efforts to provide malaria vaccine(s) to fight against malaria [70, 71], chemotherapy

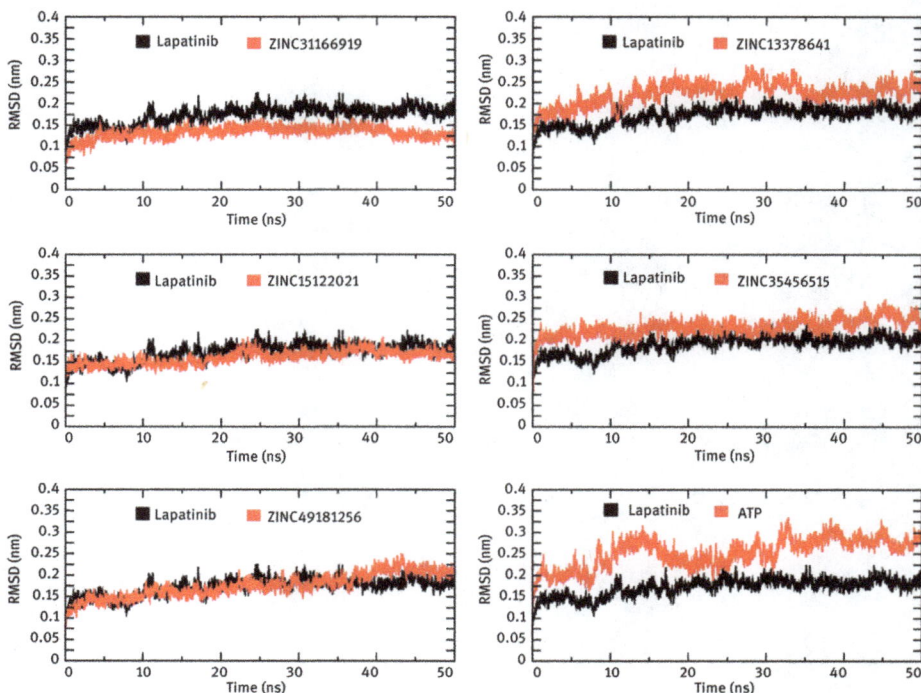

Figure 11.3: The rmsd of the heavy atoms of HER2-ligand systems over 50 ns MD simulation runtime for some of the selected compounds compared to lapatinib (FAD approved) [53]. Figure reproduced with permission.

Table 11.1: IC_{50} values (µM) of selected natural compounds against HER2 kinase and HER2-overexpressing SKBR3 and BT474 cell lines [53]. This table was adapted with permission.

Compounds	Enzymatic IC_{50} (µM)	Cell Inhibition IC_{50} (µM)	
		SKBR3	BT474
ZINC31166919	2.63 ± 0.03	8.61 ± 0.45	6.78 ± 0.68
ZINC15122021	0.18 ± 0.002	1.22 ± 0.05	4.11 ± 0.95
ZINC49181256	9.18 ± 0.01	>50	>50
ZINC13378641	3.71 ± 0.03	26.48 ± 1.62	18.55 ± 2.06
ZINC35456515	>10	>50	>50
lapatinib	0.06 ± 0.001	0.38 ± 0.02	0.45 ± 0.03

still remains the main choice of treatment considering its protective efficacy and targeted population/ethnic group. The work of Musyoka et al. had as a focus the disruption of the *Plasmodium* hemoglobin degradation pathway using as hit/lead molecules. This pathway represents the most integral process for the growth and replication of

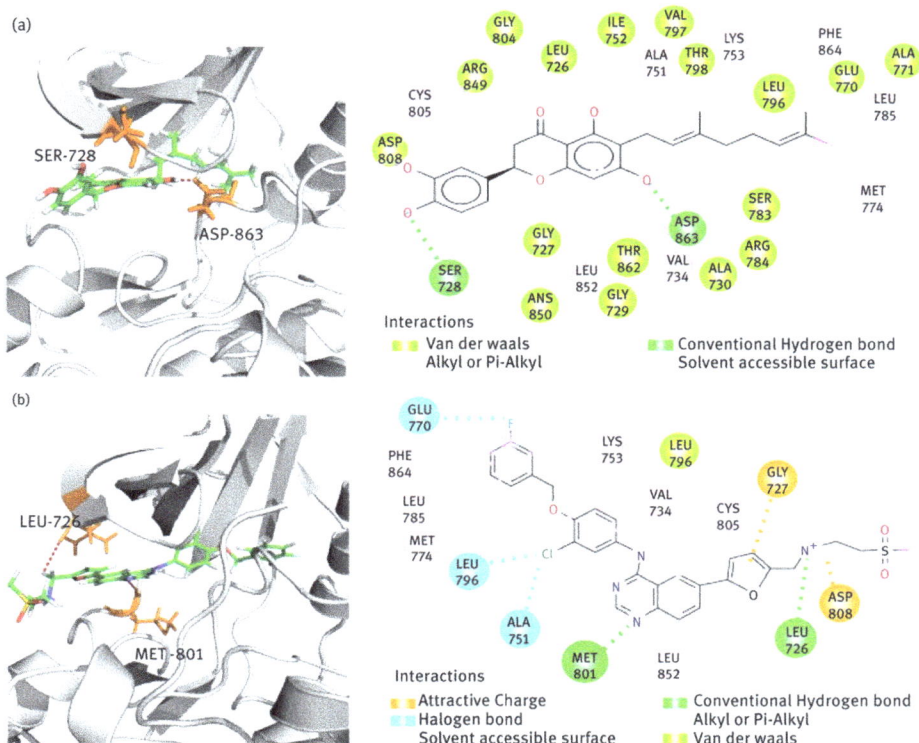

Figure 11.4: Comparison of binding models of lapatinib and ZINC15122021 against HER2 at the atomic level. (a) The interaction of ZINC15122021 with HER2; (b) the interaction of lapatinib with HER2. The 2D diagram interactions were shown as dashed lines between receptor residues and ligand atoms [53]. Figure reproduced with permission.

Plasmodium parasites within the host's erythrocytes. Thus, focusing their search for efficient and potent antimalarial hit molecules using integrated structure-based CADD method to target this pathway (Figure 11.5) [64, 72–75].

In the reported study, Musyoka et al. docked 23 SA non-peptidic natural compounds (identified from literature; Figure 11.6) into 11 selected target proteins (falcipain (FP) proteins, namely FP-1, FP-2, FP-2′ and FP-3 of *P. falciparum*; vivapain 2 and 3 (VP-2 and VP-3) of *P. vivax*, knowlesipain 2 and 3 (KP-2 and KP-3) of *P. knowlesi*, bergheipain 2 (BP-2) of *P. berghei*, chabaudipain 2 (CP-2) of *P. chabaudi* and yoelipain 2 (YP-2) of *P. yoelii*) using AutoDock4.2 [76–82]. The docking outcomes showed that 22 out of 23 of the SA natural compounds exhibited poor binding affinities to all the proteases used. This was in accordance with the nature of the "trench-like" binding pocket of protease which could not accommodate the long carbon chain or a circularized nucleus of the poor binders.

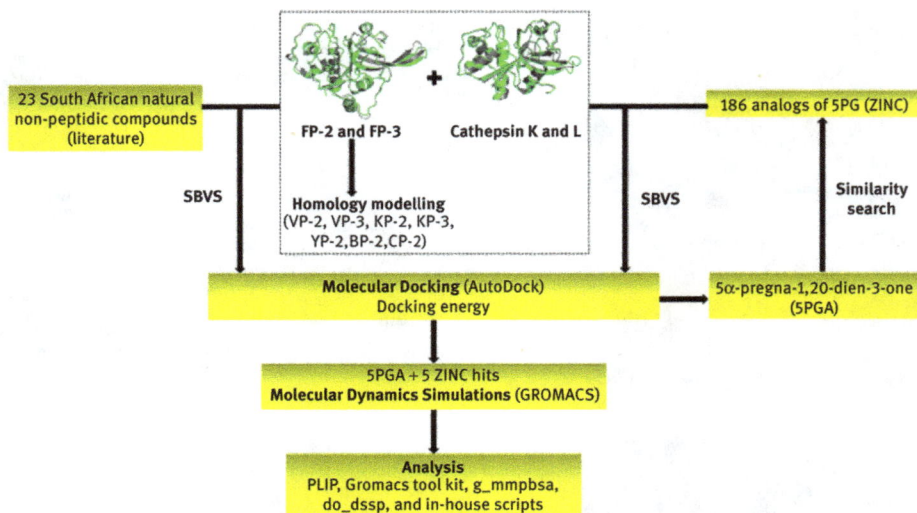

Figure 11.5: Summary of the workflow used by Musyoka et al. [64]. Figure reproduced with permission.

Nevertheless, 5α-pregna-1,20-dien-3-one (5PGA) (identifier number SANC00146) isolated from *Capnella thyrsoidea* was identified as a potential hit against most plasmodial proteases amongst other reported activities [83]. The extended S2 subsite of all plasmodial proteases accommodated 5PGA perfectly and had predictive inhibitory value in nanomolar levels in FP-2, VP-2 and KP-3 (Figure 11.7). The antimalarial activity of 5PGA had not been reported before, but, was shown to elicit an inflammatory response in a different context through the release of superoxide ions in neutrophils from rabbit cells [81].

Musyoka et al. further performed a structural similarity search based on the obtained binding affinity and interaction analyses results of 5PGA as a starting template. This guided the identification of 186 compounds (analogous to 5PGA) from the ZINC database which were all docked against the 11 target proteins. This led to the selection of five compounds (ZINC36371307, ZINC03869631, ZINC04532950, ZINC04579000, and ZINC052477224; Figure 11.8) together with 5PGA for further MD simulations (evaluation of their stability) and binding free energy (BFE) calculations based on their estimated inhibitory profiles. Docking poses for the different molecules revealed additional interactions with other residues within the S2 subsite binding pocket when compared to 5PGA due to the extended carbon chains of varied lengths (Figure 11.9). Generally, the rmsd of all the protein-ligand complexes remained stable after 5 ns of the MD simulation run.

Analysis of the ligand-binding poses and binding free energy results revealed the residues necessary for the interaction. An example is the case of 5PGA and the best ZINC hits (ZINC03869631, ZINC04532950, and ZINC05247724) exhibiting different

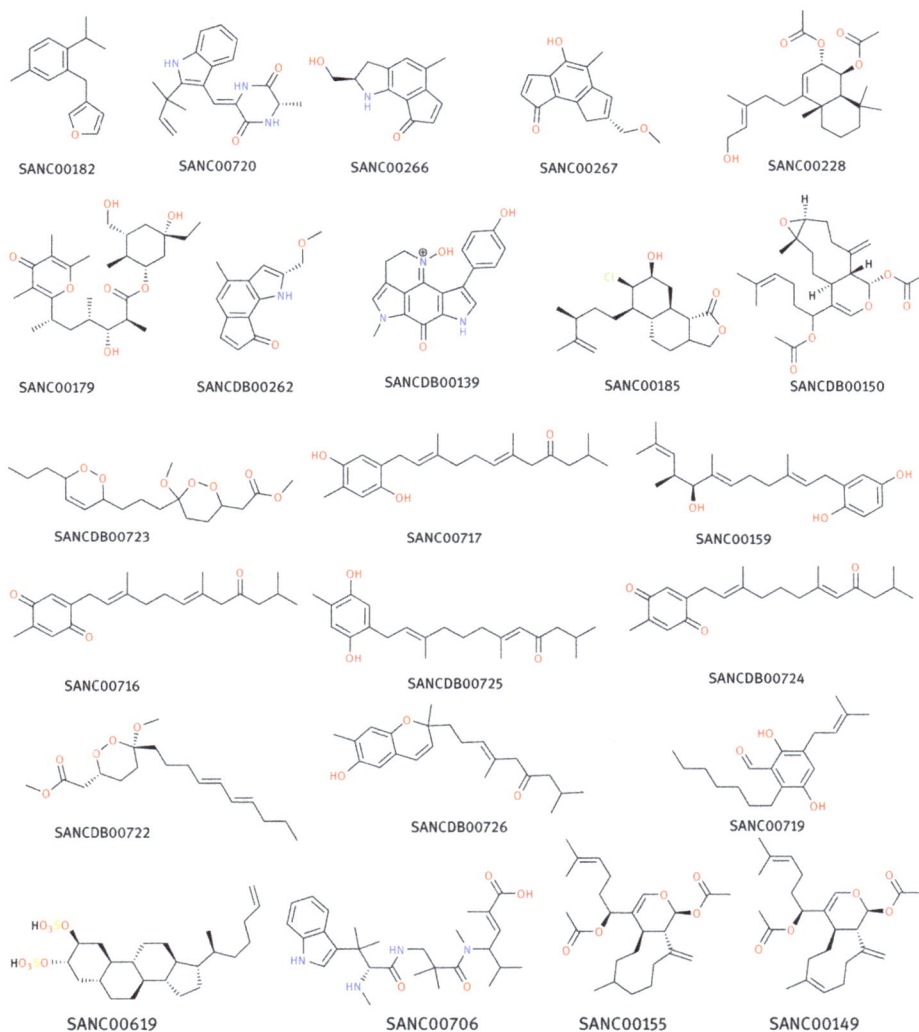

Figure 11.6: Non-peptidic natural compounds from SA with their identity numbers in the SANCDB [64]. Figure adapted with permission.

pose between human Cat L and the FPs (Figure 11.9). In Cat L, the ligands all adopted different docking poses while a similar docking pose was observed for the ligands in complex with the FPs (the case of FP-2:5PGA complex was an exception). The reported results by Musyoka et al. [64], thus, serve as a guide that can be explored through chemical modifications to search for some more drug-like derivatives and potent plasmodial cysteine protease inhibitors.

Estimated inhibition constant (Ki) of 5PGA

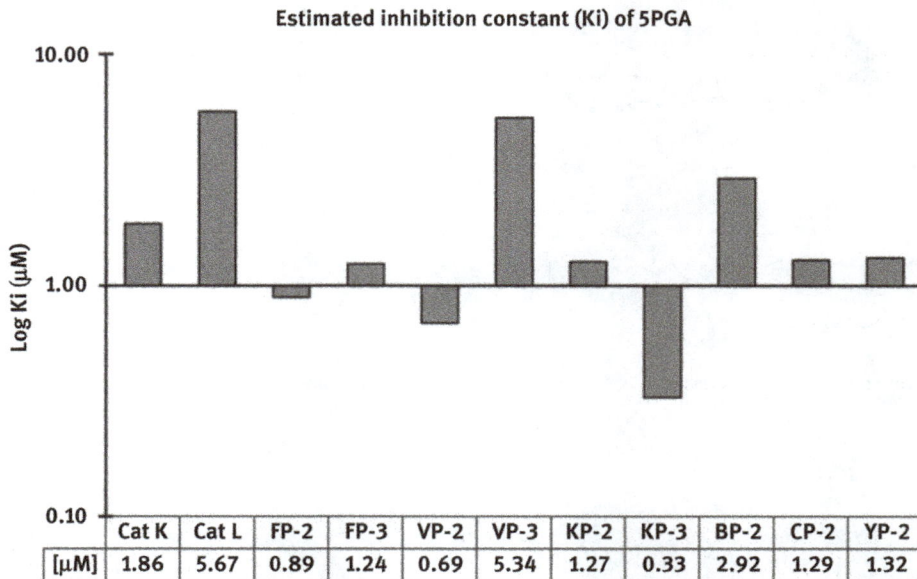

	Cat K	Cat L	FP-2	FP-3	VP-2	VP-3	KP-2	KP-3	BP-2	CP-2	YP-2
[µM]	1.86	5.67	0.89	1.24	0.69	5.34	1.27	0.33	2.92	1.29	1.32

Figure 11.7: 5PGA-protein predicted inhibitor constants as determined by AutoDock software [64]. Figure reproduced with permission.

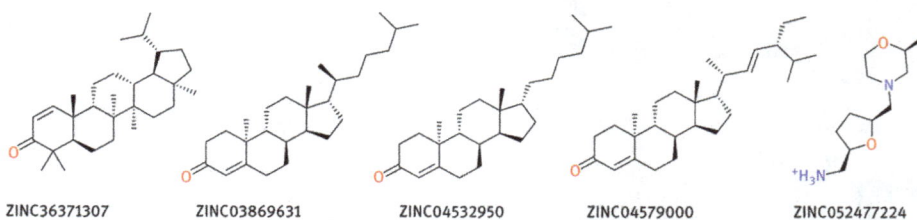

| ZINC36371307 | ZINC03869631 | ZINC04532950 | ZINC04579000 | ZINC052477224 |

Figure 11.8: Selected hits from the ZINC database after docking against human cathepsins and plasmodial cysteine proteases [64]. The figure was adapted with permission.

11.3.2 Ligand-based CADD methods

Ligand-based methods typically aim to identify those structural features that are important and responsible for the observed biological activity of a molecule by considering sets of known active and inactive and/or decoy molecules. Ligand-based strategies for the discovery of lead compounds include similarity searches in compound databases, structural alignment of molecules for the identification of pharmacophores and virtual screening, and machine learning (ML) algorithms in order to establish quantitative structure-activity relationships (QSARs) to predict properties of candidates and guide the design and/or optimization of new molecules.

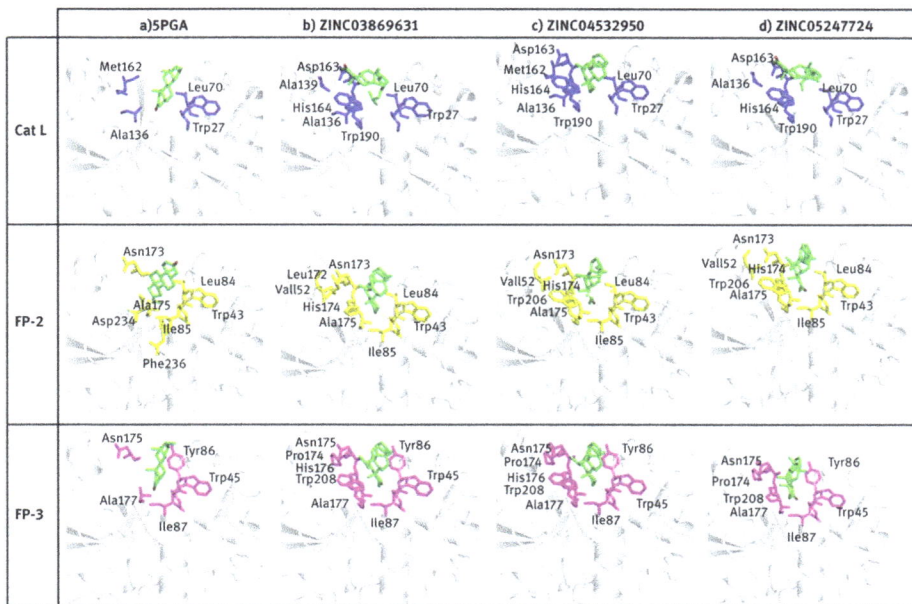

Figure 11.9: Binding pocket amino acid residue interactions patterns of bound 5PGA, ZINC03869631, ZINC04532950, and ZINC05247724 with Cat L (blue), FP-2 (yellow) and FP-3 (magenta) [64].
Figure reproduced with permission.

In order to search for novel and structurally diverse Topoisomerase I (Top1) inhibitors, Pal et al. [84] performed virtual screening based on a ligand-based-pharmacophore model generated from the chemical features of 29 camptothecin (CPT) derivatives. The cytotoxic quinoline alkaloid was isolated from the stem and bark of *Camptotheca acuminate*. *C. acuminate* is used in ancient Chinese traditional medicine for the treatment of common colds, psoriasis, liver problems and digestive problems. After further investigations, this compound is now being used to treat cancer [85]. CPT and several analogs of CPT (Figure 11.10) are clinically used as Top1 poison in cancer therapies. Top1 represents one of the important targets for cancer chemotherapy because it is essential for resolving the topological problems associated with DNA supercoiling during replication and transcription [86]. While Top1 suppressor reduces the proper functioning of topoisomerase by interacting with the protein, Top1 poisons, on the other hand, stabilize the DNA-Top1 ternary complex (DNA-Top1-poison), thus preventing the relegation of DNA strand and eventually resulting in the cancer cell death. However, limitations such as instability of the hydroxy lactone ring, dose-limiting side effect, multidrug resistance, solubility, and severe side effects are some of the challenges faced with CPT and its analogs as Top1 poisons. Therefore, there is a need to search for novel and structurally diverse Top1 inhibitors.

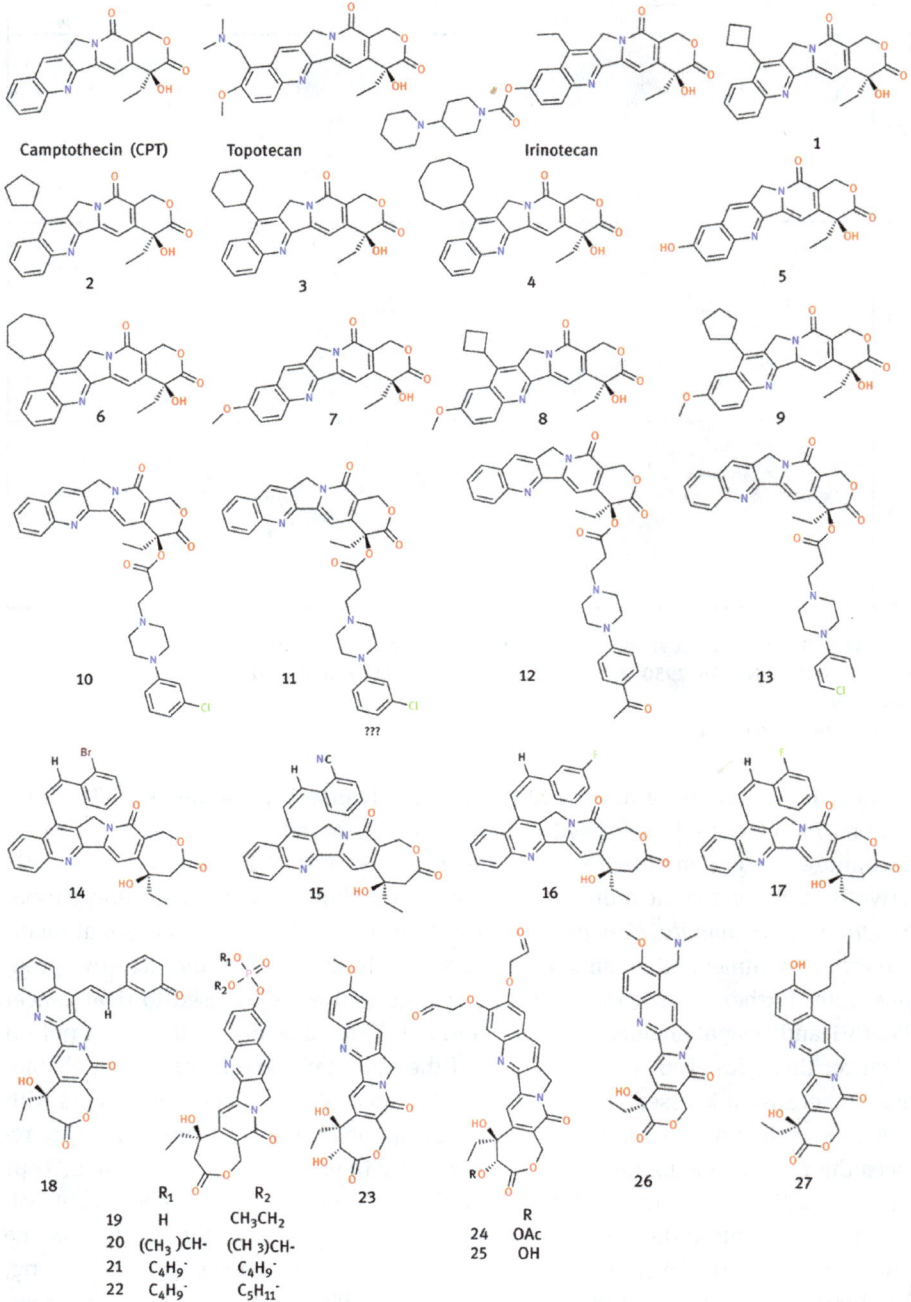

Figure 11.10: CPT and analogs that were used as training set molecules [84]. The figure was adapted with permission.

In the quest to search for new Top1 inhibitors, Pal et al. summarized the structural features of seven different classes of CPT derivatives (with corresponding IC_{50} values ranging from 0.003 to 11.4 µM) to develop a pharmacophore-based 3D-QSAR (using HypoGen algorithm) technique to screen the ZINC Database containing 1,087,724 molecules [84]. Furthermore, filtration of the retrieved hits using Lipinski's rule of five (for drug-like molecules) and SMART protocol (for filtering out the molecules with unwanted functional groups) was performed. Additional filtering using estimated activity < 1.0 µM to reduce the range of the activity scale provided the authors with 212,670 molecules. The best-validated pharmacophore model (Hypo1) had the following features; hydrogen bond acceptor (HBA), hydrogen bond donor (HBD), and ring aromaticity (RA) (Figure 11.11). Hypo1 was utilized like a 3D query for screening the ZINC Database for possible "hit molecules" that should fit with all the possible features of the query pharmacophore.

Topoisomerase I (PDB ID: 1T8I) co-crystalized with the reference inhibitor CPT (in a planar geometry) was used to check if the filtered hit molecules adopt similar docking poses as observed with CPT. The molecules adopting similar docking pose to CPT while showing interactions with the important active site residues (Arg364, Asp533 amino acids, and nucleotide DC112, DA113, DT10) were selected as more desirable hit molecules (Figure 11.12). The selected molecules were further subjected to ADMET and toxicity evaluation leading to the selection of three virtual hit molecules (ZINC68997780, ZINC38550809, and ZINC15018994) for MD simulation studies (Figure 11.13).

An all-atom MD simulation for 10 ns using GROMACS 5.1.5 was performed on the docking poses of the three selected molecules. The preferred orientations of the docked ligands similar to the co-crystallized ligand were taken as the initial structure for the MD studies and root-mean-square fluctuations (rmsf) and rmsd amongst others were used to evaluate the stability of the molecules. It was concluded that over the MD simulation period, the main protein backbone conformation did not change. This was an indication that without the ligand, the fluctuations were similar in those regions. Thus, there were no significant conformational changes observed for all three lead compounds before and after bindings, showing that binding poses of the individual ligands were maintained throughout the whole 10 ns dynamics trajectories (Figure 11.14).

1.3.3 Machine Learning

The continuous and remarkable growth in data in the world is estimated to ~ 35 trillion gigabytes by 2020 [87]. How to efficiently mine the large-scale chemistry data is becoming a crucial problem for drug discovery. ML techniques are advancing as efficient strategies to mine, manage, and analyze this growing amount of data [88–92]. Applications of ML techniques have also emerged as a means to increase

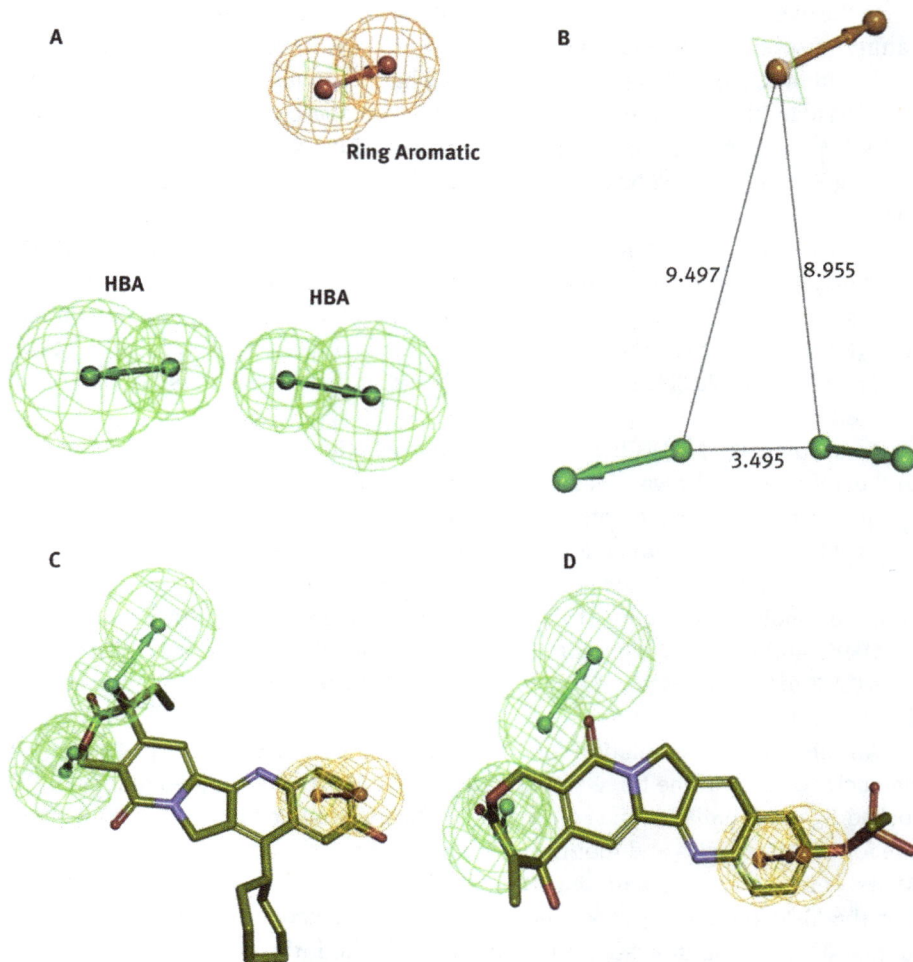

Figure 11.11: The best HypoGen pharmacophore model (Hypo1). (A) Chemical features present in Hypo 1 (B) 3D spatial arrangement and the distance constraints between the chemical features. HBA (Green), ring aromatic (Brown). (C) Hypo1 mapped on to the most active molecule (compound 4) in the training set (D) Hypo 1 mapped on to the least active molecule (compound 19) in the training set [84]. The figure was adapted with permission.

the chance of pharmaceutical companies to find new active molecules from the large-scale chemistry and biomedical data [93]. Larger data volumes in combination with increased automation technologies have promoted the further use of ML. An example is demonstrated by Grisoni et al., where the authors designed fully automated ML models leading to the suggestion of two compounds that act as mimetics of the NP (−)-galantamine for the treatment of Alzheimer's disease (AD) via a virtual screening protocol [94].

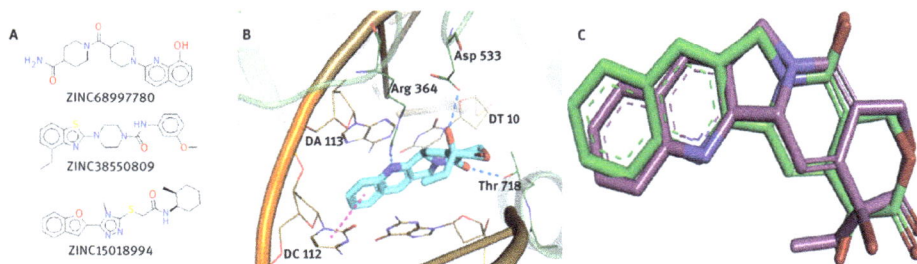

Figure 11.12: (A) Selected compounds, (B) Binding interaction of the native ligand within the active site of 1T8I. Hydrogen bond (blue dotted line) and π-π hydrophobic (magenta dotted line) interactions are shown. (C) Molecular overlay of the experimental binding pose of CPT and predicted binding pose of CPT after docking [84]. The figure was adapted with permission.

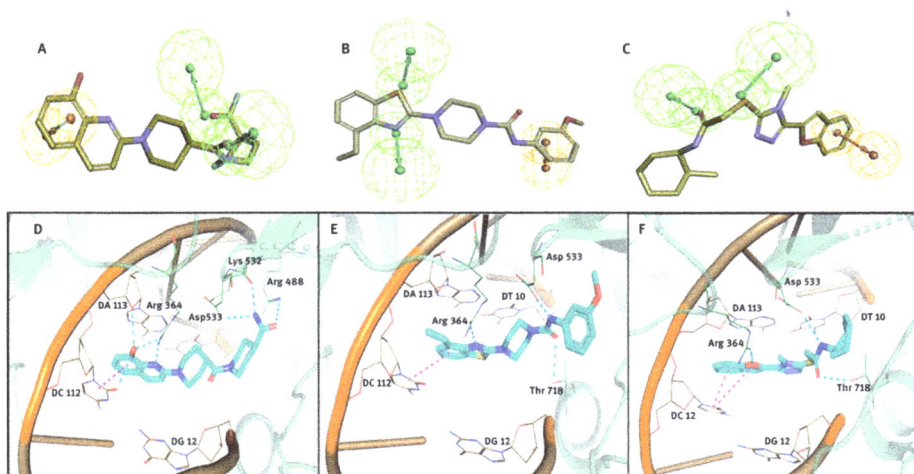

Figure 11.13: Overlay of lead molecules on the pharmacophore Hypo1. (A) ZINC68997780, (B) ZINC38550809, (C) ZINC15018994, Binding interaction of the hit compounds, (D) ZINC68997780, (E) ZINC38550809, (F) ZINC15018994 in the active site of human Top1 protein. Blue dotted line indicates the hydrogen bond and magenta dotted line indicates the π-π hydrophobic interaction (For interpretation of the references to color in this figure legend, the reader is referred to the web version of this article [84]). The figure was adapted with permission.

Increased interest in the simultaneous modulation of multiple targets and/or multi-treatment therapy for AD has been accepted as an efficient therapy strategy to treat AD [95–97]. Galantamine has been proven to act as a cholinesterase inhibitor; acting as a multitarget inhibitor on both nicotinic and muscarinic acetylcholine receptors [98–101]. However, the high cost associated with its isolation (~ US$ 50,000 per kg) coupled with the low concentration of galantamine in nature led to the design of total synthesis methods for this compound [101]. In order to overcome some of these

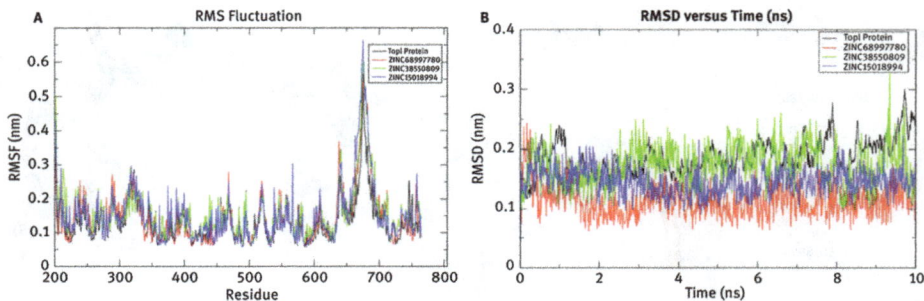

Figure 11.14: (A) Root-mean-square fluctuations of protein and selected leads for 10 ns production run. (B) Root-mean-square deviations of protein and selected leads for 10 ns production run [84]. Figure reproduced with permission.

challenges, the search for better synthetic mimics with the desired galantamine polypharmacology to manage AD and other neurological disorders have received a lot of attention. Thus, Grisoni et al. used computer-assisted ML techniques to develop automated workflow/models for compound screening and target profiling in order to explore new chemical space (Figure 11.15) [94].

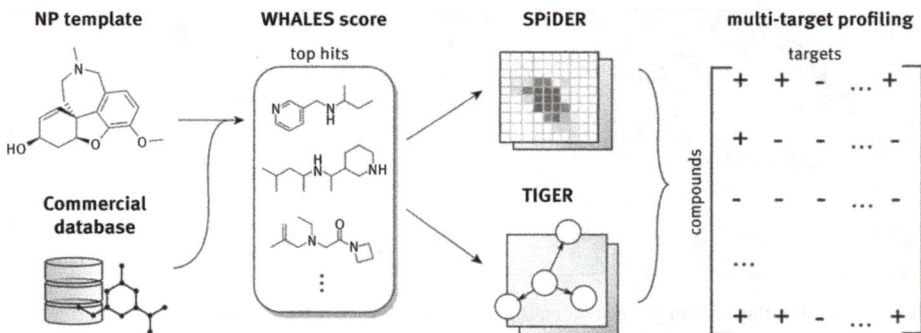

Figure 11.15: Graphical representation of the automated computer-assisted screening program [94]. The figure was reproduced with permission.

In the study, Grisoni et al. developed fully automated ML models using (–)-galantamine as the starting molecule to computationally screen a library of 3,383,942 commercially available compounds. The goal was to identify NP-inspired small molecules with multi-target activity on enzymes and receptors related to AD. The screening library was assembled from Asinex, ChemBridge screening compound collection, Enamine advanced and HTS collections, and Specs screening compounds. Twenty top-scoring compounds (with three-dimensional pharmacophore similarity to (–)-galantamine) were selected by two ML models (using SPiDER (SOM-based Prediction of Drug Equivalence

Relationships) and TIGER (Target Inference GEneratoR) as macromolecular targets inhibitors [102, 103]. While SPiDER is based on self-organizing maps, TIGER, on the other hand, is an extension of the SPiDER concept by an ensemble information scoring. Macromolecular target predictions were considered relevant in the study if they had:

1. high SPiDER confidence ($p < 0.025$),
2. high TIGER score ($S > 20$), or
3. both SPiDER and TIGER predictions agreed (conditions: SPiDER $p < 0.05$ and TIGER $S > 10$).

In vitro characterization of the 20 selected hits led to the identification of eight molecules (See Figure 11.16; five of them shared one or more targets with (−)-galantamine) with activity on at least one of the selected targets as well as a target inhibition or stimulation ranging from 28% to 100% at 50 μM concentration.

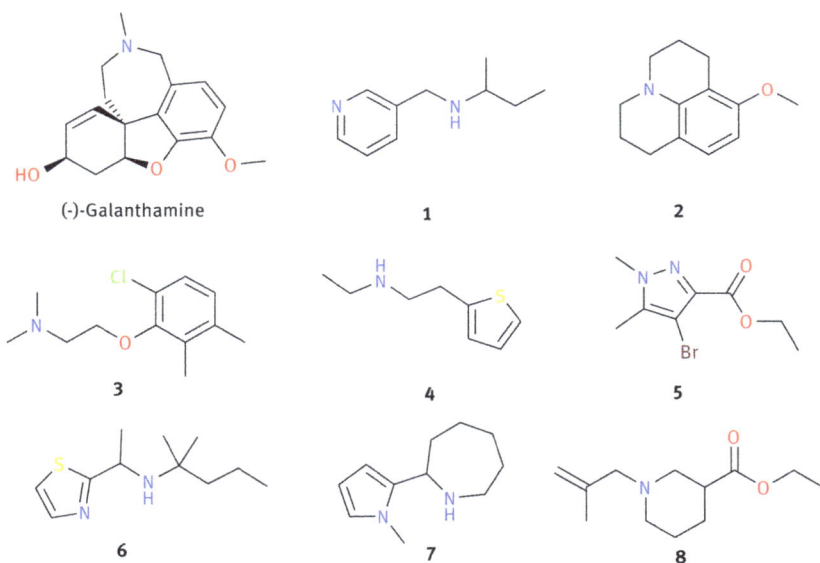

Figure 11.16: Virtual screening hits [94]. The figure was adapted with permission.

Of the reported compounds, **1** uniquely showed 60.3% *in vitro* binding activity on nAChR, while **2**–**5** and **7** were identified as novel 5-HT$_{2B}$ modulators. Compounds **2**, **4**, **5**, and **7** demonstrated 5-HT$_{2B}$ modulation activity (antagonistic effect) ranging from 47% to 90%. Compound **3**, on the other hand, demonstrated 5-HT$_{2B}$ modulation activity (agonistic effect) of 100%. Thus, the suggested molecules **1** (nAChR), **3** (5-HT2B), **4** (5-HT2B), **5** (5-HT2B), **6** (mAChR), and **8** (mAChR) can be explored to develop novel modulators of acetylcholine and serotonin signaling. Additionally, an observed dual activity of compound **3** (on AChE and 5-HT2B) and compound **8** (expanded to include mAChR M1 antagonism) identified from the ML approaches could act as a potentially superior treatment strategy to treat AD. Furthermore, analysis from the docking studies (using MOE v. 2018.01 (Chemical Computing Group ULC, Molecular Operating Environment, Montreal, QC, Canada)) of compounds **3** and **8** showed that the compounds had similar binding poses to the reference compound (−)-galantamine (Figure 11.17). Due to the fragment-like size of **3** and **8**, these compounds can be further explored to maximize the entire binding site regions and improve their binding affinity to AChE, 5-HT$_{2B}$, and M1.

11.4 Conclusion

Historically, medicinal plants have been used in the treatment of several ailments. Investigations of such species yielded very helpful compounds that have served as new drugs, as starting scaffolds for modifications or supplements. The growing interest in NPs has resulted in the development of modern methods to facilitate the identification, isolation, and characterization of novel NPs. This is continuously leading to an increased arena of diverse chemical molecules. Technical and financial drawbacks in a typical rational drug discovery process using experimental procedures have given birth to CADD methods as more effective ways to explore the diverse and complicated nature of NPs as potential drug/scaffold candidates. Some of these CADD discovery approaches (e. g. structure-based, ligand-based, and ML) discussed herein are contributing enormously to the drug discovery pipeline. The application of CADD methods has led to the proposal of several molecules with NP-based origin as potential molecules that can be developed to treat several ailments including cancer, AD, and malaria.

Acknowledgements: CVS acknowledges financial support from the German Academic Exchange Services (DAAD) through a Ph.D. student fellowship. FNK benefitted from an Alexander von Humboldt return fellowship and equipment subsidy.

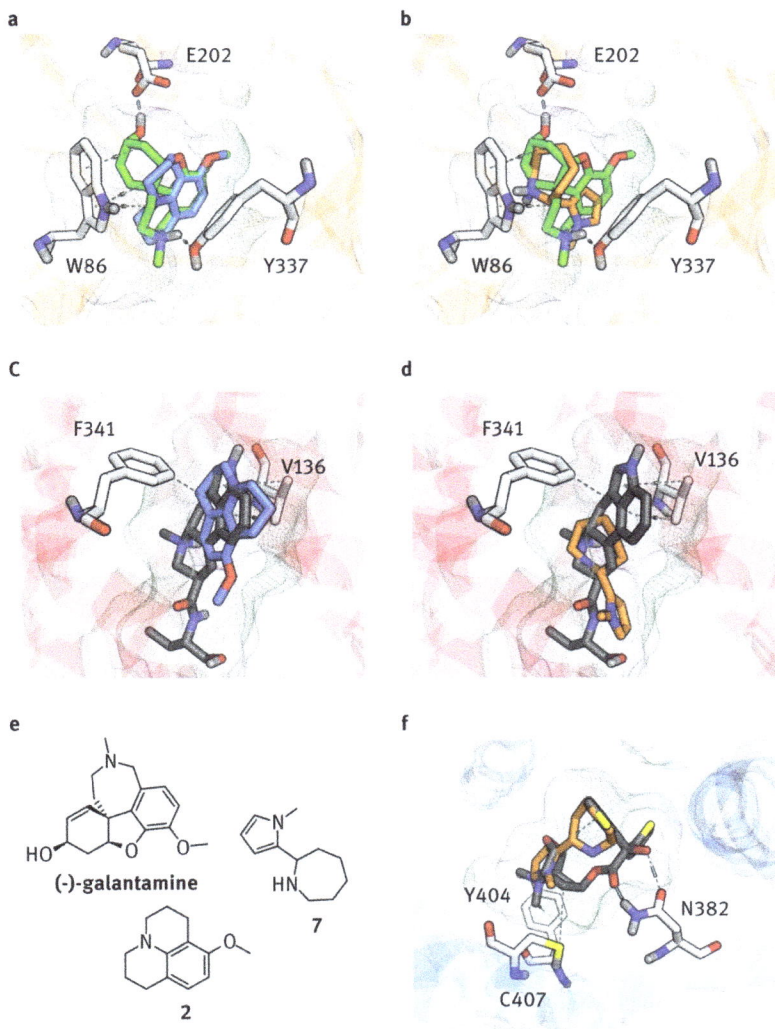

Figure 11.17: Predicted binding poses of compounds **2** (blue) and **7** (orange) in the binding sites of AChE (PDB ID: 4EY6), 5-HT2B (PDB ID: 6DRY), and mAChR M1 (PDB ID: 5CXV only for **7**).
(a) Compound **2** (blue) and the crystallized ligand, (–)-galantamine (green) in the binding site of AChE. (b) Compound **7** (orange) and the crystallized ligand (–)-galantamine (green) in the binding site of AChE. (c) Compound **2** (blue) and the crystallized ligand (H8D, grey) in the binding site of 5-HT2B. (d) Compound **7** (orange) and the crystallized ligand (H8D, grey) in the binding site of 5-HT2B. (e) Chemical structures of galantamine in comparison with compounds **2** and **7**.
(f) Compound **7** (orange) in comparison with the crystallized ligand (0HK, grey) in the binding site of mAChR [94]. The figure was adapted with permission.

References

[1]   Nosengo N. Can you teach old drugs new tricks? Nature. 2016;534:314–16.
[2]   Scannell JW, Blanckley A, Boldon H, Warrington B. Diagnosing the decline in pharmaceutical R&D efficiency. Nat Rev Drug Discov. 2012;11:191–200.
[3]   DiMasi JA, Grabowski HG, Hansen RW. Innovation in the pharmaceutical industry: new estimates of R&D costs. J Health Econ. 2016;47:20–33.
[4]   Gupta S, Basu A, Jaiswal A, Mishra N. Computational methods in drug discovery. Int J Pharm Sci Res. 2018;9:4102–8.
[5]   Heifetz A, Southey M, Morao I, Townsend-Nicholson A, Bodkin MJ. Computational methods used in hit-to-lead and lead optimization stages of structure-based drug discovery. In: Heifetz A, editor. Computational methods for GPCR drug discovery. Methods in molecular biology, vol. 1705. New York, NY: Humana Press, 2018:375–94.
[6]   Bottegoni G, Cavalli A. 9-Computational methods in multitarget drug discovery. In: Decker M, Maximilian J, editors. Design of hybrid molecules for drug development. Würzburg, Germany: University of Würzburg, 2017:239–58.
[7]   Gasteiger J. Chemoinformatics: achievements and challenges, a personal view. Molecules. 2016;21:151.
[8]   Leelananda SP, Lindert S. Computational methods in drug discovery. Beilstein J Org Chem. 2016;12:2694–718.
[9]   Cho A. Combining density functional theory with cheminformatics for development of a new-paradigm ligand screening method in computational drug discovery. Abstr Pap Am Chem S. 2016;251. PubMed PMID: WOS:000431903804869. Available at: http://bulletin.acscinf.org/PDFs/251nm/2016_spring_CINF_062.pdf. Accessed: 30 Sept 2019.
[10] Nantasenamat C, Prachayasittikul V. Maximizing computational tools for successful drug discovery. Expert Opin Drug Discov. 2015;10:321–9.
[11] Ekins S, Freundlich JS, Hobrath JV, White EL, Reynolds RC. Combining computational methods for hit to lead optimization in mycobacterium tuberculosis drug discovery. Pharm Res. 2014;31:414–35.
[12] Sliwoski G, Kothiwale S, Meiler J, Lowe EW. Computational methods in drug discovery. Pharmacol Rev. 2014;66:334–95.
[13] Talele TT, Khedkar SA, Rigby AC. Successful applications of computer aided drug discovery: moving drugs from concept to the clinic. Curr Top Med Chem. 2010;10:127–41.
[14] Van Drie JH. Computer-aided drug design: the next 20 years. J Comput Aid Mol Des. 2007;21:591–601.
[15] Clark DE. What has computer-aided molecular design ever done for drug discovery? Expert Opin Drug Dis. 2006;1:103–10.
[16] Wright GD. Unlocking the potential of natural products in drug discovery. Microb Biotechnol. 2019;12:55–7.
[17] Nishad VM, Anu S, Sundar AS. A review on natural products in drug discovery. Indo Am J Pharm Sci. 2018;5:11556–63.
[18] Thomford NE, Senthebane DA, Rowe A, Munro D, Seele P, Maroyi A, et al. Natural products for drug discovery in the twenty-first Century: innovations for novel drug discovery. Int J Mol Sci. 2018;19:1578.
[19] Sun D, Gündisch D. Editorial (Thematic issue: privileged scaffolds in natural products and drug discovery). Curr Top Med Chem. 2016;16:1199.
[20] Gaudêncio SP, Pereira F. Dereplication: racing to speed up the natural products discovery process. Nat Prod Rep. 2015;32:779–810.

[21] Khazir J, Mir BA, Mir SA, Cowan D. Natural products as lead compounds in drug discovery. J Asian Nat Prod Res. 2013;15:764–88.

[22] Ma DL, Chan DS, Leung CH. Molecular docking for virtual screening of natural product databases. Chem Sci. 2011;2:1656–65.

[23] Shen J, Xu X, Cheng F, Liu H, Luo X, Shen J, et al. Virtual screening on natural products for discovering active compounds and target information. Curr Med Chem. 2003;10:2327–42.

[24] Pereira F, Aires-de-sousa J. Computational methodologies in the exploration of marine natural product leads. Mar Drugs. 2018;16:7.

[25] Rodrigues T. Harnessing the potential of natural products in drug discovery from a cheminformatics vantage point. Org Biomol Chem. 2017;15:9275–82.

[26] Lal UR, Singh A. Edited by Atta-ur-Rahman Chapter 8- Recent developments in natural product-based drug discovery in tropical diseases. Stud Nat Prod Chem. 2016;48: 263–85.

[27] Sherman DH. Re-emergence of natural products in drug discovery and development. Abstr Pap Am Chem S. 2015;250. MEDI: 11 PubMed PMID: WOS:000432475701174. Available at: https://www.acsmedchem.org/ama/orig/abstracts/mediabstractf2015.pdf. Accessed: 30 Sept 2019.

[28] Joshi P, Stamos D, Binch H, Alvi K. Natural product inspired libraries for hit generation in drug discovery. Abstr Pap Am Chem S. 2014;248. MEDI:189. PubMed PMID: WOS:000349167402004. Available at: https://www.acsmedchem.org/ama/orig/abstracts/mediabstractf2014.pdf. Accessed: 30 Sept 2019.

[29] Zeng X, Zhang P, He W, Qin C, Chen S, Tao L, et al. NPASS: natural product activity and species source database for natural product research, discovery and tool development. Nucleic Acids Res. 2018;46:D1217–D1222.

[30] Wackett LP. Natural product databases: an annotated selection of World Wide Web sites relevant to the topics in microbial biotechnology. Microb Biotechnol. 2018;11:797–8.

[31] Choi H, Cho SY, Pak HJ, Kim Y, Choi J-Y, Lee YJ, et al. NPCARE: database of natural products and fractional extracts for cancer regulation. J Cheminform. 2017;9:2.

[32] Xie T, Song S, Li S, Ouyang L, Xia L, Huang J. Review of natural product databases. Cell Proliferat. 2015;48:398–404.

[33] Banerjee P, Erehman J, Gohlke BO, Wilhelm T, Preissner R, Dunkel M. Super Natural II-a database of natural products. Nucleic Acids Res. 2015;43:D935–D939.

[34] Saldivar F, Franco JL. Natural product databases: chemical space, diversity and suitability of virtual screening. Abstr Pap Am Chem S. 2018;256. CINF: 21 http://apps.webofknowledge.com/full_record.do?product=WOS&search_mode=DaisyOneClickSearch&qid=3&SID=D12Vje1Xx4JX8nbmoVT&page=1&doc=2. Accessed: 30 Sept 2019.

[35] Gurevich A, Mikheenko A, Shlemov A, Korobeynikov A, Mohimani H, Pevzner PA. Increased diversity of peptidic natural products revealed by modification-tolerant database search of mass spectra. Nat Microbiol. 2018;3:319–27.

[36] Chen Y, de Lomana MG, Friedrich NO, Kirchmair J. Characterization of the chemical space of known and readily obtainable natural products. J Chem Inf Model. 2018;58:1518–32.

[37] Borchert E, Jackson SA, O'Gara F, Dobson AD. Diversity of natural product biosynthetic genes in the microbiome of the deep sea sponges *Inflatella pellicula, Poecilla stracompressa,* and *Stelletta normani.* Front Microbiol. 2016;7:1027.

[38] Ntie-Kang F, Zofou D, Babiaka SB, Meudom R, Scharfe M, Lifongo LL, et al. AfroDb: a select highly potent and diverse natural product library from african medicinal plants. PLoS One. 2013;8:e78085.

[39] Chen Y, Kops CD, Kirchmair J. Data resources for the computer-guided discovery of bioactive natural products. J Chem Inf Model. 2017;57:2099–111.

[40] Wang SZ, Fang K, Dong GQ, Chen SQ, Liu N, Miao ZY, et al. Scaffold diversity inspired by the natural product evodiamine: discovery of highly potent and multitargeting antitumor agents. J Med Chem. 2015;58:6678–96.

[41] Ntie-Kang F, Ed. Chemoinformatics of natural products: fundamental concepts, Vol. 1. Berlin: De Gruyter, 2019. ISBN 978-3-11-057935-2.

[42] Newman DJ. From natural products to drugs. Phys Sci Rev. 2018;4:20180111.

[43] Pascolutti M, Quinn RJ. Natural products as lead structures: chemical transformations to create lead-like libraries. Drug Discov Today. 2014;19:215–21.

[44] Cragg GM, Newman DJ. Natural products: a continuing source of novel drug leads. Biochim Biophys Acta. 2013;1830:3670–95.

[45] Bardelle C, Coleman T, Cross D, Davenport S, Kettle JG, Ko EJ, et al. Inhibitors of the tyrosine kinase EphB4. Part 2: structure-based discovery and optimisation of 3,5-bis substituted anilinopyrimidines. Bioorg Med Chem Lett. 2008;18:5717–21.

[46] Brvar M, Perdih A, Renko M, Anderluh G, Turk D, Solmajer T. Structure-based discovery of substituted 4,5′-bithiazoles as novel DNA gyrase inhibitors. J Med Chem. 2012;55:6413–26.

[47] Labadie S, Dragovich PS, Barrett K, Blair WS, Bergeron P, Chang C, et al. Structure-based discovery of C-2 substituted imidazo-pyrrolopyridine JAK1 inhibitors with improved selectivity over JAK2. Bioorg Med Chem Lett. 2012;22:7627–33.

[48] Li X, Huang B, Zhou Z, Gao P, Pannecouque C, Daelemans D, et al. Arylazolyl(azinyl) thioacetanilides: Part 19: Discovery of novel substituted imidazo[4,5-b]pyridin-2-ylthioacetanilides as potent HIV NNRTIs via a structure-based bioisosterism approach. Chem Biol Drug Des. 2016;88:241–53.

[49] Liu L-J, Leung KH, Chan DS, Wang Y-T, Ma D-L, Leung CH. Identification of a natural product-like STAT3 dimerization inhibitor by structure-based virtual screening. Cell Death Dis. 2014;5: e1293.

[50] Paneth A, Węglińska L, Bekier A, Stefaniszyn E, Wujec M, Trotsko N, et al. Discovery of potent and selective halogen-substituted imidazole-thiosemicarbazides for inhibition of *Toxoplasma gondii* growth in vitro via structure-based design. Molecules. 2019;24:1618.

[51] Schuster D, Wolber G. Identification of bioactive natural products by pharmacophore-based virtual screening. Curr Pharm Des. 2010;16:1666–81.

[52] Tanaka T, Sugawara H, Maruoka H, Imajo S, Muto T. Discovery of novel series of 6-benzyl substituted 4-aminocarbonyl-1,4-diazepane-2,5-diones as human chymase inhibitors using structure-based drug design. Bioorg Med Chem. 2013;21:4233–49.

[53] Li JZ, Wang HY, Li JJ, Bao JK, Wu CF. Discovery of a potential HER2 inhibitor from natural products for the treatment of HER2-positive breast cancer. Int J Mol Sci. 2016;17:1055.

[54] Bw S, Wild CP, editors. World cancer report. Geneva, Switzerland: WHO Press; 2014.

[55] Diéras V, Miles D, Verma S, Pegram M, Welslau M, Baselga J, et al. Trastuzumab emtansine versus capecitabine plus lapatinib in patients with previously treated HER2-positive advanced breast cancer (EMILIA): a descriptive analysis of final overall survival results from a randomised, open-label, phase 3 trial. Lancet Oncol. 2017;18:732–42.

[56] Konecny GE, Venkatesan N, Yang G, Dering J, Ginther C, Finn R, et al. Activity of lapatinib a novel HER2 and EGFR dual kinase inhibitor in human endometrial cancer cells. Br J Cancer. 2008;98:1076–84.

[57] Montemurro F, Valabrega G, Aglietta M. Lapatinib: a dual inhibitor of EGFR and HER2 tyrosine kinase activity. Expert Opin Biol Ther. 2007;7:257–68.

[58] Piwko C, Prady C, Yunger S, Pollex E, Moser A. Safety profile and costs of related adverse events of trastuzumab emtansine for the treatment of HER2-positive locally advanced or metastatic breast cancer compared to capecitabine plus lapatinib from the perspective of the canadian health-care system. Clin Drug Investig. 2015;35:487–93.

[59] Xu B-H, Jiang Z-F, Chua D, Shao Z-M, Luo R-C, Wang X-J, et al. Lapatinib plus capecitabine in treating HER2-positive advanced breast cancer: efficacy, safety, and biomarker results from Chinese patients. Chin J Cancer. 2011;30:327–35.

[60] AnalytiCon Discovery NP. Available at: zinc.docking.org/catalogs/acdiscnp. (Accessed on: 29 June 2016).

[61] Lang PT, Brozell SR, Mukherjee S, Pettersen EF, Meng EC, Thomas V, et al. DOCK 6: combining techniques to model RNA-small molecule complexes. RNA. 2009;15:1219–30.

[62] Trott O, Olson AJ. AutoDockVina: improving the speed and accuracy of docking with a new scoring function, efficient optimization, and multithreading. J Comput Chem. 2010;31: 455–61.

[63] Pronk S, Pall S, Schulz R, Larsson P, Bjelkmar P, Apostolov R, et al. GROMACS 4.5: a high-throughput and highly parallel open source molecular simulation toolkit. Bioinformatics. 2013;29:845–54.

[64] Musyoka TM, Kanzi AM, Lobb KA, Tastan Bishop O. Structure based docking and molecular dynamic studies of plasmodial cysteine proteases against a south african natural compound and its analogs. Sci Rep. 2016;6:23690.

[65] World Health Organization (WHO). Q&A on artemisinin resistance (issued:2013/ Updated:2019). 2019, Available at: http://who.int/malaria/media/artemisinin_resistance_ qa/en/. Accessed: 30 Sept 2019).

[66] World Health Organization (WHO). World Malaria Report 2013. 2013. Available at: http:// www.who.int/malaria/publications/world_malaria_report_2013/wmr2013_no_profiles.pdf? ua=1. Accessed: 26th July 2015.

[67] Bhattarai A, Ali AS, Kachur SP, Martensson A, Abbas AK, Khatib R, et al. Impact of artemisinin-based combination therapy and insecticide-treated nets on malaria burden in Zanzibar. PLoS Med. 2007;4:e309.

[68] Griffin JT, Hollingsworth TD, Okell LC, Churcher TS, White M, Hinsley W, et al. Reducing *Plasmodium falciparum* malaria transmission in Africa: a model-based evaluation of intervention strategies. PLoS Med. 2010;7:e1000324.

[69] Severini C, Menegon M. Resistance to antimalarial drugs: An endless world war against *Plasmodium* that we risk losing. J Glob Antimicrob Resist. 2015;3:58–63.

[70] Pance A. How elusive can a malaria vaccine be? Nat Rev Microbiol. 2019;17:129.

[71] GlaxoSmithKline (GSK). GSK's malaria candidate vaccine, MosquirixTM (RTS,S), receives positive opinion from European regulators for the prevention of malaria in young children in sub-Saharan Africa. Available at: https://au.gsk.com/en-au/media/press-releases/2015/ gsk-s-malaria-candidate-vaccine-mosquirixtm-rtss-receives-positive-opinion-from-european-regulators-for-the-prevention-of-malaria-in-young-children-in-sub-saharan-africa/. Accessed: 30 Sept 2019.

[72] Goldberg DE. Plasmodial hemoglobin degradation: an ordered pathway in a specialized organelle. Infect Agents Dis. 1992;1:207–11.

[73] Pandey KC, Dixit R. Structure-function of falcipains: malarial cysteine proteases. J Trop Med. 2012;2012:345195.

[74] Rosenthal PJ. Falcipains and Other Cysteine Proteases of Malaria Parasites. In: Robinson MW, Dalton JP, editors. Cysteine proteases of pathogenic organisms. Adv Exp Med Bio, vol. 712. Boston, MA: Springer, 2011.

[75] Shah F, Mukherjee P, Gut J, Legac J, Rosenthal PJ, Tekwani BL, et al. Identification of novel malarial cysteine protease inhibitors using structure-based virtual screening of a focused cysteine protease inhibitor library. J Chem Inf Model. 2011;51:852–64.

[76] Butler MS. Natural products to drugs: natural product derived compounds in clinical trials. Nat Prod Rep. 2005;22:162–95.

[77]   Butler MS. Natural products to drugs: natural product-derived compounds in clinical trials.
       Nat Prod Rep. 2008;25:475–516.
[78]   Butler MS, Robertson AA, Cooper MA. Natural product and natural product derived drugs in
       clinical trials. Nat Prod Rep. 2014;31:1612–61.
[79]   Irwin JJ, Sterling T, Mysinger MM, Bolstad ES, Coleman RG. ZINC: a free tool to discover
       chemistry for biology. J Chem Inf Model. 2012;52:1757–68.
[80]   Musyoka TM, Kanzi AM, Lobb KA, Tastan Bishop O. Analysis of non-peptidic compounds as
       potential malarial inhibitors against *Plasmodial* cysteine proteases via integrated virtual
       screening workflow. J Biomol Struct Dyn. 2016;34:2084–101.
[81]   Davies-Coleman MT, Beukes DR. Ten years of marine natural products research at Rhodes
       University. S Afr J Sci. 2004;100:539–44.
[82]   Morris GM, Huey R, Lindstrom W, Sanner MF, Belew RK, Goodsell DS, et al. AutoDock4 and
       AutoDockTools4: Automated docking with selective receptor flexibility. J Comput Chem.
       2009;30:2785–91.
[83]   Williams DR, Walsh MJ, Miller NA. Studies for the synthesis of xenicane diterpenes.
       A stereocontrolled total synthesis of 4-hydroxydictyolactone. J Am Chem Soc.
       2009;131:9038–45.
[84]   Pal S, Kumar V, Kundu B, Bhattacharya D, Preethy N, Reddy MP, et al. Ligand-based
       pharmacophore modeling, virtual screening and molecular docking studies for discovery of
       potential Topoisomerase I inhibitors. Comput Struct Biotechnol J. 2019;17:291–310.
[85]   Yang SY. Pharmacophore modeling and applications in drug discovery: challenges and
       recent advances. Drug Discov Today. 2010;15:444–50.
[86]   Keszthelyi A, Minchell NE, Baxter J. The causes and consequences of topological stress
       during DNA replication. Genes (Basel). 2016;7:134.
[87]   Gantz J, Reinsel D. The digital universe decade – are you ready? (2010) Available at: https://
       www.emc.com/collateral/analyst-reports/idc-digital-universe-are-you-ready.pdf Accessed:
       30 Sept 2019.
[88]   Cortes C, Vapnik V. Support-Vector Networks. Mach Learn. 1995;20:273–97.
[89]   LeCun Y, Bengio Y, Hinton G. Deep learning. Nature. 2015;521:436–44.
[90]   LeCun Y. Deep learning & convolutional networks. IEEE Hot Chip Symp. 2015;2015:7477328.
[91]   Saxena AK, Schaper KJ. QSAR analysis of the time- and dose-dependent anti-inflammatory *in
       vivo* activity of substituted imidazo[1,2-a]pyridines using artificial neural networks. QSAR
       Comb Sci. 2006;25:590–7.
[92]   Salt DW, Yildiz N, Livingstone DJ, Tinsley CJ. The use of artificial neural networks in QSAR.
       Pestic Sci. 1992;36:161–70.
[93]   Gilson MK, Liu T, Baitaluk M, Nicola G, Hwang L, Chong J. BindingDB in 2015: A public
       database for medicinal chemistry, computational chemistry and systems pharmacology.
       Nucleic Acids Res. 2016;44:D1045–D1053.
[94]   Grisoni F, Merk D, Friedrich L, Schneider G. Design of natural-product-inspired multitarget
       ligands by machine learning. Chem Med Chem. 2019;14:1129–34.
[95]   Freret T, Lelong-Boulouard V, Lecouflet P, Hamidouche K, Dauphin F, Boulouard M.
       Co-modulation of an allosteric modulator of nicotinic receptor-cholinesterase inhibitor
       (galantamine) and a 5-HT4 receptor agonist (RS-67333): effect on scopolamine-induced
       memory deficit in the mouse. Psychopharmacology. 2017;234:2365–74.
[96]   Cavalli A, Bolognesi ML, Minarini A, Rosini M, Tumiatti V, Recanatini M, et al. Multi-target-
       directed ligands to combat neurodegenerative diseases. J Med Chem. 2008;51:347–72.
[97]   Hughes RE, Nikolic K, Ramsay RR. One for all? Hitting Multiple alzheimer's disease targets
       with one drug. Front Neurosci. 2016;10:177.

[98]   Kita Y, Ago Y, Higashino K, Asada K, Takano E, Takuma K, et al. Galantamine promotes adult hippocampal neurogenesis via M_1 muscarinic and $\alpha 7$ nicotinic receptors in mice. Int J Neuropsychopharmacol. 2014;17:1957–68.

[99]   Woodruff-Pak DS, Vogel RW, 3rd, Wenk GL. Galantamine: effect on nicotinic receptor binding, acetylcholinesterase inhibition, and learning. Proc Natl Acad Sci USA. 2001;98:2089–94.

[100] Almasieh M, Zhou Y, Kelly ME, Casanova C, Di Polo A. Structural and functional neuroprotection in glaucoma: role of galantamine-mediated activation of muscarinic acetylcholine receptors. Cell Death Dis. 2010;1:e27.

[101] Thomsen T, Kewitz H. Selective inhibition of human acetylcholinesterase by galanthamine in vitro and in vivo. Life Sci. 1990;46:1553–8.

[102] Schneider P, Schneider G. A computational method for unveiling the target promiscuity of pharmacologically active compounds. Angew Chem Int Ed Engl. 2017;56:11520–4.

[103] Reker D, Rodrigues T, Schneider P, Schneider G. Identifying the macromolecular targets of de novo-designed chemical entities through self-organizing map consensus. Proc Natl Acad Sci USA. 2014;111:4067–72.

Lucas Paul, Celestin N. Mudogo, Kelvin M. Mtei,
Revocatus L. Machunda, Fidele Ntie-Kang

12 A computer-based approach for developing linamarase inhibitory agents

Abstract: Cassava is a strategic crop, especially for developing countries. However, the presence of cyanogenic compounds in cassava products limits the proper nutrients utilization. Due to the poor availability of structure discovery and elucidation in the Protein Data Bank is limiting the full understanding of the enzyme, how to inhibit it and applications in different fields. There is a need to solve the three-dimensional structure (3-D) of linamarase from cassava. The structural elucidation will allow the development of a competitive inhibitor and various industrial applications of the enzyme. The goal of this review is to summarize and present the available 3-D modeling structure of linamarase enzyme using different computational strategies. This approach could help in determining the structure of linamarase and later guide the structure elucidation *in silico* and experimentally.

Keywords: linamarase, cassava, structural determination, computational strategies

12.1 Introduction

Cassava (*Manihot esculenta* Crantz) (Euphorbiaceae), also known as mandioca, yucca, tapioca or manioc. It is the leading supplier of energy ranked after rice and corn [1]. It is the most grown crop in the tropics and subtropics regions. The tuber is the primary source of carbohydrate while leaves provide protein as a vegetable. About 105 countries grow cassava as a strategical crop against famine since it can sustain and produces in drought and poor soil, can stay within the farm and be harvested at the time of demand [2]. The roots are the primary source of carbohydrate while leaves provide vitamins, minerals, and protein, as well as a vegetable that is available throughout the year. The leaves have high crude protein, and the amino acids, which are well balanced and its amount is beyond the minimal amount recommended by the Food and Agriculture Organization [2]. Leaves also have various minerals like iron, zinc, manganese, magnesium and calcium, vitamin B1, B2, C and carotenoids [2, 3], The combination of cassava roots and leaves can provide a meal with almost all essential dietary needs.

This article has previously been published in the journal *Physical Sciences Reviews*. Please cite as: A computer-based approach for developing linamarase inhibitory agents. Physical Sciences Reviews [Online] 2020 DOI: 10.1515/psr-2019-0098

https://doi.org/10.1515/9783110668896-012

All body parts of the cassava except seeds contain cyanogenic glucosides compounds known as linamarin and lotaustralin. Linamarase hydrolyses these compounds to hydrogen cyanide as the main product [4]. There are about 5,000 varieties of cassava; all of them are known to contain cyanogenic glucosides which range from 10 to 500 mg HCN/kg. Based on the amount of hydrogen cyanide (HCN) released, cassavas are classified into three groups: group one those with greater than 100 mg HCN/kg are called very bitter and very toxic, group two those with between 50 to 100 mg HCN/kg are regarded as moderate bitter and moderate toxic while those with less than 50 mg HCN/kg are sweet cassava [5].

Cyanogenic glucosides are mainly composed of Linamarin (95%) and Lotaustralin (5%) [6], enzymatic hydrolysis of cyanogenic glucosides by linamarase is initiated by any physical damage of cassava tissue this allows interaction between enzyme from cell wall which is physically separated from substrate found in cell vacuole so are not compartmentalized [7]. Interaction between enzymes and substrate (mainly Linamarin) start by the release of glucose and acetone cyanohydrin at pH > 4 and temperature >35 °C is converted to hydrogen cyanide [8], Figure 12.1. The presence of hydroxynitrile lyase (HNL) helps to complete the reaction of acetone cyanohydrin to cyanide. This enzyme is highly available in cassava leaves and very little in the roots [9]. The study by [10] has reported that processed cassava flours contain high levels of acetone cyanohydrin but little linamarin or HCN, this is due to the little amount of HNL in the roots, which brings about this accumulation.

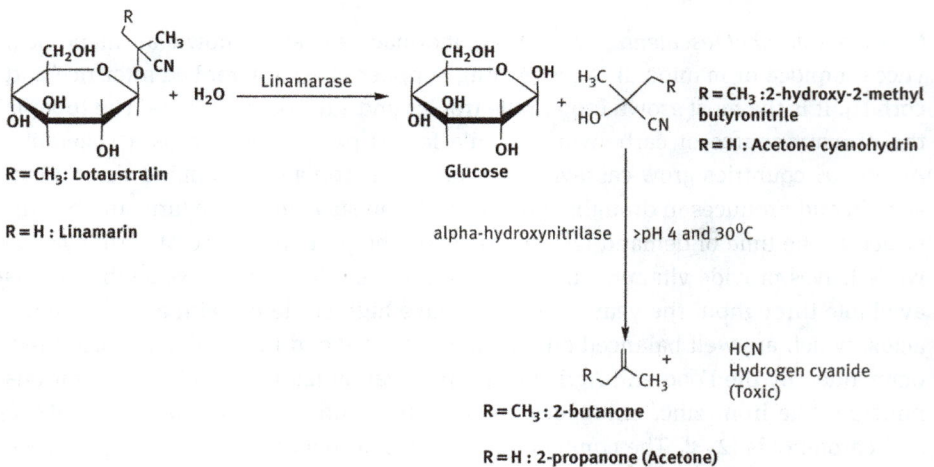

Figure 12.1: A complete enzymatic reaction between linamarase and linamarin.

Linamarin is a chemically stable compound, soluble in water and resists boiling in acid. Acetone cyanohydrin is also soluble in water and has a boiling point of 82 °C. But, HCN is a volatile compound, which evaporates only at 27 °C so volatilize at ambient temperature. So for effective processing techniques to be useful, we should reduce cyanogens to a safe level. The methods should maximize cassava tissue rupture to ensure effective enzyme-substrate interaction, to release acetone cyanohydrin and finally, volatile cyanide [11, 12].

The potential concentration of HCN determines the toxicity of the cyanogenic product consumed. If inadequately processed food is ingested, the HCN concentration is expected to be high within the body. For the toxicity of cyanogens depends on the following factors:

(a) Unsufficient processing of plant which causes linamarin or HCN to remain in the food.

(b) When raw cassava is consumed or insufficiently processed cassava product.

HCN, when released continuously until when low pH value from the stomach, deactivates the enzyme (linamarase). Cyanide ingested into the body always follows the metabolic pathway of detoxification, whereby rhodanese works by converting it to thiocyanide which later excreted in the urine [13]. When HCN ingested in the body gets absorbed quickly into the blood and combines with all forms of iron (methemoglobin and hemoglobin) which are present in erythrocytes [14]. The body eliminates the toxic cyanide by using the enzyme rhodanese, which contains an active disulfide group. It works by reacting with thiosulphate and cyanide, which converts cyanide to excretable thiocyanate, for this process to be complete sulfur donors that usually is provided by dietary sulfur amino acids are highly required [15, 16]. HCN binds to the Fe^{3+}/Fe^{2+} present in cytochrome and inactivates its activity.

This HCN inhibits the oxygen uptake and then causes glucose and lactic acid accumulation and deficiency of Adenosine triphosphate (ATP)/adenosine diphosphate (ADP), which brings the body to anaerobic instead of aerobic respiration [17]. HCN in the body inhibits many enzymatic reactions; if they contain iron, copper or molybdenum and its effect is highly and immediately appreciated in the respiratory system and heart. The amount recommended for cassava products, should not exceed 10 mg HCN/kg [4]. Any consumption of cassava products beyond the recommended amount can cause the following health problems vomiting, nausea, diarrhoea, dizziness, headache, stomach pains and sometimes death.

Linamarase is among the β-glucosidase belonging to the GH1 family which can convert glucosyl group from a glycoside (nonreducing) or carbohydrate by hydrolysis resulting in water or by transglycosylation gives alcohol. It has the $(\beta/\alpha)_8$ barrel structure, the properties of acid-base catalysis and the nucleophilic are contributed by the two carboxylic acid residues at β-strands 4 and 7 [18]. One most crucial property of linamarase which differentiate it from other GH1 family is the ability to effectively catalyze the transglycosylation using primary, secondary and tertiary alcohol as acceptors [19]. However, cannot synthesize oligosaccharides and glycosides by reverse hydrolysis [20].

The detailed crystal structure of enzyme linamarase is still lacking [21]. The only effort has been done is to obtain function active-site amino acid residues, which has been modeled using homology modeling by the MODELLER9v4 program (Figure 12.2) [22]. In the study by [21], they modeled the residues which are likely to be involved in the activity of dalcochinase in an effort to identify the amino acids which bring about enzyme specificity as compared to linamarase which have 47% sequence similarities.

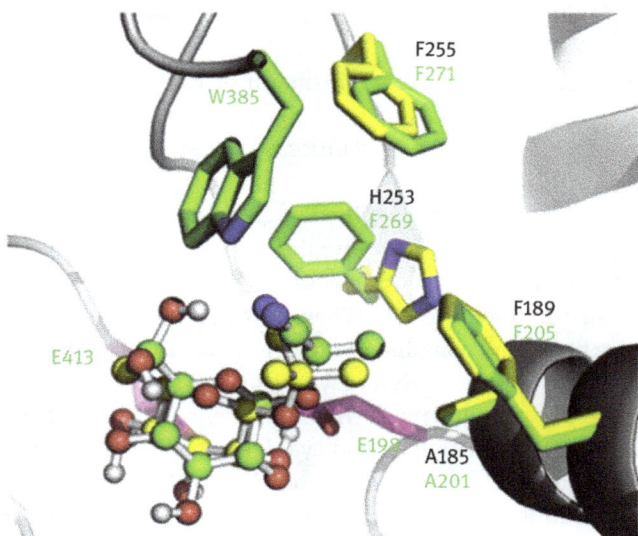

Figure 12.2: The generated three-dimensional models by MODELLER 9v4 of dalcochinase and Wild-type linamarase mutant's (1185A/N189F/V255F) (adapted from reference [22]).

There is the inclusion of linamarin in the 1185A/N189F/V255F mutant of dalcochinase as well as linamarase models. The catalytic acid/base (E198) and nucleophile (E413) of linamarase are shown in pink its docked linamarin is in yellow. The residues of linamarase and its docked linamarin are shown in green adopted from [22].

This review describes the computational approach toward the development of linamarase inhibitor, which is more competitive than natural substrate linamarin. For the development of linamarase inhibitor, we need first to develop *in silico* model structure of the enzyme (linamarase) by homology modeling. Then steps to discover the best and stable inhibitor, this will involve virtual screening (VS), molecular docking and finally, molecular dynamics. The details of these steps are analyzed below.

12.2 Determination of the three-dimension linamarase structure prediction

Generally, there are several computational methodologies and algorithms which are currently used to solve the problems of three-dimensional (3-D) structure of the protein which have not yet experimentally determined. The only available information is the sequence of amino acid, these provide essential information that relates to the 3-D macromolecules structures which are obtained by the experimental method like Protein Crystallography (X-ray diffraction), electron microscopy or nuclear magnetic resonance (NMR) [23]. There are four main methods that can be used as an approach of obtaining the linamarase 3-D structure these include the following.

12.2.1 Method without database information

The method uses the *ab initio* method which uses the concept from thermodynamics assuming all native protein structure always corresponds to the global minimum free energy [24–26]. Here, it does not use the structural templates from a database like Protein Data Bank (PDB). It mainly considers the potential energy functions in integrating the parameters of all atoms. The general goal is to obtain a global minimum free energy that corresponds to the native protein [26–28]. Using this approach, we can predict the new folds, since it is not limited to template from PDB. This principle uses the following simulation package; AMBER (Assisted Model Building with Energy Refinement) [29, 30], CHARMM (Chemistry at HARvard Molecular Mechanics) [31], UNRES [32], GROMACS (Groningen Machine for Chemical Simulation) [33], TINKER (Software tools for Molecular Design) [34].

12.2.2 Methods with database information

The starting point here is the 3-D protein structure is obtained from the database, and it compares the fragments of the target sequences to that of known protein. The short amino acid sub-sequence of the target structure against the known protein's structure fragments [35]. The newly discovered protein structure will be composed of a similar structure of motifs like the known protein. Therefore, this method is based on fragments of amino acid sequence with a different motif which, when combined, they form the 3-D protein structure [26]. The homology fragments are used for finding the structures which are achieved through scoring functions and algorithm optimization to get the structure with the lowest potential energy [36]. The fragment-based approach always needs to look for a criterion that exists between the fragments so that the final fragment will have a high chance of being inserted at the final structure predicted as summarized in illustration (Figure 12.3) [37].

This method is similar to *ab initio* when it finds polypeptide structures with the lowest energy, but the main difference is that it uses the database to predict the structure of polypeptide [38].

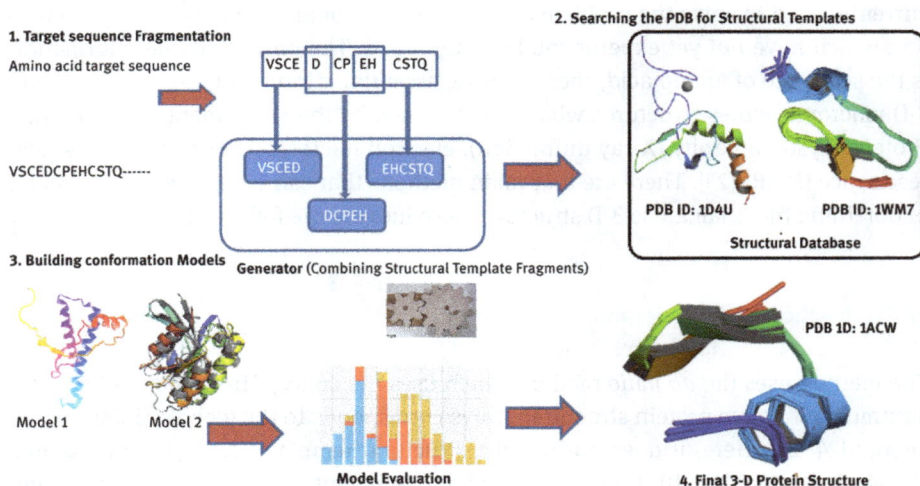

Figure 12.3: Schematic representation of the method based on fragments (adapted from reference [23]).

12.2.3 Fold recognition and threading methods

The method involves using the amino acid sequence and evaluates how well it fits the 3-D structure of the protein of the known. This approach is used because structures are more evolutionary preserved than the sequence [35, 39–41]. The sequential order is followed in placing the target amino acid sequences and is governed by two procedures; first searching the correct replacement between the target sequence versus the model which is in the space of possible sequence-structure alignment. The threads tried to find the templates with a similar fold that have or don't have direct evolutionary relations (analogue).

12.2.4 Comparative modeling method and sequence alignment strategies

This method use target protein's amino acid sequences to align against know protein's amino acid sequence (used as a template). The information of a known protein is experimentally determined and deposited in PDB [42]. If there are high similarities between the two amino acid sequences, then the structural information of the known (template) can be used to modal the target protein of interest [43, 44]. The homology protein with full information obtained experimentally are the ones to be used to model the target protein, and their amino acid residue is similar as they occupy the same position in the homology protein and have similar physic-chemical properties. Currently, comparative modeling is highly used because it is useful in protein structure prediction, which has more impact in the field of drug discovery [45]. The sequence alignment can either be pairwise which is the sequence–sequence comparison or multiple sequence comparison. The first approach uses the target sequence to compare with sequences in the database independently [46]. It uses methods like FASTA [47, 48], PSI-BLAST [49] and BLAST [50], while multiple sequence comparison allows multiple sequence alignments whereby the sensitivity of the search is maximized [51–53]. The methods used here include CLUSTALW [54], PSI-BLAST [49] and T-COFFEE [55].

In the study by [21]identify the specific residues with similarities but have different catalytic properties. So eight amino acid residues in the glycine binding pocket of dalcochinase were replaced with respective residues of linamarase. Since the crystal structure of both enzymes is unavailable, then homology modeling of dalcochinase which has 47% amino acid sequence identity with linamarase, were performed by using ClustalW 2.0 with similar procedures as reported by [54]. The 3-D structure of wild-type dalcochinase was obtained using the template with a 45% identity to dalcochinase from Maize β-glucosidase 1 (Figure 12.4) with its substrate DIMBOA-β-D-glucoside with PDB code 1E56A as also reported by [56].

The structure was built using the Sybyl 7.2 molecular modeling package, the overall structure of the model was checked by the PROCHECK, ProSA, Verify-3D and WHATIF programs [57–60]. Whereby PROCHECK showed 97% of the residues in the homology were located in the most favorable regions of the Ramachandran plot, whereby only 1.5% was in the rejected region when compared with the template with 98.5% and 0%. PROCHECK was also used to obtain G-factor, which was 0.22 above −0.5 which shows the model was reliable as compared to 0.34 of a template.

The ProSA Z-scores is explained well by its application to check the errors of the 3-D models. It can be used to determine errors of the experimentally and theoretically determined structures. It uses the coordinates then the structure's energy is evaluated by using a distance-based pair potential as well as the potential that relates to solvent exposure. The Z-scores validates the quality of the model and measures the total energy of the structure with regards to energy distribution derived from random conformations. The obtained Z-score which appears outside a range property of native proteins verifies erroneous structure [58]. In general positive value correspond to problematic or erroneous parts of a model. For example the model showedProSA Z-score of −8.15 while template −9.68 this is within the acceptable range [58]. The compatibility of the residues with the surrounding environment was done by verify-3D which scored above 0.2 (91.2%) shows reliable as compared with 93.9% for the template. WHATIF program managed to bring the confidence of the packing quality which scored above −5.0 and the picture of the 3-D

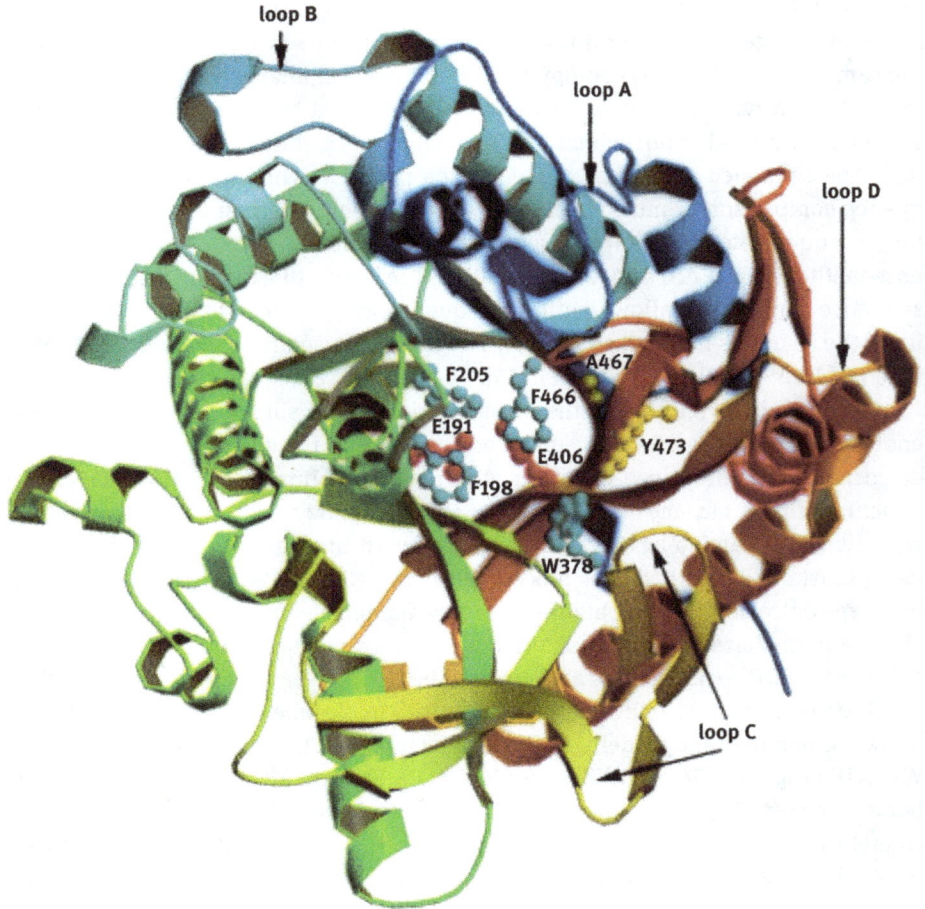

Figure 12.4: The three-dimensional structure of Maize β-glucosidase (from N terminus dark blue to C terminus dark red) (adapted from reference [56]).

model of the active site pocket of dalcochinase was generated by using PyMOL version 0.99.

Another study by [22] used the report of single mutation brought by replacing eight amino acid residues in the dalcochinase's binding pocket using residues from linamarase. The mutants namely 1185A, N189F, V255F which have been identified to contribute to the hydrolytic and transglycosylation specificity of dalcochinase. The 1185A and V255F mutants have a low contribution to natural (Dal-Glc) substrate while all three mutants have more significant transglycosylation activities by using primary and secondary alcohol as acceptors, but none of these mutants demonstrated linamarase activities of transglycosylation; glucose to tertial alcohol and hydrolyzing linamarin. So in the standing study by [22], it brought an intention of further mutating the residues of dalcochinase in order to attain the linamarase specificities (transglycosylation and hydrolysis reaction). So the 3-D model of the previously reported mutants (1185A, N189F, V255F) of dalcochinase [61] and wild-type linamarase [18] were created using MODELLER9v4 [62], using the template of cyanogenic β-glucosidase from *Trifolium repens* L with (PDB code 1CBG), which have a similarity of 60% to dalcochinase and 51% to wild-type linamarase [56]. The model's quality was checked using PROCHECK, ProSA and Verify-3D programs [58, 59] where the active site was defined as 15 Å which is at the center of residues E182 and E396 of dalcochinase.

12.3 Identification and validation of linamarase inhibitors

The approach of identifying the inhibitor of a specific enzyme uses the same approach of drug discovery which targets to obtain a small molecule, known as the entity that can preferentially interact with the valid target. The target which has identified to cause or have a link with the disease or biological effect and need to be inhibited [63].

12.3.1 Virtual Screening for the identification of linamarase inhibitors

This is described as the step by step approach of searching novel compounds that referred to as hit and led with potential biological effects is achieved by filtering and narrowing down until the lead is obtained as an alternative to the natural ligand. Depending on the intended application the databases for VS consist of up to about 10 million compounds and they can be obtained from compound libraries that are provided from commercial venders, public and commercial databases. The application of VS depends mainly on the availability of the validated structural target (3-D structure). VS is categorized into either structural-based virtual screening (SBVS) or ligand-based virtual screening (LBVS).

12.3.1.1 Structure-based virtual screening
This approach is used in identifying the best ligand through searching to the chemical library for identifying its interaction with drug target, and it uses the 3-D structure of the protein which obtained either experimentally using X-ray crystallography, NMR or computational modeling [64]. Where the candidate is docked then ranked based on the binding affinity to the binding site [65].

12.3.1.2 Ligand-based virtual screening
The approach uses information obtained from the known ligand rather than structural protein for led identification as well as optimization, and it typically applied

when there is no 3-D structure of the protein. It depends on the pharmacophores and relies on the knowledge of the ligand that will bind the active site of the biological target. The primary goal is to come up with the structure which retains the physico-chemical properties. The approach is based on the principle that structurally similar molecules will always have similar properties [66].

12.3.2 Molecular docking

A method is an essential tool in the field of drug discovery and design. The main objective is to predict the best ligand's conformation in a target binding site/protein of known 3-D structure [67]. It concentrates much on either accuracy of the struc-ture or correct prediction activity. The algorithms and scoring function allows the evaluation of the interaction of compounds and potential targets. It starts from sim-ple, then advances to its complicated stage of the scoring function. It depends mainly on electrostatic and van der Waals of the interaction of solvation or entropic effects.

There are basic ways of representing the protein and ligands during the dock-ing process these ensure evaluation of their methods used, which include; atomic, surface, and grid [68]. Atomic representation is mainly for evaluating pair-wise atomic interaction which brings the complexity so it uses potential energy function [69]. Surface-based is used to minimize the angles of the opposite or different inter-acting molecules [70, 71]. Rigid body access the energy contribution of receptor spe-cifically on grid points which are used in ligand scoring and stores electrostatic and van der Waals (potentials) [72].

12.3.2.1 Searching methods for ligands

These are methods that allow molecular flexibility by focusing on the algorithms which treat ligands flexible but in a few cases, protein. These methods are divided into (a) systematic approach, (b) random or stochastic method and (c) shape matching.

12.3.2.1.1 Systematic searching

A method is used for flexible ligand docking whereby all number of possible confor-mations for ligand binding to the active site is measured, visiting the degrees of freedom of ligand. It can be considered in three approaches; Exhaustive, incremen-tal/fragment, and assembling of conformation. The exhaustive is done by rotating all bonds of the ligand at a specific time allocated. They generate good conformation. To avoid the exhaustive search, the screening is done initially for different poses, fil-tered and then optimized. It uses Glide [73, 74] and FRED [75] for the sampling method. The fragment method is used to avoid the combinatorial explosion. So the algorithm uses the fragments which may be generated by three steps.

(i) Core fragment selection (ii) Core fragment ligand placement (iii) Incremental ligand placement.

By incrementally growing of the ligand into the binding site at a specific time, it is covalently linked, the programs used include Dock [76], LUDI [77], FlexX [77], ADAM [78] and eHiTs [79]. The conformational ensemble methods search for the pregenerated ligand confirmation with the libraries. The binding mode is compared by ranking them by considering the scoring energy. Programs used include FLOG [80], DOCK 3.5 [81], PhDock [82], MS-DOCK [83], MDock [84, 85] and Q-Dock [86].

12.3.2.1.2 Random search (stochastic algorithms)

The algorithms consider the conformation of ligands at the active site, whether individually or populated. It examines the translational/rotational space and conformational space of ligand. The approach includes;

The first is the Monte Carlo method (MC); the algorithms allows the generation of random conformation with translation and rotation at an initial stage of ligands docking at the active site. The new configuration is generated after scoring of the initial one and the probability of being accepted is achieved by considering the Boltzmann probability function,

$$P \sim exp\left[\frac{-(E_1 - E_0)}{k_B T}\right] \tag{12.1}$$

where E_O and E_1 are the energy scores of the ligand before and after the random change.

Respectively, k_B is the Boltzmann constant, and T is the absolute temperature of the system. MC uses programs like DockVision [87], ICM [88], Prodock [89] and MCDOCK [90].

The second approach is genetic/evolutionary algorithms, and this uses the approach of biological competition and population dynamics. The varying parameters are included in the chromosome and randomly varied. The result produced is evaluated by its fitness. The chromosomes that produce optimal characteristics are crossed to produce the next generation. This uses GOLD [91, 92], AutoDock [93], DIVALI [94], DARWIN [95], MolDock [96], PSI-DOCK [97], FLIPDock [98], Lead finder [99] and EADock [100].

The third approach is Tabu search algorithms, and this considers the Tabu (those rejected conformation). It operates by making random changes on available conformations then each change is ranked. The Tabu is determined, and their changes that have the lower value are going to be accepted even if it was in tabu otherwise the non-tabu change is accepted. The program uses PRO_LEADS [101] and PSI-DOCK [97].

12.3.2.1.3 Shape matching

This method is used at the initial stage of docking, and it is the simplest approach. It places the ligand in the position that its molecular surface is in complementary with the surface of the binding site of the protein involves translation and rotational which allow different orientations of ligands at the binding site. So, it mainly looks at which the ligands will be easily placed at the binding site as quickly as possible. Programs used include DOCK [102], FRED [103], FLOG [104], LigandFit [105], Surflex [106] MS-Dock [67] and MDock [107, 108].

12.3.2.2 Scoring functions

For determining the accuracy of the docking algorithm, it is important to consider the scoring function. So it is the fundamental element for the protein-ligand algorithm. It looks at the reliability and efficiency of any algorithm. The scoring functions can be grouped into three categories: Force field, empirical and knowledge-based scoring functions.

12.3.2.2.1 Force field scoring function

It uses the molecular mechanics' force field to consider the interaction between the ligand and receptor. Basically, the Van der Waals energies, the bond bending/stretching/torsional energies, are used at this strategy. The program which can be used here include AMBER [109] or CHARMM [110, 111]. The effect of water for the force field is accounted for by the inclusion of distance-dependent dielectric constant $\varepsilon(r_{ij})$, which uses a program like DOCK [112] and implements eq. (12.2) below.

$$E = \sum_i \sum_j \left(\frac{A_{ij}}{r_{ij}^{12}} - \frac{B_{ij}}{r_{ij}^6} + \frac{q_i q_j}{\varepsilon(r_{ij}) r_{ij}} \right) \tag{12.2}$$

where r_{ij} stands for the distance between protein atom i and ligand atom j, A_{ij} and B_{ij} are the VDW parameters, and q_i and q_j are the atomic charges. $\varepsilon(r_{ij})$ is usually set to $4r_{ij}$, reflecting the screening effect of water on electrostatic interactions.

12.3.2.2.2 Empirical scoring function

This method is useful for reproducing the experimental data by summation of the binding energy of ligand to a receptor such as VDW, electrostatic, hydrogen bonding, desolvation, entropy and hydrophobicity represented by eq. (12.3) below:

$$\Delta G = \sum_i W_i \cdot \Delta G_i \tag{12.3}$$

where $\{\Delta G_i\}$ is for individual empirical energy and $\{W_i\}$ produced by the binding affinity after training the complex ligand-protein from the known 3-D structures. So, this method mainly depends on the information from the crystal structure of different protein-ligand complex whose binding affinities are known. It uses programs like Glide Score [74, 113], LUDI [114–117], SCORE [118–120] X-SCORE [121], ChemScore [122, 123], SFscore [124, 125].

12.3.2.2.3 Knowledge-based scoring function

These are scoring functions which aim at reproducing the experimental structures of protein-ligand complexes. It uses the potential of mean force which is given by inverse Boltzmann eq. (12.4).

$$w(r) = -k_B T \ln\left[\rho(r)/\rho^*(r)\right] \tag{12.4}$$

where k_B is the Boltzmann constant, T is the absolute temperature of the system, $\rho(r)$ is the number density of the protein-ligand atom pair at distance r in the training set, and $\rho^*(r)$ is the pair density in a reference state where the interatomic interactions are zero. Different from the field and empirical scoring function have a right balance between the accuracy and speed because the potentials from eq. (4) above are obtained from a larger number of structures rather than generating the known affinity, it uses a programs like PMF [126, 127], DrugScore [128], SMOG [129].

12.3.2.2.4 Consensus scoring function

The approach is not a specific type of scoring function; it combines and balances the scoring information by removing/minimizing errors of each scoring function. So it makes the true binder to be distinguished from the others [130, 131]. Programs used include X-Score and MultiScore [132, 133].

12.3.2.2.5 Clustering and entropy-based scoring methods

This method also used to improve the performance of different scoring functions. It includes the entropic effect by taking the ligand binding modes then divided into various clusters [134, 135] The impact of entropic contribution in each of the clusters is considered, and it uses AutoDock [135, 136].

A case study for docking was reported by [21], after homology modeling and identification of the structure of Thai rosewood dalcochinase which have high similarity to linamarase. The natural substrate (Dal-Glc) of dalcochinase were docked into the active site pocket of dalcochinase obtained using homology modeling, the docking program used was GOLD 3.1 [137]. Maize β-glucosidase was used by docking with DIMBOA-β-D-glucoside, which achieved a fitness score of 69.15, and when as compared with the fitness of the score for docking Dal-Glc into the dalcochinase it gave score 61.41. The interaction of the substrate positioned it by stacking π–π interaction between the phenyl ring of substrate and indole ring of the conserved Trp368. The distance between amino acid residues and Dal-Glc (substrate) specifically in the binding pocket of dalcochinase model was predicted by molecular docking as summarized in Table 12.1.

Table 1.1: The binding pocket of dalcochinase model showing the distances between Dal-Glc and amino acid residues obtained by molecular docking, (adapted from reference [21]).

Dal-Glc position	Amino acid position	Distance (Å)
Sugar ring		
O2	Y325-HH	2.49
O3	N181-HD22	2.89
O4	W453-He1	1.48
O6	W445-He1	2.16
H at O2	E396-Oe1	1.72
C		
Methylene	H235-He1	2.34
Methylene	W368-Hz3	2.51
Ring-A		
Ring-B		π–π (3.52–4.09)
Ring-C		
OH	W368-He1	2.35

In different studies [22] and [21], the authors considered the single mutation, whereby residues Ile185, Asn189, and Val255 in dalcochinase binding pocket were replaced by Ala201, Phe205, and Phe271 from linamarase consider Figure 12.2. This affected dalcochinase by increasing transglycosylation, mainly primary and secondary alcohol. However, this replacement of single mutation did not change dalcochinase activities to be similar to that of linamarase specifically transglycosylation to tertiary alcohol.

To bring dalcochinase specificity to linamarase multiple mutations of the identified key amino acid residues. The homology models of the 1185A/N189F/V255F which considered a triple mutation of the respective enzyme were obtained and docked to linamarin which its structure was driven from PubChem (CID 11128) using the AutoDock version 4.2 [138]. The analysis of the docked conformation was analyzed by Accelrys DsVisualizer 3.0 consider Figure 12.2.

The results of docking of these multiple mutations show that for both enzymes when were docked with linamarin the covalent bond glycosidic and the C1 atom are proposed to be at the same location.

12.4 Conclusion

The enzymatic reaction of linamarase is associated with application in biotechnology like transglycosylation and hydrolysis by using acceptors. The most important and which has more concern is in cassava which the hydrolysis of cyanogenic glucoside and ultimately leads to the release of toxic product hydrogen cyanide. To get the insight of linamarase's potential and how can be used to different productive reaction, there is a need to elucidate the structure experimentally. However, due to the improvement of computational approaches in carrying out structural modeling, this work has reviewed and analyzed the approaches to be used predict the structure of linamarase. Homology modeling is mainly used then other techniques can be applied to identify the competitive inhibitor against the natural substrate and inhibits the enzymatic reaction. These are mainly molecular docking and molecular dynamics. In homology modeling of dalcochinase, it has proved that the enzyme shares 47% amino acid sequence with dalcochinase, which its structure is now available. So it can be used as a template to generate the 3-D structure of linamarase.

Regardless, of these similarities the enzymes they have the different catalytic ability as linamarase works by hydrolyzing the natural substrate linamarin and dalcochinase the substrate dalcochinin-8-O-β-D-glucoside, but not the reverse. Their distinct also is that dalcochinase can catalyze the transglycosylation of primary and secondary alcohol but not tertiary while linamarse can work effectively on both. Additionally, linamarase cannot synthesize oligosaccharides and glycosides by reverse hydrolysis while dalcochinase can do.

Therefore, regardless of the availability of amino acid sequences similar to linamarase, there is a great need to find out the crystal structure experimentally. This will unlock more scientific understanding, research, and application of linamarase especially in food processing like products with glycosides residues retained during the processing.

Acknowledgements: The authors are indebted to the Nelson Mandela African Institutional of Science and Technology (NM-AIST) through African Development Bank (AfDB) project for financial support. FNK acknowledges a return fellowship and equipment subsidy from the Alexander von Humboldt foundation, Germany.

References

[1]   Andama M, Oloya B. Effectiveness of traditional processing techniques in reducing cyanide levels in selected cassava varieties in Zombo District, Uganda. Int J Food Sci Biotechnol. 2017;2:121–5.
[2]   Montagnac JA, Davis CR, Tanumihardjo SA. Processing techniques to reduce toxicity and antinutrients of Cassava for use as a staple food. Compr Rev Food Sci Food Saf. 2009;8:17–27.
[3]   Burns AE, Bradbury JH, Cavagnaro TR, Gleadow RM. Journal of food composition and analysis total cyanide content of cassava food products in Australia. J Food Compos Anal. 2012;25:79–82.

[4]   Nicolau AI. Safety of fermented cassava products. In: Prakash V, Martín-Belloso O, Keener L, Astley S, Braun S, McMahon H, Lelieveld H, editors. Regulating safety of traditional and ethnic foods. London, Oxford, Boston, New York und San Diego: Academic Press, 2016:319–35.

[5]   AttahDaniel BE, Ebisike K, Adeeyinwo CE, Ojumu TV, Olusunle SO, Adewoye OO. Towards arresting the harmful effect of cyanogenic potential of cassava to man in the environment. Int J Eng Sci. 2013;2:100–4.

[6]   Nartey F. Manihot Esculemta (Cassava): cyanogenesis, ultrastructure and seed germination. Australia: Copenhagen, Munksgaard, 1978.

[7]   Mcmahon JM, White WL, Sayre RT. Review article: cyanogenesis in cassava (manihot esculenta crantz). J Exp Bot. 1995;46:731–41.

[8]   Mkpong OE, Yan H, Chism G, Sayre RT. Purification, characterization, and localization of linamarase in cassava. Plant Physiol. 1990;93:176–81.

[9]   White WL, Arias-Garzon DI, McMahon JM, Sayre RT. Cyanogenesis in cassava: the role of hydroxynitrile lyase in root cyanide production. Plant Physiol. 1998;116:1219–25.

[10]   Tylleskar T, Banea M, Bikangi N, Cooke RD, Poulter NH, Rosling H. Cassava cyanogens and konzo, an upper motoneuron disease found in Africa. Lancet. 1992;339:208–11.

[11]   Cooke RD. An enzymatic assay for the total cyanide content of cassava (manihot esculenta crantz). J Sci Food Agric. 1978;29:345–52.

[12]   Brien GM, Taylor AJ, Poulter NH. Improved enzymic assay for cyanogens in fresh and processed cassava. J Sci Food Agric. 1991;56:277–89.

[13]   Frankenberg L. Enzyme therapy in cyanide poisoning: effect of rhodanese and sulfur compounds. Arch Toxicol. 1980;45:315–23.

[14]   Food Standards Australia New Zealand, Cassava and bamboo shoots; a human health risk assessment. 2005.

[15]   Cardoso AP, Ernesto M, Cliff J, Egan SV, Bradbury JH. Cyanogenic potential of cassava flour: field trial in Mozambique of a simple kit. Int J Food Sci Nutr. 1998;49:93–9.

[16]   Rosling H. Measuring effects in humans of dietary cyanide exposure from cassava. International Society for Horticultural Science (ISHS), Nov. 1994.

[17]   Solomonson LP. Cyanide as a metabolic inhibitor. Cyanide Biol. 1981;2013:11–28.

[18]   Keresztessy Z, Kiss L, Hughes MA. Investigation of the active site of the cyanogenic β-D-glucosidase (linamarase) from Manihot esculenta Crantz (cassava). II. Identification of Glu-198 as an active site carboxylate group with acid catalytic function. Arch Biochem Biophys. 1994;315:323–30.

[19]   Svasti J, Phongsak T, Sarnthima R. Transglucosylation of tertiary alcohols using cassava β-glucosidase. Biochem Biophys Res Commun. 2003;305:470–5.

[20]   Srisomsap C, Subhasitanont P, Techasakul S, Surarit R, Svasti J. Synthesis of homo- and hetero-oligosaccharides by Thai rosewood β- glucosidase. Biotechnol Lett. 1999;21:947–51.

[21]   Kongsaeree PT, Ratananikom K, Choengpanya K, Tongtubtima N, Sujiwattanarat P, Porncharoennop C, et al. Substrate specificity in hydrolysis and transglucosylation by family 1 β-glucosidases from cassava and Thai rosewood. J Mol Catal B Enzym. 2010;67:257–65.

[22]   Tongtubtim N, Thenchartanan P, Ratananikom K, Choengpanya K, Svasti J, Kongsaeree PT. Multiple mutations in the aglycone binding pocket to convert the substrate specificity of dalcochinase to linamarase. Biochem Biophys Res Commun. 2018;504:647–53.

[23]   Dorn M, Silva MB, Buriol LS, Lamb LC. Three-dimensional protein structure prediction: methods and computational strategies. Comput Biol Chem. 2014;53:251–76.

[24]   Anfinsen CB, Haber E, Sela M, White FH. The kinetics of formation of native ribonuclease during oxidation of the reduced polypeptide chain. Proc Natl Acad Sci USA. 1961;47:1309–14.

[25]   Anfinsen CB. Principles that govern protein folding Publication;. Science. 1973;181:223–30.

[26]   Büssow KA. Protein structure prediction. Concepts and applications. Anal Bioanal Chem. 2006;386:1579–80.

[27]   Bujnicki JM. Protein-structure prediction by recombination of fragments. ChemBioChem. 2006;7:19–27 1579–80.

[28]   Osguthorpe DJ. Ab initio protein folding. Curr Opin Struct Biol. 2000;10:146–52.

[29]   Case DA, Cheatham III TE, Darden T, Gohlke H, Luo R, Merz Jr KM, et al. The Amber biomolecular simulation programs. J Comput Chem. 2005;26:1668–88.

[30]   Pearlman DA, Case DA, Caldwell JW, Ross WS, Cheatham III TE, DeBolt S, et al. AMBER, a package of computer programs for applying molecular mechanics, normal mode analysis, molecular dynamics and free energy calculations to simulate the structural and energetic properties of molecules. Comput Phys Commun. 1995;91:1–41.

[31]   Best RB, Zhu X, Shim J, Lopes PE, Mittal J, Feig M, et al. Optimization of the additive CHARMM all-atom protein force field targeting improved sampling of the backbone ϕ, ψ and side-chain χ 1 and χ 2 dihedral Angles. J Chem Theory Comput. 2012;8:3257–3273.

[32]   Liwo A, Kaźmierkiewicz R, Czaplewski C, Groth M, Ołdziej S, Wawak RJ, et al. United-residue force field for off-lattice protein-structure simulations: III. Origin of backbone hydrogen-bonding cooperativity in united-residue potentials. J Comput Chem. 1998;19:259–76.

[33]   Pronk S, Páll S, Schulz R, Larsson P, Bjelkmar P, Apostolov R, et al. GROMACS 4.5: A high-throughput and highly parallel open source molecular simulation toolkit. Bioinformatics. 2013;29:845–54.

[34]   Kundrot CE, Ponder JW, Richards FM. Algorithms for calculating excluded volume and its derivatives as a function of molecular conformation and their use in energy minimization. J Comput Chem. 1991;12:402–9.

[35]   Floudas CA, Fung HK, McAllister SR, Mönnigmann M, Rajgaria R. Advances in protein structure prediction and de novo protein design: a review. Chem Eng Sci. 2006;61:966–88.

[36]   Simons KT, Kooperberg C, Huang E, Baker D. Assembly of protein tertiary structures from fragments with similar local sequences using simulated annealing and Bayesian scoring functions. J Mol Biol. 1997;268:209–25.

[37]   Zhang Y, Skolnick J. SPICKER: A clustering approach to identify near-native protein folds. J Comput Chem. 2004;25:865–71.

[38]   Moult J, Fidelis K, Kryshtafovych A, Schwede T, Tramontano A. Critical assessment of methods of protein structure prediction (CASP)—Round XII. Proteins Struct Funct Bioinforma. 2018;86:7–15.

[39]   Finkelstein AV, Ptitsyn OB. Why do globular proteins fit the limited set of folding patterns? Prog Biophys Mol Biol. 1987;50:171–90.

[40]   Levitt M, Chothia C. Structural patterns in globular proteins. Nature. 1976;261:552–8.

[41]   Setubal JC, Meidanis J. Introduction to computational molecular biology. Boston: PWS Pub, 1997.

[42]   Berman HM, Battistuz T, Bhat TN, Bluhm WF, Bourne PE, et al. The protein data bank. Acta Crystallogr Sect D Biol Crystallogr. 2002;58:899–907.

[43]   Bajorath J, Stenkamp R, Aruffo A. Knowledge-based model building of proteins: concepts and examples. Protein Sci. 1993;2:1798–810.

[44]   Sternberg MJ, Thornton JM, Blundell TL, Sibanda BL. Knowledge-based prediction of protein structures and the design of novel molecules. Nat Int J Sci. 1987;326:347–52.

[45]   Kopp J, Schwede T. Automated protein structure homology modeling: a progress report. Pharmacogenomics. 2004;5:405–16.

[46]   Apostolico A, Giancarlo R. Sequence alignment in molecular biology. J Comput Biol. 1998;5:173–96.

[47]   Pearson WR, Lipman DJ. Improved tools for biological sequence comparison. Proc Natl Acad Sci USA. 1988;85:2444–8.

[48]   Lipman DJ, Pearson WR. Rapid and sensitive protein similarity searches. Science (80-.). 1985;227:1435–41.

[49]   Altschul SF, Madden TL, Schäffer AA, Zhang J, Zhang Z, Miller W, et al. Gapped BLAST and PSI-BLAST: a new generation of protein database search programs. Nucleic Acids Res. 1997;25:3389–402.

[50]   Altschup SF, Gish W, Pennsylvania T, Park U. Basic local alignmen. J Mol Biol. 1990;215:403–10.

[51]   Notredame C. Recent progresses in multiple sequence alignment: a survey. Pharmacogenomics. 2002;3:131–44.

[52]   Notredame C. Recent evolutions of multiple sequence alignment algorithms. PLoS Comput Biol. 2007;3:1405–8.

[53]   Thompson JD, Plewniak F, Poch O. A comprehensive comparison of multiple sequence alignment programs. Nucleic Acids Res. 1999;27:2682–90.

[54]   Thompson JD, Higgins DG. CLUSTAL W: improving the sensitivity of progressive multiple sequence alignment through sequence weighting, position-specific gap penalties and weight matrix choice. Nucleic Acids Res. 1994;22:4673–80.

[55]   Notredame C, Higgins DG, Heringa J. T-Coffee: a novel method for fast and accurate multiple sequence alignment. J Mol Biol. 2000;302:205–17.

[56]   Czjzek M, Cicek M, Zamboni V, Bevan DR, Henrissat B, Esen A. The mechanism of substrate (aglycone) specificity in β-glucosidases is revealed by crystal structures of mutant maize β-glucosidase-DIMBOA, -DIMBOAGlc, and -dhurrin complexes. Proc Natl Acad Sci USA. 2000;97:13555–60.

[57]   Laskowski RA, MacArthur MW, Moss DS, Thornton JM. PROCHECK: a program to check the stereochemical quality of protein structures. J Appl Crystallogr. 1993;26:283–91.

[58]   Wiederstein M, Sippl MJ. ProSA-web: interactive web service for the recognition of errors in three-dimensional structures of proteins. Nucleic Acids Res. 2007;35:407–10.

[59]   Kresge JS, Leonowicz CT, Roth ME, Vartuli WJ, Beck JC. Assessment of protein models with three-dimensional profiles. Nature. 1992;359:710–13.

[60]   Vriend G. WHAT IF: a molecular modeling and drug design program. J Mol Graph. 1990;8:52–6.

[61]   Ketudat Cairns JR, Champattanachai V, Srisomsap C, Wittman-Liebold B, Thiede B, Svasti J. Sequence and expression of Thai Rosewood beta-glucosidase/beta-fucosidase, a family 1 glycosyl hydrolase glycoprotein. J Biochem. 2000;128:999–1008.

[62]   Eswar N, Webb B, Marti-Renom MA, Madhusudan MS, Eramian D, Shen M, et al. Comparative protein structure modeling using Modeller. Curr Protoc BioinformaticsCurr. 2006;15:5.6.1–5.6.30.

[63]   Lavecchia A, Giovanni C. Virtual screening strategies in drug discovery: a critical review. Curr Med Chem. 2013;20:2839–60.

[64]   Li Q, Shah S. Structure-based virtual screening. Methods Mol Biol. 2017;1558:111–24.

[65]   Dror O, Shulman-peleg A, Nussinov R, Wolfson HJ. Predicting molecular interactions. Curr Med Chem. Curr. 2004;11:71–90.

[66]   Hamza A, Wei NN, Zhan CG. Ligand-based virtual screening approach using a new scoring function. J Chem Inf Model. 2012;52:963–74.

[67]   Kukol A. Molecular docking. In: Kahl G, editor(s). The dictionary of genomics, transcriptomics and proteomics, 5th ed. Vol. 443. Mannheim: Wiley-Blackwell, 2015:1–1.978-3-527-32852-9.

[68]   Halperin I, Ma B, Wolfson H, Nussinov R. Principles of docking: an overview of search algorithms and a guide to scoring functions. Proteins Struct Funct Genet. 2002;47:409–43.

[69] Taylor JS, Burnett RM. DARWIN: A program for docking flexible molecules. Proteins Struct Funct Genet. 2000;41:173–91.

[70] Norel R, Lin SL, Wolfson HJ, Nussinov R. Shape complementarity at protein–protein interfaces. Biopolymers. 1994;34:933–40.

[71] Norel R, Petrey D, Wolfson HJ, Nussinov R. Examination of shape complementarity in docking of unbound proteins. Proteins. 1999;36:307–17.

[72] Goodford PJ. A computational procedure for determining energetically favorable binding sites on biologically important macromolecules. J Med Chem. 1985;28:849–57.

[73] Friesner RA, Banks JL, Murphy RB, Halgren TA, Klicic JJ, Mainz DT, et al. Glide: A new approach for rapid, accurate docking and scoring. 1. Method and assessment of docking accuracy. J Med Chem. 2004;47:1739–49.

[74] Halgren TA, Murphy RB, Friesner RA, Beard HS, Frye LL, Pollard WT, et al. Glide: A new approach for rapid, accurate docking and scoring. 2. Enrichment factors in database screening. J Med Chem. 2004;47:1750–9.

[75] McGann MR, Almond HR, Nicholls A, Grant JA, Brown FK. Gaussian docking functions. Biopolymers. 2003;68:76–90.

[76] Ewing TJ, Kuntz ID. Critical evaluation of search algorithms for automated molecular docking and database screening. J Comput Chem. 1997;18:1175–89.

[77] Bohm HJ. The computer program LUDI: a new method for the de novo design of enzyme inhibitors. J Comput Aided Mol Des. 1992;6:61–78.

[78] Mizutani MY, Tomioka N, Itai A. Rational automatic search method for stable docking models of protein and ligand. J Mol Biol. 1994;243:310–26.

[79] Zsoldos Z, Reid D, Simon A, Sadjad BS, Johnson AP. eHiTS: An innovative approach to the docking and scoring function problems. Curr Protein Pept Sci. 2006;7:421–35.

[80] Miller MD, Kearsley SK, Underwood DJ, Sheridan RP. FLOG: a system to select 'quasi-flexible' ligands complementary to a receptor of known three-dimensional structure. J Comput Aided Mol Des. 1994;8:153–74.

[81] Lorber DM, Shoichet BK. Flexible ligand docking using conformational ensembles Despite important successes. Protein Sci. 1998;7:938–50.

[82] Joseph-McCarthy D, Thomas BE, Belmarsh M, Moustakas D, Alvarez JC. Pharmacophore-based molecular docking to account for ligand flexibility. Proteins Struct Funct Genet. 2003;51:172–88.

[83] Sauton N, Lagorce D, Villoutreix BO, Miteva MA. MS-DOCK: accurate multiple conformation generator and rigid docking protocol for multi-step virtual ligand screening. BMC Bioinformatics. 2008;9:1–12.

[84] Di Costanzo L, Jr LV, Christianson DW. Ensemble docking of multiple protein structures: considering protein structural variations in molecular docking. Proteins. 2006;642:637–42.

[85] Huang SY, Zou X. Efficient molecular docking of NMR structures: application to HIV-1 protease. Protein Sci. 2006;16:43–51.

[86] Brylinski M, Skolnick J. Q-dock: low-resolution flexible ligand docking with pocket-specific threading restraints. J Comput Chem. 2008;29:1574–88.

[87] Thomsen R, Christensen MH. MolDock: a new technique for high-accuracy molecular docking. J Med Chem. 2006;49:3315–21.

[88] Benigni R, Bossa C. Mechanisms of chemical carcinogenicity and mutagenicity: a review with implications for predictive toxicology. Chem Rev. 2011;111:2507–36.

[89] McGann M. FRED and HYBRID docking performance on standardized datasets. J Comput Aided Mol Des. 2012;26:897–906.

[90] Chen R, Li L, Weng Z. ZDOCK: an initial-stage protein-docking algorithm. Proteins Struct Funct Genet. 2003;52:80–7.

[91] Kitchen DB, Decornez H, Furr JR, Bajorath J. Docking and scoring in virtual screening for drug discovery: methods and applications. Nat Rev Drug Discov. 2004;3:935–49.

[92] Brooijmans N, Kuntz ID. Molecular recognition and docking algorithms. Annu Rev Biophys Biomol Struct. 2003;32:335–73.

[93] Rishton GM. Reactive compounds and in vitro false positives in HTS. Drug Discov Today. 1997;2:382–4.

[94] Moitessier N, Englebienne P, Lee D, Lawandi J, Corbeil CR. Towards the development of universal, fast and highly accurate docking/scoring methods: a long way to go. Br J Pharmacol. 2008;153:7–26.

[95] Huang SY, Grinter SZ, Zou X. Scoring functions and their evaluation methods for protein-ligand docking: recent advances and future directions. Phys Chem Chem Phys. 2010;12:12899–908.

[96] Krovat E, Steindl T, Langer T. Recent advances in docking and scoring. Curr Comput Aided-Drug Des. 2006;1:93–102.

[97] Jain AN. Scoring functions for protein-ligand docking. Curr Protein Pept Sci. 2006;7:407–20.

[98] Seiler KP, George GA, Happ MP, Bodycombe NE, Carrinski HA, Norton S, et al. ChemBank: a small-molecule screening and cheminformatics resource database. Nucleic Acids Res. 2008;36:351–9.

[99] Meng EC, Shoichet BK, Kuntz ID. Automated docking with grid-based energy evaluation. J Comput Chem. 1992;13:505–24.

[100] Goodsell DS, Olson AJ. Automated docking of substrates to proteins by simulated annealing. Proteins Struct Funct Bioinforma. 1990;8:195–202.

[101] Shoichet BK, Leach AR, Kuntz ID. Ligand solvation in molecular docking. Proteins. 1999;34:4–16.

[102] Song CM, Lim SJ, Tong JC. Recent advances in computer-aided drug design. Brief Bioinform. 2009;10:579–91.

[103] Knox C, Law W, Jewison T, Liu P, Ly S, Frokis A, et al. DrugBank 3.0: A comprehensive resource for 'Omics' research on drugs. Nucleic Acids Res. 2011;39:1035–41.

[104] Gaulton A, Bellis LJ, Bento AP, Chambers J, Davies M, Hersey A, et al. ChEMBL: a large-scale bioactivity database for drug discovery. Nucleic Acids Res. 2012;40:1100–7.

[105] Del Rio A, Barbosa AJ, Caporuscio F, Mangiatordi GF. CoCoCo: a free suite of multiconformational chemical databases for high-throughput virtual screening purposes. Mol Biosyst. 2010;6:2122–8.

[106] Bissantz C, Folkers G, Rognan D. Protein-based virtual screening of chemical databases. 1. Evaluation of different docking/scoring combinations. J Med Chem. 2000;43:4759–67.

[107] Baell JB. Observations on screening-based research and some concerning trends in the literature. Future Med Chem. 2010;2:1529–46.

[108] Lagorce D, Maupetit J, Baell J, Sperandio O, Tufféry P, Miteva MA, et al. The FAF-Drugs2 server: a multistep engine to prepare electronic chemical compound collections. Bioinformatics. 2011;27:2018–20.

[109] Seifert MH. Robust optimization of scoring functions for a target class. J Comput Aided Mol Des. 2009;23:633–44.

[110] Charifson PS, Corkery JJ, Murcko MA, Walters WP. Consensus scoring: a method for obtaining improved hit rates from docking databases of three-dimensional structures into proteins. J Med Chem. 1999;42:5100–9.

[111] Feher M. Consensus scoring for protein-ligand interactions. Drug Discov Today. 2006;11:421–8.

[112] Raub S, Steffen A, Kämper A, Marian CM. AIScore - Chemically diverse empirical scoring function employing quantum chemical binding energies of hydrogen-bonded complexes. J Chem Inf Model. 2008;48:1492–510.

[113] Warren GL, Andrews CW, Capelli AM, Clarke M, LaLonde J, Lambert MH, et al. A critical assessment of docking programs and scoring functions. J Med Chem. 2006;49:5912–31.

[114] Willett P, Barnard JM, Downs GM. Chemical similarity searching. J Chem Inf Comput Sci. 1998;38:983–96.

[115] Ma X, Jia J, Zhu F, Xue Y, Li Z, Chen Y. Comparative analysis of machine learning methods in ligand-based virtual screening of large compound libraries. Comb Chem High Throughput Screen. 2009;12:344–57.

[116] Böhm HJ. The development of a simple empirical scoring function to estimate the binding constant for a protein-ligand complex of known three-dimensional structure. J Comput Aided Mol Des. 1994;8:243–56.

[117] Boehm HJ. Prediction of binding constants of protein ligands: a fast method for the prioritization of hits obtained from de novo design or 3D database search programs. J Comput Aided Mol Des. 1998;12:309–23.

[118] Melville JL, Burke EK, Hirst JD. Machine learning in virtual screening. Comb Chem High Throughput Screen. 2009;12:332–43.

[119] Wang R, Liu L, Lai L, Tang Y. SCORE: a new empirical method for estimating the binding affinity of a protein-ligand complex. J Mol Model. 1998;4:379–94.

[120] Tao P, Lai L. Protein ligand docking based on empirical method for binding affinity estimation. J Comput Aided Mol Des. 2001;15:429–46.

[121] Wang R, Lai L, Wang S. Further development and validation of empirical scoring functions for structure-based binding affinity prediction. J Comput Aided Mol Des. 2002;16:11–26.

[122] Eckert H, Vogt I, Bajorath J. Mapping algorithms for molecular similarity analysis and ligand-based virtual screening: design of DynaMAD and comparison with MAD and DMC. J Chem Inf Model. 2006;46:1623–34.

[123] Eldridge MD, Murray CW, Auton TR, Paolini GV, Mee RP. Empirical scoring functions: I. The development of a fast empirical scoring function to estimate the binding affinity of ligands in receptor complexes. J Comput Aided Mol Des. 1997;11:425–45.

[124] Barreiro G, Guimarães CR, Tubert-Brohman I, Lyons TM, Tirado-Rives J, Jorgensen WL. Search for non-nucleoside inhibitors of HIV-1 reverse transcriptase using chemical similarity, molecular docking, and MM-GB/SA scoring. J Chem Inf Model. 2007;47:2416–28.

[125] Rarey M, Kramer B, Lengauer T, Klebe G. A fast flexible docking method using an incremental construction algorithm. J Mol Biol. 1996;261:470–89.

[126] Muegge I. A knowledge-based scoring function for protein-ligand interactions: probing the reference state. Perspect Drug Discov Des. 2000;20:99–114.

[127] Muegge I, Martin YC. A general and fast scoring function for protein-ligand interactions: a simplified potential approach. J Med Chem. 1999;42:791–804.

[128] Gohlke H, Hendlich M, Klebe G. Knowledge-based scoring function to predict protein-ligand interactions. J Mol Biol. 2000;295:337–56.

[129] DeWitte RS, Ishchenko AV, Shakhnovich EI. SMoG: de novo design method based on simple, fast, and accurate free energy estimates. 2. Case studies in molecular design. J Am Chem Soc. 1997;119:4608–17.

[130] Smith RD, Hu L, Falkner JA, Benson ML, Nerothin JP, Carlson HA. Exploring protein-ligand recognition with binding MOAD. J Mol Graph Model. 2006;24:414–25.

[131] Okuno Y, Tamon A, Yabuuchi H, Niijima S, Minowa Y, Tonomura K, et al. GLIDA: GPCR - Ligand database for chemical genomics drug discovery - Database and tools update. Nucleic Acids Res. 2008;36:907–12.

[132] Mcgaughey GB, Sheridan RP, Bayly CI, Culberson JC, Kreatsoulas C, Lindsley S, et al. Comparison of topological, shape, and docking methods in virtual screening. J Chem Inf ModelJ . 2007;47:1504–19.

[133] Krejsa CM, Horvath D, Rogalski SL, Penzotti JE, Mao B, Barbosa F, et al. Predicting ADME properties and side effects: the BioPrint approach. Curr Opin Drug Discov Devel. 2003;6:470–80.

[134] Lin X, Huang XP, Chen G, Whaley R, Peng S, Wang Y, et al. Life beyond kinases: structure-based discovery of sorafenib as nanomolar antagonist of 5-HT receptors. J Med Chem. 2012;55:5749–59.

[135] Corbeil CR, Englebienne P, Moitessier N. Docking ligands into flexible and solvated macromolecules. 1. Development and validation of FITTED 1.0. J Chem Inf Model. 2007;47:435–49.

[136] Wang R, Fang X, Lu Y, Yang CY, Wang S. The PDBbind database: methodologies and updates. J Med Chem. 2005;48:4111–19.

[137] Verdonk ML, Cole JC, Hartshorn MJ, Murray CW, Taylor RD. Improved protein-ligand docking using GOLD. Proteins. 2003;52:609–23.

[138] Morris GM, Huey R, Lindstrom W, Sanner MF, Belew RK, Goodsell DS, et al. AutoDock4 and AutoDockTools4: automated docking with selective receptor flexibility. J Comput Chem. 2009;30:2785–91.

Part V: **Tutorials and Glossary**

Fidele Ntie-Kang, Kiran K. Telukunta, Linda Jahn
and Jutta Ludwig-Müller

13 Searching natural product databases

Abstract: This article was originally conceived as a set of tutorial exercises for undergraduate biology majors for a natural products course at the Institute of Botany, Technische Universität Dresden (2020/21 academic year). The overall goals of these exercises are to identify different compound databases (open access and commercial) from which compounds could be searched for academic reasons. The reader will be Initiated to basic tools for querying compound databases (including searching for molecular properties, similarity searching and sub-structure searching), the concept of chemical scaffolds and the familiarity with the major chemical scaffolds encountered in the natural products lectures. In addition, the read should be able to identify two secondary metabolite databases and use the web to search them, use of the web browser for querying compound databases (including searching for molecular properties, similarity searching and sub-structure searching). Application of the concept of chemical scaffolds and the familiarity with the major chemical scaffolds encountered in natural products lectures is expected. The last set of tutorial exercises are intended to identify databases containing taxonomic information on plants and animal species, use a web browser to search the contents of two open access databases that help to establish the taxonomy of medicinal plants. This includes establishing a clear definition of chemotaxonomy by use of examples. e.g. employ the ANPDB to establish the relationship between a selected list of plant families and their chemical content.

Keywords: chemotaxonomy, databases, drug discovery, natural products, scaffolds

13.1 Introduction

In order to get familiar with the theoretical background of the material in this article, the reader is advised to consult previous chapters of this book series, e.g. chapter 1 of volume I. This tutorial chapter is work estimated for one full day. The reader is, however, free to organize their time flexibly. The times set on each session should only serve as guidelines. It begins with a literature assignment, which the reader is expected to complete by reading through the referenced article and summarize in 2 pages maximum (A4 size, Times New Roman font 12 double spacing).

https://doi.org/10.1515/9783110668896-013

13.1.1 Literature assignment

Read and summarize the following article:

Ntie-Kang F, Telukunta KK, Döring K, Simoben CV, Moumbock AF, Malange YI, et al. NANPDB: A resource for natural products from Northern African sources. J Nat Prod. 2017;80:2067–2076. DOI: 10.1021/acs.jnatprod.7b00283.

URL: https://pubmed.ncbi.nlm.nih.gov/28641017/

13.1.2 Plan of the day

09:00–09:30 General Introduction, FAQ session
09:30–10:30 Session I (Compound Databases: ZINC, PubChem)
10:30–12:30 Session II (Secondary Metabolite Databases: SuperNatural II, ANPDB)
12:30–13:30 Lunch Break
13:30–15:30 Session III (Taxonomy Databases: NCBI Taxonomy, Tropicos)
15:30–17:00 Session IV (Chemotaxonomy: ANPDB)

13.2 Session I (Compound databases) 1 hour

13.2.1 Objectives

- Identify different compound databases (open access and commercial) from which compounds could be searched for academic reasons.
- Initiation to basic tools for querying compound databases (including searching for molecular properties, similarity searching and sub-structure searching).
- The concept of chemical scaffolds and the familiarity with the major chemical scaffolds encountered in the natural products lectures.

13.1.1 ZINC database

1. The most used and most open access compound database (including both organic and inorganic molecules) is Zinc Is Not Commercial (simply ZINC), accessible via the internet browser on the link (https://zinc15.docking.org/).
2. Take 5 minutes to read by clicking "Get Started", then "About ZINC 15 Resources".
3. Answer the following questions:
   a. What is the most current version of ZINC?
   b. About how many substances have been deposited in ZINC15?
   c. About how many genes have been annotated with ZINC15 compounds?

    d. About how many types of rings have been included in the compounds/substances in ZINC15?

    e. Which are the three most abundant rings?

    f. What percentage of Quinolines are included in all biogenic substances in ZINC15?

    g. Repeat the previous exercise (f) with all two member rings.

    h. Search for substances that include the Quinoline scaffold in ZINC15.
         Procedure: Go to the "Substances" menu. Sketch the Quinoline structure on the panel that appears (otherwise, you could simply copy the SMILES pattern from the previous exercise (f) and paste on this panel, i.e. from http://zinc15.docking.org/rings/, you copy the Quinoline SMILE, which is c1ccc2ncccc2c1 and paste on the https://zinc15.docking.org/substances/home/ panel, under "Search using one"). If you press "Enter" on your keyboard, a list of suppliers of this metabolite and its target class and biological activities appears.

4. To carry out a sub-structure search, return to the previous page (with Quinoline scaffold on the "substances" panel. Make sure, that you mean Quinone? scaffold appears on this panel). Under the "Search with" drop down menu beneath this panel, choose "Substructure".

    a. How many substances in ZINC15 include the Quinoline scaffold?

    b. Which other scaffold would you classify as a "privileged" scaffold?

13.2.1.2 PubChem database

1. On an internet browser, search PubChem (https://pubchem.ncbi.nlm.nih.gov/).

2. Search some basic information about the molecules Muscarin and Vocamine.

3. On an internet browser, search PubMed (http://www.ncbi.nlm.nih.gov/pubmed).

4. Search some basic information about the molecule Muscarin and Vocamine.

5. What is the basic difference between the information available from PubChem and that from PubMed?

6. Select the first article from PubMed on Vocamine and state the following information: journal name, title, volume, issue, page numbers, year of publication, month of publication, DOI, PMID.

7. Do the same thing for Quinoline on PubChem and PubMed.

8. Carry out a substructure search of substances in PubMed that include the Quinoline scaffold.
    Procedure: Go to the PubChem page and click on "Draw Structure". Draw the Quinoline structure or import the SMILES from the ZINC15 exercise and paste beside the SMILES drop down menu and press enter. Now click on the "Search this Structure" button beneath. A new page opens. On this new page, select the "Substructure" menu (click on it).

9. How many substances in PubChem include the Quinoline substructure?
10. One of these is hydroxychloroquine. What is the most important use of this compound?
11. Now, select hydroxychloroquine (PubChem CID: 3652). Browse through the physical properties, spectral data, vendor information and clinical data on this compound.
12. Under which chemical class would you place hydroxychloroquine?
   Further tutorials are available here (https://www.ncbi.nlm.nih.gov/guide/train ing-tutorials/).

13.3 Session II (Secondary metabolite databases) 2 hours

13.3.1 Objectives

- Identify two secondary metabolite databases and use the web to search them.
- The use of the web browser for querying compound databases (including searching for molecular properties, similarity searching and sub-structure searching).
- Application of the concept of chemical scaffolds and the familiarity with the major chemical scaffolds encountered in the natural products lectures

13.3.1.1 SuperNatural database

1. Open the main page of the SuperNatural II database on a new browser (http://bioinf-applied.charite.de/supernatural_new/index.php).
2. Click in the "Compounds" top menu. On the left menu, search for purchasable Quinolines in this database (under compound name type "Quinoline" and check that the "Purchasable" box is selected. Now click "Find"). How many of these exist?
3. Would you accept the statement that Quinoline is a "privileged scaffold"?
4. Carry out a similar exercise using the right menu (you can use the SMILES in session I or re-draw the structure).
5. Now, carry out a similarity search on the same structure. What major difference do you observe between the compounds in the similarity search and those of the substructure search outputs?
6. Identify one class of compounds that was encountered during the lectures. What is the scaffold of this chemical class?
7. Now draw this scaffold in ZINC, PubChem and SuperNatural II and identify the metabolites that contain the scaffold and those which are similar to it.

13.3.1.2 African natural products database

Open the main page of the African Natural Products Database (ANPDB) a new browser (http://african-compounds.org/anpdb/). Data currently uploaded come from two regions in Africa (Northern Africa and Eastern Africa).

The following sets of tutorials will help you browse the Northern African Natural Products Database (NANPDB), which is a subset of this database:

Outline
- Searches
  - Compounds/Bioactivity
  - Compound ID
  - Similarity
  - Substructure
  - Keyword

- Lists
  - Compounds
  - Families
  - Species

- Cards
  - Compound
  - Species

Searches

Compounds/Bioactivity

If you know the name of the compound of interest just type in the name or synonym.
1. In this field either a compound name, its synonym or a bioactivity can be given.
2. These are compounds which are similar to the given compound name or compounds which contain the queried bioactivity.
3. The number of compounds on the list indicates how many compounds or synonyms are found for the corresponding search. In order to view the output, you can click on the compound name on Field (2).
4. The number on Field (4) indicates the number of source species containing the corresponding molecule or synonym.

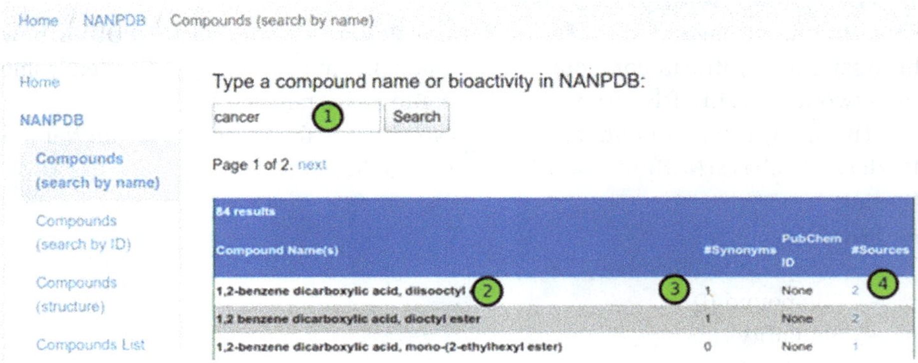

Figure 13.1: ANPDB search menus.

Compound ID search

1. After getting the compound ID from the method given in using downloads (link), it can be used in Field (1)

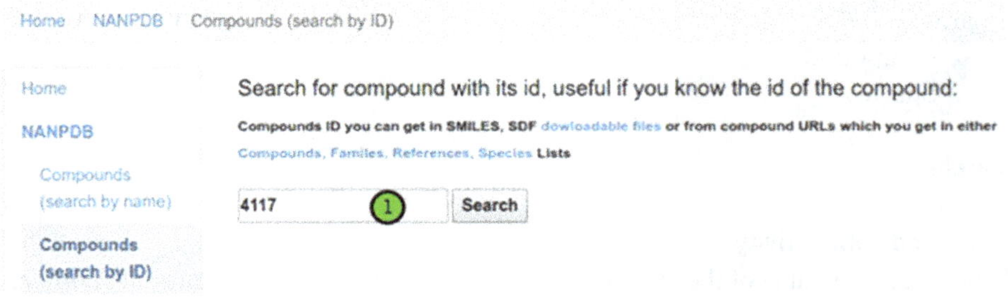

Figure 13.2: ANPDB search by compound ID.

Similarity search

1. Here you can draw the structure for which you want to do either similarity or the substructure search. The structure drawn on the Field (1) window is 2-hydroxy-benzoic acid. It can be drawn by picking the benzene ring on the top panel, followed by the single bond, the oxygen atom. The carboxylic group can be drawn by clicking on the single bond, then attaching the single bond, double bond and oxygen atoms, placing each one at its appropriate positions.
2. This automatically generates the SMILES string for the drawn molecule on Field (2).
3. In Field (3), you can give the preferred Tanimoto filter value, which is useful in similarity search.

4. In this list box Field (4) you can select either similarity or substructure search after drawing the desired molecule.
5. You then click "Search" and wait for the results.

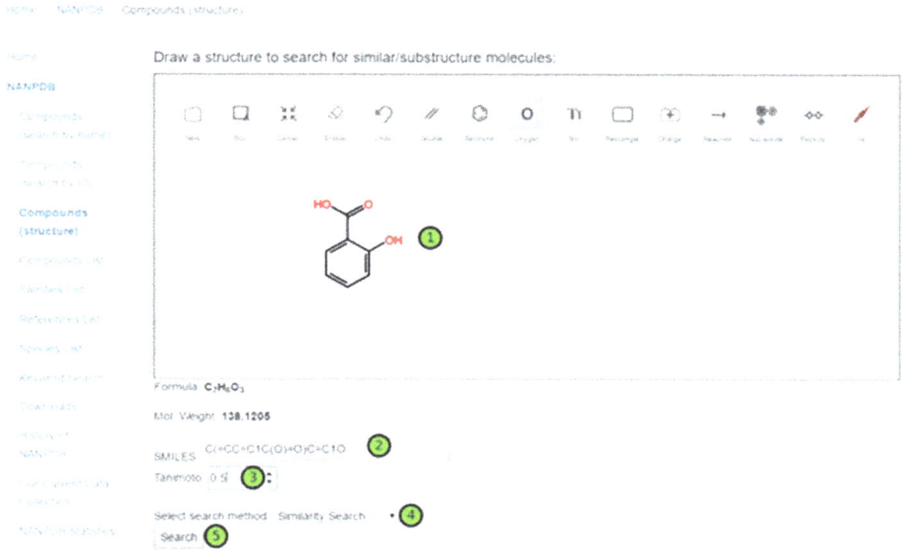

Figure 13.3: ANPDB search by similarity and substructure.

Substructure search

In the **Substructure search** module, you are able to search for molecules having a particular substructure or the fragment of a molecule. You may want to know all NANPDB molecules containing the chemical substructure. *Substructure search* has to be done in a similar way as the *Similarity search*. The only difference is that at the step 4 of the above search, you choose *Substructure search* instead of *Similarity search*.

Keyword search

1. In this field any keyword can be given to be searched in Field (1).
2. By clicking a compound in the column of Field (2), you are automatically linked to the compound card (link) of the corresponding compound.
3. The number on Field (3) indicates how many synonyms exist for the corresponding molecule. In order to view those synonyms click the number.

4. The number on Field (4) indicates from how many sources corresponding molecule or their synonyms are obtained. In order to view those sources click the number

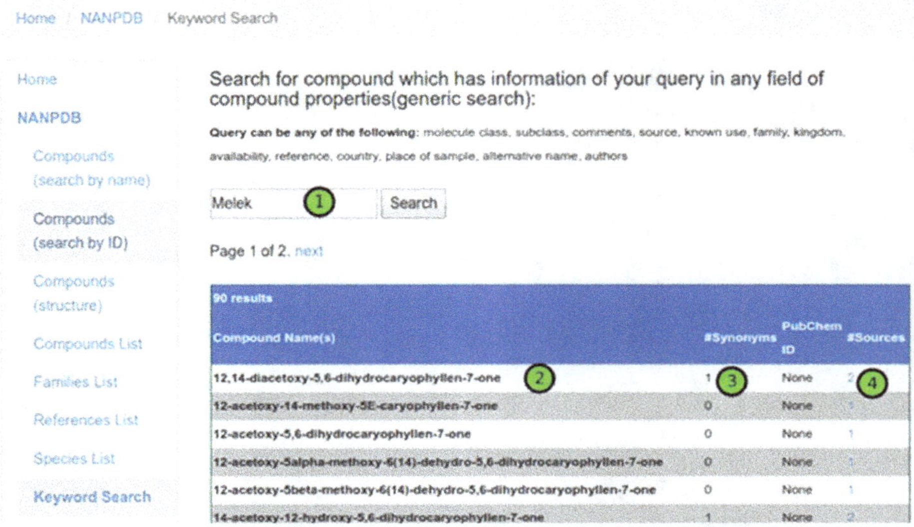

Figure 13.4: ANPDB fields in the compound search.

Lists

Compounds list

1. The NANPDB compounds have been arranged alphabetically and numerically, Field (1). By clicking any of the letters (A-Z) or numbers (1–9), compounds whose names begin with the corresponding letter or number are shown on the list.
2. You can now select or search for the compound of interest on the list in Field (2).
3. The number on Field (3) indicates how many synonyms are there for the corresponding molecule. This can be viewed by clicking on the number in Field (3).
4. If a PubChem ID of the corresponding compound is available it is shown on Field (4). PubChem reference can viewed by clicking number in Field(4)
5. The number on Field (5) indicates the number of source species containing a particular molecule or their synonyms.

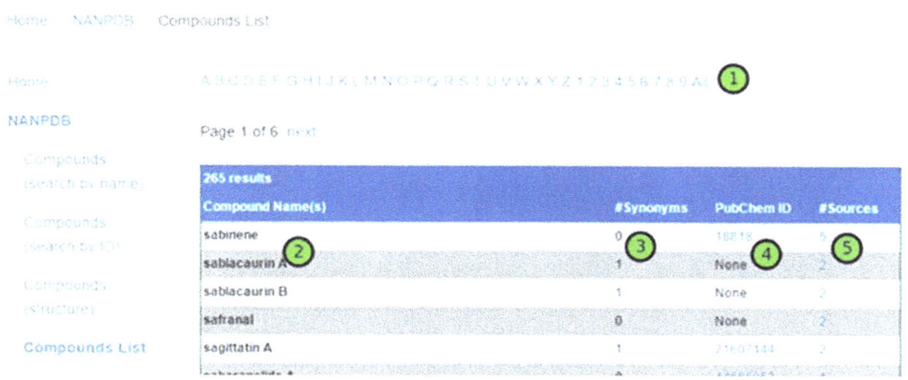

Figure 13.5: ANPDB compounds list.

Families list

1. The NANPDB species have been arranged alphabetically into their respective families, Field (1). By clicking any of the letters *(A-Z)* or *All.*
2. By clicking any of the first letter of the respective family names, you get an output, Field (2). Families for letter A are shown on the picture. By clicking on a particular family, all its corresponding species are shown.
3. Field (3) corresponds to the kingdom of the corresponding family.
4. The number on Field (4) indicates number of species in a particular family.
5. The number on Field (5) indicates number of compounds found in all source species of the corresponding family in the row.

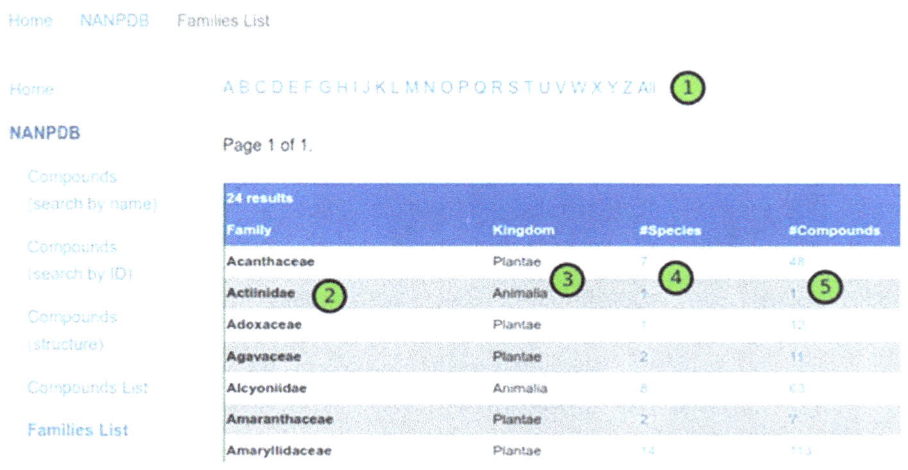

Figure 13.6: ANPDB families list.

Species list

1. The NANPDB species have also been arranged alphabetically, Field (1). By clicking any of the letters *(A-Z)* or *All*, species whose names begin with that letter will be shown on the results.
2. Species names of the selected letter of the alphabet are listed on Field (2). Upon clicking the species name, you will be provided with detailed information of the species in species card (link).
3. The respective kingdoms of the corresponding species are displayed on Field (3).
4. The respective families of the corresponding species are displayed on Field (4).
5. Field (5) shows the number of compounds contained in the corresponding species.
6. Field (6) provides the link to the wiki page describing the species, if available.

Figure 13.7: ANPDB species list.

Cards

Compound Card

1. The chemical structure of the respective compound is shown at the top of the compound card. This is followed by the SMILES string and the compound synonyms, Field (1). You will see that the number of synonyms of the given compound are displayed along with their source species information. You can click on either the highlighted synonym name or its species names (in blue) to access the additional information.

Home NANPDB

Home

NANPDB

Details of quercetagetin-3,6-dimethyl ether

Image:

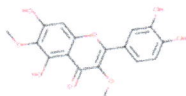

SMILES: COc1c(O)cc2c(c1O)c(=O)c(c(o2)c1ccc(c(c1)O)O)OC

Synonyms: 5,7,3',4'-tetrahydroxy-3,6-dimethoxyflavone from *Cleome species*
axillarin from *Madia divaricata*

PubChem:

Properties	MW: 346.293	cLogP: 1.382	HBD: 3.0
	HBA: 6.0	NRB: 6.0	logBB: -1.828
	logKp: -4.368	logHERG: -4.81	#metab: 6
	TPSA: 220.305	logS: -3.382	logKhsa: -0.145
	Lipinski Violations: 0	Comment: No comment	Molecule Class: Flavonoid
	Molecule Subclass: flavonol aglycone		

Source Species Information
- **Source:** *Francoeuria crispa*
- **Known use:** None reported
- **Family:** Compositae-Asteraceae
- **Kingdom:** Plantae
- **Availability of source sample:** Herbarium of Botany Department, Faculty of Science, Mansoura University (Egypt)
- **Reference:** no reference
- **Country:** Egypt
- **Place of sample collection:** Elkuraiemat-Elzafarana road
- **GPS Location:** Unavailable
- **Collected on:** April 1, 1987
- **Taxonomy:** link to taxonomic data
- **Wikipedia:** None
- **Additional Information:** Reported in literature source as April 1987

Predicted toxicity from pkCSM
- **AMES toxicity:** No (Yes/No)
- **Max. tolerated dose (human):** 0.627 (log mg/kg/day)
- **hERG I inhibitor:** No (Yes/No)
- **hERG II inhibitor:** Yes (Yes/No)
- **Oral Rat Acute Toxicity (LD50):** 2.27 (mol/kg)
- **Oral Rat Chronic Toxicity (LOAEL):** 1.546 (mg mg/kg_bw/day)
- **Hepatotoxicity:** No (Yes/No)
- **Skin Sensitisation:** No (Yes/No)
- **T.Pyriformis toxicity:** 0.288 (mg ug/L)
- **Minnow toxicity:** 2.113 (log mM)

Reference information
- **Type:** Journal article
- **Reference:** Phytochemistry 1990,29(8),2581-2584
- **Title:** Sesquiterpene lactones and kaurane glycosides from *Francoeuria crispa*
- **PubMed:** None
- **Link:** Link to reference

Authors information
- **Author(s):** Abdel-Mogib M., Jakupovic J., Dawidar AM., Metwally MA., Abou-Elzahab M.
- **Curator(s):** Moumbock AFA, Ntie-Kang F, Yong JN

Figure 13.8: ANPDB compound card.

2. Field (2) shows the PubChem ID of the compound. Upon clicking this link, you will be redirected to the PubChem page, where additional information about this compound could be accessed. This is followed by computed physicochemical information (from *QikProp*), source species information, the predicted toxicity data (from *pkCSM*) and the reference information.
3. Field (3) shows the information about the authors of the article from which taken compound is obtained, upon clicking the author's name it redirects to author's references list(link)

Species Card
1. Field (1) gives the name of the selected species.
2. Field (2) gives the names of all compounds identified from the selected species. Each compound card can be accessed by simply clicking on the highlighted compound links.
3. Field (3) shows the names of the authors of the article where information about the selected species is taken from. Upon clicking you can see Author's reference(link).

13.4 Session III (Taxonomy databases) 2 hours

13.4.1 Objectives

– Identify databases containing taxonomic information on plants and animal species.
– Use a web browser to search the contents of two open access databases that help to establish the taxonomy of medicinal plants.

13.4.1.1 NCBI taxonomy database
1. Open the NCBI Taxonomy database on a new web browser (https://www.ncbi.nlm.nih.gov/taxonomy).
2. We are interested in two common plants, a food crop (rice plant) and the lover plant that produces the rose flower. ***You might later want to reproduce this study with the plant that produces your favorite foodstuff.***
3. Search for the scientific name of these plants (e.g. from previous knowledge, Google search, Wikipedia, etc.).
4. Search the scientific name of each plant in the NCBI Taxonomy database.
5. Provide a full classification of each species.
6. Identify the Taxonomy ID (txid) of each species. What is the significance of this number?
7. Of what use is the genetic information provided for each species?

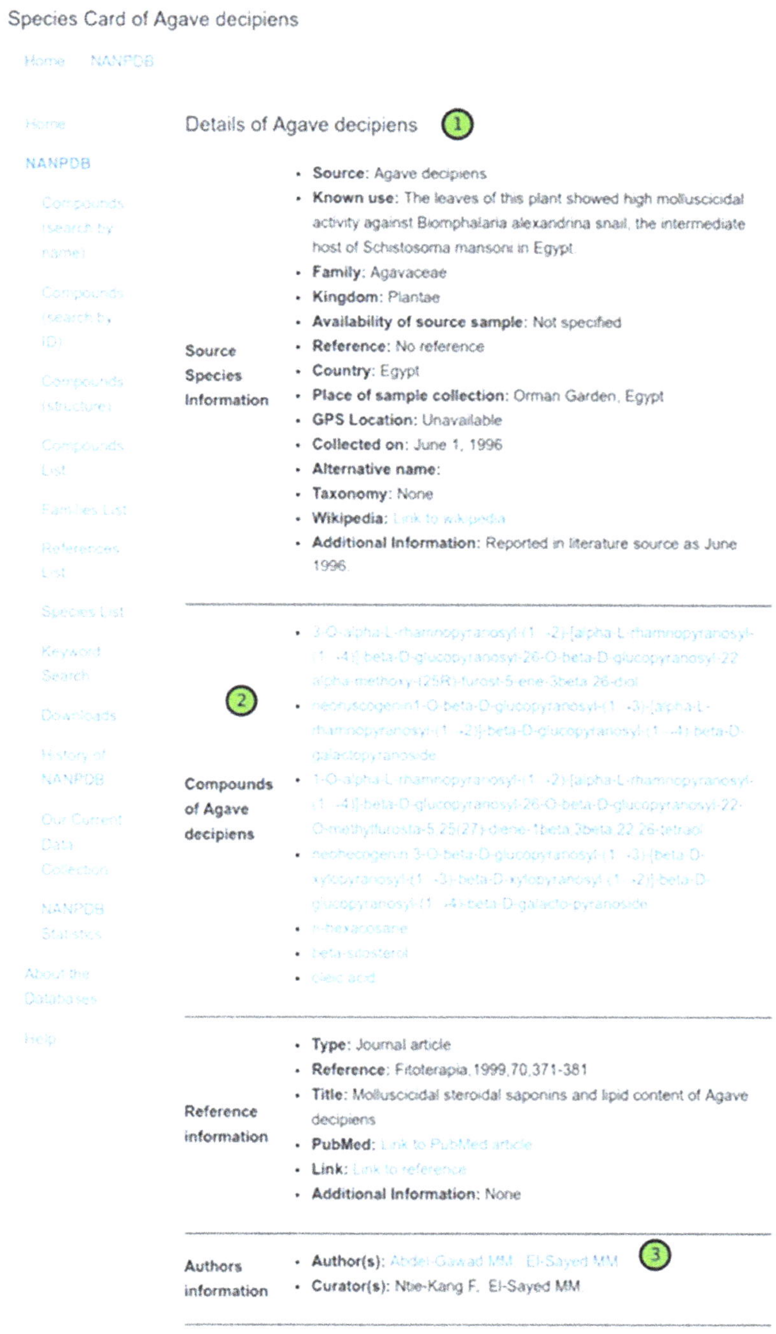

Species Card of Agave decipiens

Home   NANPDB

Home   Details of Agave decipiens   ①

NANPDB
                                    • **Source:** Agave decipiens
Compounds   • **Known use:** The leaves of this plant showed high molluscicidal
(search by   activity against Biomphalaria alexandrina snail, the intermediate
name)   host of Schistosoma mansoni in Egypt.
                                    • **Family:** Agavaceae
Compounds   • **Kingdom:** Plantae
(search by   • **Availability of source sample:** Not specified
ID)   Source   • **Reference:** No reference
                     Species   • **Country:** Egypt
Compounds   Information   • **Place of sample collection:** Orman Garden, Egypt
(structure)   • **GPS Location:** Unavailable
                                    • **Collected on:** June 1, 1996
Compounds   • **Alternative name:**
List   • **Taxonomy:** None
                                    • **Wikipedia:** Link to wikipedia
Families List   • **Additional Information:** Reported in literature source as June
References   1996.
List

Species List
                                    • 3-O-alpha-L-rhamnopyranosyl-(1→2)-[alpha-L-rhamnopyranosyl-
Keyword   (1→4)]-beta-D-glucopyranosyl-26-O-beta-D-glucopyranosyl-22-
Search   alpha-methoxy-(25R)-furost-5-ene-3beta,26-diol
                     ②   • neoruscogenin1-O-beta-D-glucopyranosyl-(1→3)-[alpha-L-
Downloads   rhamnopyranosyl-(1→2)]-beta-D-glucopyranosyl-(1→4)-beta-D-
                                      galactopyranoside
History of
NANPDB   Compounds   • 1-O-alpha-L-rhamnopyranosyl-(1→2)-[alpha-L-rhamnopyranosyl-
                     of Agave   (1→4)]-beta-D-glucopyranosyl-26-O-beta-D-glucopyranosyl-22-
Our Current   decipiens   O-methylfurosta-5,25(27)-diene-1beta,3beta,22,26-tetraol
Data   • neohecogenin-3-O-beta-D-glucopyranosyl-(1→3)-[beta-D-
Collection   xylopyranosyl-(1→3)-beta-D-xylopyranosyl-(1→2)]-beta-D-
                                      glucopyranosyl-(1→4)-beta-D-galactopyranoside
NANPDB   • n-hexacosane
Statistics   • beta-sitosterol
                                    • oleic acid
About the
Databases

Help   • **Type:** Journal article
                                    • **Reference:** Fitoterapia,1999,70,371-381
                                    • **Title:** Molluscicidal steroidal saponins and lipid content of Agave
                     Reference   decipiens
                     information   • **PubMed:** Link to PubMed article
                                    • **Link:** Link to reference
                                    • **Additional Information:** None

                     Authors   • **Author(s):** Abdel-Gawad MM, El-Sayed MM   ③
                     information   • **Curator(s):** Ntie-Kang F, El-Sayed MM

Figure 13.9: ANPDB species card.

8. Distinguish between the information provided for the Genetic code, the Mitochondrial genetic code, and the Plastid genetic code.
9. Which other features of this database would you find useful?
10. Click the link beneath to the Tropicos database. Comment of the various subspecies (if any) of each species studied.
11. Continue browsing the NCBI database of move to III.2 (Tropicos Database).

13.4.1.2 Tropicos database

1. Open the Tropicos database on a new web browser (https://www.tropicos.org/home).
2. Quickly look through each menu in the database.
3. Using the "Tools" menu of Tropic, query a rare (preferably medicinal) plant species of your choice (provide a full classification of your species!). On the "Tools" menu, select "Functions.
4. Attempt a DNA specimen search for your species.
5. What is the ethnobotanical use of your species?
6. Use the "GeoLocate" function to attempt to find a possible place of collection of your species.
7. Repeat the exercise using the "Specimen Geographic Search" function this time. How widely distributed is your species? Compare your results with querying the plant that produces your favorite foodstuff.

13.5 Session IV (Chemotaxonomy) 1.5 hours

13.5.1 Objectives

– Establish a clear definition of chemotaxonomy by use of examples.
– In this exercise, we shall employ the ANPDB to establish the relationship between a selected list of plant families and their chemical content.

1. Open the ANPDB database on a new browser.
2. Using the entire database (both NANPDB and EANPDB), search the following plant families:
   A = Ceasalpiniaceae
   B = Cactaceae
   C = Rubiaceae
   D = Verbenaceae
3. For each plant family, tabulate the species represented.
4. For each species, identify the compound classes present.

5. Represent the distribution of compound classes per family in pie charts.
6. In terms of percentage compounds of number of compounds per class and in terms of absolute numbers, represent the entire data in a histogram.
7. Compare your results with the conclusions of the results published in Figure 40 of the article (https://pubmed.ncbi.nlm.nih.gov/32423415/).

Funding

Financial support is acknowledged from the German Academic Exchange Services (DAAD) to FN-K through the guest professorship program.

Fidele Ntie-Kang, Linda Jahn and Jutta Ludwig-Müller

14 From genes to secondary metabolites

Abstract: This article was originally conceived as a set of tutorial exercises for undergraduate biology majors for a natural products course at the Institute of Botany, Technische Universität Dresden (2020/21 academic year). The overall goals of these exercises are to identify different genome-based databases from which plant-based genomic data could be searched for academic reasons, initiate the reader to basic tools for querying genome-based databases for studying the relationships between genes and metabolites and to introduce reader to the relationship between genes and secondary metabolites. The reader is also expected to identify a gene cluster database for plant-based secondary metabolites and use the web browser for querying gene clusters for plant secondary metabolite biosynthesis. The reader is further expected to identify online tools for the analysis of biosynthetic gene clusters (BGCs) and secondary metabolites in plants, fungi and microbes, use these from a web browser to submit jobs on a web server and retrieve the results and analyze the results and attempt to establish a relationship between a secondary metabolite and its BGC.

Keywords: biosynthesis, gene clusters, genomics, secondary metabolites, specialized metabolism

14.1 Introduction

In order to get familiar with the theoretical background of the material in this article, the reader is advised to consult previous chapters of this book, e.g. in part III. This tutorial article is work estimated for one full day. The reader is, however, free to organize their time flexibly. The times set on each session should only serve as guidelines. It begins with a literature assignment, which the reader is expected to complete by reading through the referenced article and summarize in two pages maximum (A4 size, Times New Roman font 12 double spacing).

14.1.1 Literature assignment

Read and summarize the following article:

Kautsar SA, Suarez Duran HG, Blin K, Osbourn A, Medema MH. plantiSMASH: automated identification, annotation and expression analysis of plant biosynthetic gene clusters. Nucleic Acids Res. 2017;45:W55–W63. DOI: 10.1093/nar/gkx305. URL: https://pubmed.ncbi.nlm.nih.gov/28453650/

https://doi.org/10.1515/9783110668896-014

14.1.2 Plan of the day

09:00–09:30 General Introduction
09:30–11:00 Session I (Genome-based databases)
11:00–12:30 Session II (Database for gene cluster analysis: antiSMASH and Phytozome)
12:30–13:30 Lunch Break
13:30–15:00 Session III (Database for gene cluster analysis: antiSMASH and Phytozome Cont'd)
15:00–17:00 Session IV (Tools for gene cluster analysis: antiSMASH and plantiSMASH)

14.2 Session I (Genome-based databases) 1.5 hours

14.2.1 Objectives

– Identify different genome-based databases from which plant-based genomic data could be searched for academic reasons.
– Initiation to basic tools for querying genome-based databases for studying the relationships between genes and metabolites.
– Introduction to the relationship between genes and secondary metabolites.

14.2.2 The medicinal plant genomics database

1. This database is accessible via the internet browser on the link (http://medicinalplantgenomics.msu.edu/).
2. Take 15 min to browse through the various menus and read the contents.
3. Identify any usefulness of this database.
4. What could be the possible limitations of this database?
5. Write down a half page description of this database.
6. 3Make a tabular comparison between this database and the NCBI Taxonomy database (seen on Day 1)
7. Check more databases containing information of gene clusters and secondary metabolites (https://www.secondarymetabolites.org/databases/).

14.3 Sessions II & III (Databases for gene cluster analysis) 3 hours

14.3.1 Objectives

- Identify a gene cluster database for plant-based secondary metabolites.
- The use of the web browser for querying gene clusters for plant secondary metabolite biosynthesis.

14.3.1.1 The antiSMASH database

1. What is a gene cluster?
2. Are plant genes usually clustered?
3. What is the general difference between genes in bacteria, fungi and plants?
4. Provide a general procedure for sequencing the DNA in plants.
5. Open the main page of the antiSMASH database on a new browser (https://antismash-db.secondarymetabolites.org). This database includes gene clusters for microbial secondary metabolites. Check the equivalent databases for fungal- and plant-based secondary metabolite gene clusters.
6. Browse the "Statistics" menu and check which secondary metabolite gene clusters are most abundantly represented in this database.
7. Click on the topmost secondary metabolite class and search for the gene clusters for this metabolite class in the organism "*Bacillus cereus* FORC087".
8. Take a screenshot of the query output and you see something like the picture beneath.

9. Now click on "Overview"
10. By clicking on the various "Regions" or numbered rectangles, e.g. 1.1, 1.2, etc., what do you notice? Which information is available? For "Region 1.1", you will observe something like this!

11. How well distributed are the biosynthetic gene clusters (BGCs) on the chromosome/contig?
12. Now go back to the "Overview" table. On which column do you find about the following?
    a. the list of identified gene clusters in an organism.
    b. the cluster type.
    c. the cluster coordinates.
    d. the secondary metabolite class
    e. the percentage BGC similarity with respect to the MIBiG database.

13. You are still on the "Overview" page. For "Region 2.1", the gene cluster type corresponds to non-ribosomal polyketide synthase (NRPS) and the cluster coordinates range from 87,556 to 132,506. This is best hit for the "tauraramide" gene cluster in the MIBiG database of characterized gene clusters. Click of number 13%, the percentage similarity of this BGC to that in the MIBiG database.

    a. On the new page that opens, select the "NRPS" gene (brown). Click on the brown arrow and the information will be displayed on the right panel. This is what you would see:

F8159_RS05670
amino acid adenylation domain-containing protein
Locus tag: F8159_RS05670
Protein ID: WP_151524627.1
Gene: None
Location: 107,566 - 112,506, (total: 4941 nt)
biosynthetic (rule-based-clusters) NRPS: AMP-binding
biosynthetic (rule-based-clusters) NRPS: Condensation
biosynthetic-additional (rule-based-clusters) PP-binding
biosynthetic-additional (rule-based-clusters) NAD_binding_4
biosynthetic-additional (smcogs) SMCOG1127:condensation domain-containing protein (Score: 368; E-value: 1.3e-111)
NCBI BlastP on this gene
View genomic context
MiBIG Hits
AA sequence: Copy to clipboard
Nucleotide sequence: Copy to clipboard

Notes
The first two lines refer to the RefSeq/GenBank annotation and is not generated by antiSMASH

    Identify the location of this cluster.
- *On line 6, you'll find information on the highly conserved enzyme HMM profiles (core-enzymes), which are indicative of a specific BGC type.*
- *On lines 9 and 10, you will find information on the smCOG classification. Each gene of the cluster is compared to a database of clusters of orthologous groups of proteins involved in secondary metabolism. This information then is used to provide an annotation of the putative function of the gene products.*
- *Following are links to NCBI BLAST search and NCBI genome viewer. These two links only work in case the genome was downloaded from NCBI.*

The next link connects you with BLAST hits to the MIBiG sequences
– *A copy of DNA or amino acid sequences are available on the last two links. These could be copied on a clipboard and pasted to other programs for searching.*

b. Link this to the NORINE database (bottom right panel) and state which core structures are predicted as products of this gene cluster.

Notes

The core structure predicted is shown on the top right window.
This is followed by the predicted monomer sequence.
The link to the NORINE NRP database is beneath the core structure (for NRPs).
Beneath are the predicted details for the PKS/NRPS domains.
For each domain, by clicking on the " + " sign, you unfold the domain.
At the bottom of the panel, there is a link which open directly to the NORINE peptide database query form.

c. Now, under detailed domain architecture, click on the domain "A". You observe something like this!

Notes

The letter "A" refers to the protein domain type responsible for the biosynthesis of the metabolite.
This is followed by the location of the domain with respect to the full length protein.
Following is the link to the NCBI BlastP.
The rest are amino acid and DNA sequence information of the selected domain for copying and pasting to other programs.

7d. How useful did you find this database?

e. For a more extended tutorial on the use of bacterial, fungal and plant BGC data-bases, please check here (https://docs.antismash.secondarymetabolites.org/PDFmanual/antiSMASH5manual.pdf).

14.3.1.2 The Phytozome database

1. Open the main page of the Phytozome database (https://phytozome.jgi.doe.gov/pz/portal.html). Data currently uploaded plant are genomes.
2. Quickly run through the embedded user guide (https://phytozome.jgi.doe.gov/pz/QuickStart.html).
3. What common features does the Phytozome database have in common with the antiSMASH database for microbial gene cluster?

14.4 Session IV (Tools for gene cluster analysis) 2 hours

14.4.1 Objectives

– Identify online tools for the analysis of BGCs and secondary metabolites in plants, fungi and microbes.
– Use these from a web browser to submit jobs on a web server and retrieve the results.
– Analyze the results and attempt to establish a relationship between a secondary metabolite and its BGC.

14.4.1.1 The antiSMASH tool

1. The tool antiSMASH (antibiotics and secondary metabolites analysis shell) is a comprehensive pipeline for the automated mining of finished or draft genome data for the presence of secondary metabolite BGCs. It runs as a server on https://antismash.secondarymetabolites.org/, which is accessible via a web browser although it can also be installed locally on your computer. The plant version of it (plantiSMASH) runs under the same principle but only came up recently, so we shall begin by running one or two jobs on the microbial version to gain some experience.
2. Data submission is on the panel accessible via the web browser (https://antismash.secondarymetabolites.org/#!/start).

Data can be uploaded in Genbank (recommended), EMBL, or plain FASTA for-mat. If the user uploads FASTA files, which do not contain any annotation or anno-tated coding sequences, putative genes are identified with Prodigal (10) (for bacterial sequences) or GlimmerHMM (11) (for fungal sequences). If the NCBI accession num-ber is known, antiSMASH can also automatically retrieve the data from NCBI. If you

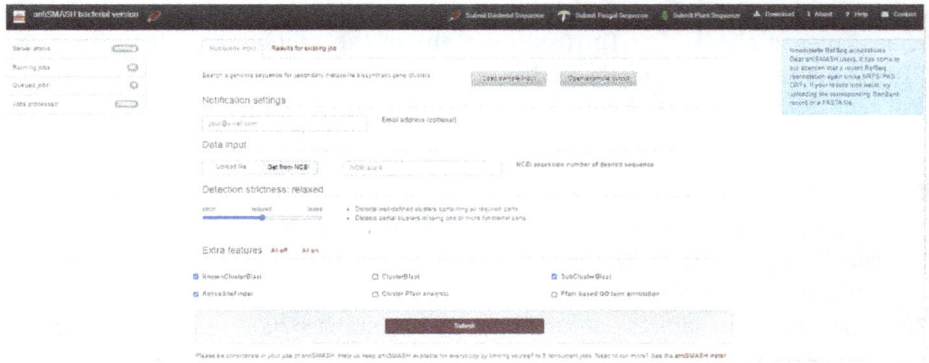

work 8with draft genome sequences, it is preferable to use scaffolded sequences containing "N" characters for the gaps, as gene clusters can only be identified if positional information is available.

14.4.2 Notes

- The quality of the prediction by antiSMASH is highly dependent on the quality of the input data. If you analyze poor quality draft genomes with many (thousands) of small contigs and low N50, antiSMASH may not be able to detect gene clusters if they are scattered over multiple contigs.
- Many of the specificity determination algorithms in antiSMASH depend on the presence/absence of conserved amino acids at specific positions in the enzyme sequence. If the sequence quality is low and thus also the amino acid of the translated proteins cannot be fully trusted all predictions have to be taken with precautions
- Fungal and plant sequences could be uploaded and searched by simply clicking and selecting the appropriate panel on the top menu button.

3. Follow the procedure outlines beneath to submit a job:
a. Select the correct type of analysis you plan to carry out. Some functions are only applicable or useful for bacterial, fungal or plant sequences (select "Submit Bacterial Sequence" from the top menu).
b. Enter your email address on the appropriate space provided. This option permits you to receive the analysis into your inbox when it is completed. If you did not provide an email address, please bookmark the link to this page or copy and paste the link somewhere (otherwise you will lose your results).
c. Enter Genbank or RefSeq accession number to directly download sequence from NCBI (you can collect the NCBI acc# of the species studied in the previous

example on the antiSMASH database, *Bacillus cereus* FORC087. In this case, accession code is NZ_CP029454.1. You get this by simply searching the organism name in the NCBI Nucleotide database.

d. Upload your sequence by using the "Upload file" button and selecting the sequence file (Fasta or GenBank format) to upload. Otherwise, you simply type the accession code about in the appropriate space. Allow all other settings by default. For details of the settings, consult page 10 of the document (https://docs.anti smash.secondarymetabolites.org/PDFmanual/antiSMASH5manual.pdf).

e. Press "Submit" button at the end of the page.

f. Relax and expect the results.

g. Once the job is completed, click on "Results". You see something like this:

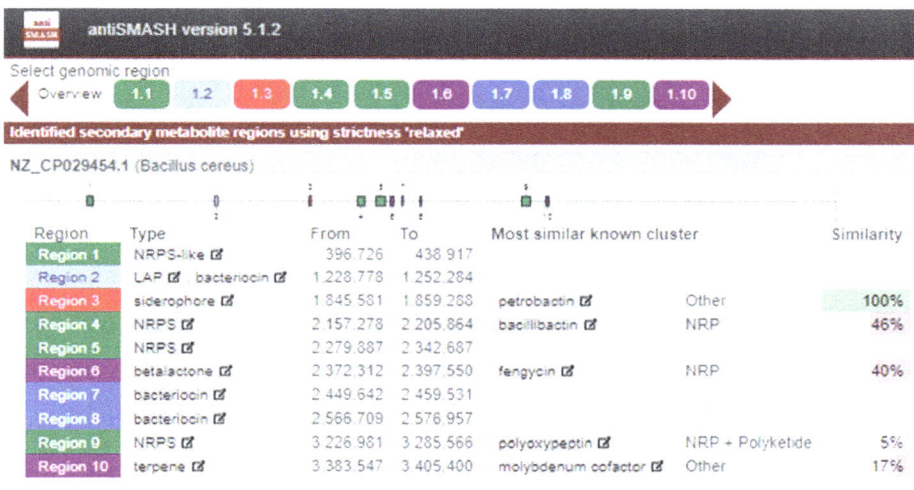

Otherwise, you get a nice email that looks like that below, with a link to access your results:

h. Proceed to interpret the results as described in Section II.1 steps 10 and 11.

− What are the various types of candidate clusters present in this organism's genome (chemical hybrid, interleaved, neighboring, single)?

− What is the largest (dominant) candidate cluster?

− antiSMASH is used to identify the genetic region as a secondary metabolite gene cluster. Where do you find such information on the results panel?

− Provide information regarding Region 1.10 (location, biosynthetic cluster, metabolites).

Your antiSMASH job bacteria-71678db9-9e68-4b12-9f8b-4ddca617b4b5 finished.

noreply@secondarymetabolites.org
Do 08.10.2020 01:47

An: Ntie-Kang, Fidele;

Dear antiSMASH user,

The antiSMASH job bacteria-71678db9-9e68-4b12-9f8b-4ddca617b4b5 you submitted on 2020-10-07 23:36:26.898473 with the filename 'NZ_CP029454.1.gbk' has finished with status done.

You can find the results on
https://antismash.secondarymetabolites.org/upload/bacteria-71678db9-9e68-4b12-9f8b-4ddca617b4b5/index.html

Results will be kept for one month and then deleted automatically.

If you found antiSMASH useful, please check out
https://antismash.secondarymetabolites.org/#!/about
for information on how to cite antiSMASH.

14.4.2.1 The plantiSMASH tool

1. Go back to the original page where you had submitted the analysis for the microbial sequence and select "Submit Plant Sequence". You will see something like this:

2. Are there any jobs currently running on the server? How many jobs are currently on the queue? For how long has the oldest job on the queue lasted already? Top left corner!

3. 11Complete the data as in antiSMASH (e.g. email address for retrieving the results). Now, check the complete sequence of *Zea mays* alpha zein gene cluster (22-kDa, 78,101 bp linear DNA) in the NCBI Nucleotide database (accession: AF031569.1). Include the NCBI ACC# AF031569.1 in the appropriate input entry.

4. Click on the "+" buttons and include additional settings (e.g. check both "Compare to plantiSMASH predicted clusters" and "Compare to registered known clusters from MIBiG Database". Leave the "Advanced options" as default. Press "Submit" button at the end of the page. The rest proceeds as previously described for antiSMASH.

5. What is your observation?

6. Now, do the same for the *Arabidopsis thaliana* chromosome 1 sequence (30,427,671 bp linear DNA), with accession # NC_003070.9. These are the expected results.

7. Analyze the gene clusters responsible for alkaloid and terpene biosynthesis.

8. Do the same for the whole genome shotgun sequence of *Solanum lycopersicum* cultivar Heinz (1706 chromosome 1, SL3.0, 98,455,869 bp linear DNA), with accession NC_015438.3

9. Which are the major secondary metabolites in this species?

10. Assign BGCs for each class of secondary metabolite.

Additional Useful Links!!!

1. https://genomevolution.org/wiki/index.php/Sequenced_plant_genomes

2. http://www.softberry.com/berry.phtml?topic=fgenesh&group=programs&subgroup=gfind&gclid=EAIaIQobChMIibnIwrL66wIVw-ntCh0HowUCEAAYBCAAEgI4APD_BwE

3. https://onlinelibrary.wiley.com/doi/full/10.1111/tpj.13485

4. https://www.biost.com/bac-libraries-arrayed-non_arrayed?gclid=Cjw-KCAjwwab7BRBAEiwAapqpTNo1WEbuANYNIHMTa6tCHHAu0xOh4VXPuVOTh9le Wuem0fA5VicMnRoC1ckQAvD_BwE
5. https://www.ncbi.nlm.nih.gov/pmc/articles/PMC5449196/

Funding

Financial support is acknowledged from the German Academic Exchange Services (DAAD) to FN-K through the guest professorship program.

Fidele Ntie-Kang, Abdurrahman Olğaç, Conrad V. Simoben,
Abdulkarim Najjar, Sergi Herve Akone, Lucas Paul,
Ramsay S. T. Kamdem, Abraham Madariaga-Mazón,
Celestin N. Mudogo, Donatus B. Eni, Boris D. Bekono,
Fatima Baldo, Dina Robaa, Daniel M. Shadrack,
Ricardo B. Hernández-Alvarado, Karina Martinez-Mayorga,
José L. Medina-Franco, Daniel Svozil and Wolfgang Sippl

Glossary of terms used in chemoinformatics of natural products: advanced concepts and applications

Note to the Reader: The following list of definitions of recurrent terms used in this book is not meant to be exhaustive. The Reader is also invited to consult the Glossary of volume 1 of this book.

Ab initio: Literally means "from first principles" or "from the beginning", implying that the only inputs into an ab initio calculation are physical constants.

Acetone cyanohydrin (ACH): An organic compound used in the production of methyl methacrylate, the monomer of the transparent plastic polymethyl methacrylate (PMMA), also known as acrylic. It liberates hydrogen cyanide easily, so it is used as a source of such. For this reason, this cyanohydrin is also highly toxic.

ADMET: Abbreviations for the following pharmacokinetic and pharmacological properties related to absorption, distribution, metabolism, excretion and toxicity.

Affinity: Affinity is a measure of the tightness with which a drug binds to the receptor. Intrinsic activity is a measure of the ability of a drug that is bound to the receptor to generate an activating stimulus and produce a change in cellular activity. Both agonists and antagonists can bind to a receptor.

Aflatoxins: Mycotoxins mainly produced by the fungi *Aspergillus flavus* and *Aspergillus parasiticus*, biosynthesized by the polyketide route.

Agonist: A compound (endogenous ligand or drug) that causes a physiological or a pharmacological action when in complex with a receptor.

Algorithm: A logical procedure for answering scientific queries, usually performed by the use of a computer.

Alkaloid: A class of naturally occurring organic compounds that contain (mostly heterocyclic) nitrogen atoms.

Amino acids: The building blocks of proteins.

Antagonist: A compound or drug that competes with the endogenous ligand in order to create a complex with a receptor.

antiSMASH: Acronym for antibiotics & Secondary Metabolite Analysis Shell, a genome mining platform, the first comprehensive pipeline capable of identifying

https://doi.org/10.1515/9783110668896-015

biosynthetic loci covering the whole range of known secondary metabolite compound classes (polyketides, non-ribosomal peptides, terpenes, aminoglycosides, aminocoumarins, indolocarbazoles, lantibiotics, bacteriocins, nucleosides, beta-lactams, butyrolactones, siderophores, melanins and others), under development since 2011, coordinated by the Weber and Medema groups. Available at https://anti smash.secondarymetabolites.org.

Applicability domain: The applicability domain of a QSAR model is the physico-chemical, structural or biological space, knowledge or information on which the training set of the model has been developed, and for which it is applicable to make predictions for new compounds.

Area Under the Curve: Two dimensional area under the entire ROC curve, which is a measurement of an algorithm's ability to classify compounds into specific categories such as active and inactive. Also see "*ROC curve*" and "*ROC analysis*".

Artificial neural network (or connectionist systems): Computing systems vaguely inspired by the biological neural networks that constitute animal brains. The neural network itself is not an algorithm, but rather a framework for many different machine learning algorithms to work together and process complex data inputs. Such systems "learn" to perform tasks by considering examples, generally without being programmed with any task-specific rules.

AUC: See "Area under the curve".

Bayesian: A branch of statistical methods, including "Bayesian inference", "Bayesian probability", Bayesian classifier, Bayesian regression, etc. Bayesian regression, for example, could be thought of as a way of performing a range of QSAR analyses between two extremes. There exists prior information in the form of a previous regression on the relevant set of previous data – the training set of complexes used to fit the original scoring function.

Bayesian classifier: A supervised learning algorithm used for classification tasks.

BCG: see "*Biosynthetic gene cluster*"

BEDROC (Boltzmann-enhanced discrimination of ROC): Contains the discrimination power of the RIE metric but incorporates the statistical significance from ROC and its well-behaved boundaries.

Beta-glucosidase: An enzyme that catalyzes the hydrolysis of the glycosidic bonds to terminal non-reducing residues in beta-D-glucosides and oligosaccharides, with release of glucose.

Binding affinity: How strongly two things bind (e. g. a protein-ligand interaction).

Binding free energy: A sum of the intermolecular interactions between the ligand and the protein and the internal steric energy of the ligand.

Binding site: This is a region on a macromolecule such as a protein that binds to another molecule with specificity.

Biological activity: Biological activity or pharmacological activity describes the beneficial or adverse effects of a drug on living matter. When a drug is a complex chemical mixture, this activity is exerted by the substance's active ingredient or pharmacophore but can be modified by the other constituents. Among the various properties of chemical compounds, pharmacological/biological activity plays a crucial role since it suggests uses of the compounds in the medical applications. However, chemical compounds may show some adverse and toxic effects which may prevent their use in medical practice (see "Toxicity").

Biosynthesis (biosynthesize): The production of a chemical compound by a living organism.

Biosynthetic gene cluster (BCG): Organized groups of genes involved in the production of specialized metabolites. Typically, one BGC is responsible for the production of one or several similar compounds with bioactivities that usually only vary in terms of strength and/or specificity.

BitterDB: A database of bitter molecules developed in the group of Prof. Masha Niv. Accessible at http://bitterdb.agri.huji.ac.il/dbbitter.php.

Bitter taste receptors (TAS2Rs or T2Rs): These belong to the superfamily of seven-transmembrane G protein–coupled receptors, which are the targets of > 50% of drugs currently on the market. Canonically, T2Rs are located in taste buds of the tongue, where they initiate bitter taste perception.

Building blocks: In the chemical sense, building blocks are small molecules that are used in the enzymatic process of assembling larger molecules like proteins, polysaccharides, polyterpenes, etc.

CAS: Acronym for Chemical Abstracts Services. The CAS Registry Number® is used to identify your substance of interest. This is universally used to provide a unique, unmistakable identifier for chemical substances. The CAS registry information includes daily updated information on literature references to the substance experimental and predicted property data (e. g. boiling and melting points, etc.), CA Index Names and synonyms, commercial availability of compounds/substances, preparative methods, spectra, regulatory information from international sources, etc. Also see "*Chemical Abstracts Services (CAS) Number*".

Cassava: The starchy tuberous root of a tropical tree, used as food in tropical countries. It has a shrubby tree from which cassava root is obtained. It is native to tropical America and cultivated throughout the tropics.

Catalyst: A substance that accelerates a chemical reaction without itself being affected.

CATS2: Chemically advanced template search. Topological pharmacophore descriptors designed for scaffold hopping, de novo design, and machine learning.

5CADD: Computers-aided drug design – a broad range of computational approaches that help in modern drug discovery.

Chemical Abstracts Services (CAS) Number: A unique code for identifying a compound in the CAS Directory. Also see "*CAS*"

Chemical scaffold: Also see "*Scaffold*".

Chemical space: A collection of molecules meaningfully related by a mathematical object. For practical applications, chemical space can be used as a "tool" that helps to find associations in complex data and rapidly exploit the increasing information available for the discovery of drugs and other research areas such as food science.

Chemoinformatics (cheminformatics or chemical informatics): A scientific branch which involves searching and managing information stemming from chemical structures. This is the science of handling, indexing, archiving, searching, and evaluating information that is specific to chemical structures and is used in data mining, information retrieval, information extraction, and machine learning.

Chemotaxonomy: The system of classifying organisms by their chemical content, i. e. a method of biological classification based on similarities and dissimilarity in the structure of certain compounds among the organisms being classified.

cLogD (see also logD): Calculated logD value at a specified pH.

cLogP (see also logP): Calculated log of the octan-1-ol/water partition coefficient.

Cluster of chemical compounds: A group of chemical compounds sharing similar properties such as structural or physicochemical properties.

Clustering of compounds: See "Cluster of chemical compounds".

Clustering algorithm: See "Cluster of chemical compounds".

Combinatorial biosynthesis: Application of genetic engineering to modify biosynthetic pathways to natural products in order to produce new and altered structures using nature's biosynthetic machinery.

Compound library (or chemical database): A collection of chemical compounds in electronic format and/or with available physical samples.

Computational toxicology: A scientific discipline that focuses on the development of mathematical and computer-based models for understanding and predicting adverse health effects caused by chemicals, such as environmental pollutants and pharmaceuticals.

Cyanide: A chemical compound that contains the group $C \equiv N$. This group, known as the cyano group, consists of a carbon atom triple-bonded to a nitrogen atom.

Cyanogenic glycosides: Natural plant toxins that are present in several plants, most of which are consumed by humans.

Cytotoxicity: The possession of such destructive action, particularly in reference to lysis of cells by immune phenomena and to antineoplastic agents that selectively kill dividing cells.

Database: A collection of objects organized according to specific characteristics and rules.

Data mining: Useful information which can be extracted from large sets of molecules using various methods.

Dataset (or data set): A collection of data. Most commonly a data set corresponds to the contents of a single database table, or a single statistical data matrix, where every column of the table represents a particular variable (e. g. chemical structure file or input in the form of SMILES entry or InChI key), and each row corresponds to a given member of the data set in question.

DEcRyPT: Drug-Target Relationship Predictor. A machine learning method built with CATS2 descriptors.

Dereplication: A strategy used for the previous identification of known compounds in a complex sample, based on comparing spectroscopic and spectrometric data with the previously reported data from known compounds.

Derivatives of Natural Products: Chemical modification of natural product structures. Examples would be all of the beta-lactams (>>30,000) synthesized from the base penicillin G and cephalosporin C.

DiaNat-DB: A database of natural products with antidiabetic properties, collected by the Martinez and Madariaga groups.

Diversity of a chemical database: Variety of chemical scaffolds, property space, and functional group space within a chemical database. The degree of the variety can be measured by various properties such as physical properties or descriptors generated by the use of scientific software.

DMNP: Abbreviation for "Dictionary of Marine Natural Products", a large commercially available collection of natural products data for compounds from marine sources. Available at: http://dmnp.chemnetbase.com.

DNA methyltransferases: Families of enzymes that attach the methyl group to DNA.

DNP: Abbreviation for "Dictionary of Natural Products", a large commercially available collection of natural products data. Available at: http://dnp.chemnetbase. com.

Docking: Computational technique used to find the binding mode of a ligand in a ligand-receptor complex by searching the conformational and orientational space of a ligand and receptor for a geometry with favorable binding energy.

Docking experiments: *In silico* experiments in order to investigate if there is a fit between a ligand and a binding site and the stability of the complex created.

Docking score: See "*Scoring function*".

Drug: According to IUPAC's definition, a drug is a compound against a human or an animal disease.

Drug discovery: The process of identifying a new drug and bringing it to market. It involves different scientific disciplines including biology, chemistry and pharmacology.

Druglike (or Drug-like): A comparison between various chemical compounds and known drugs in order to estimate the degree of similarity based on different properties. A drug-like compound is defined as a compound with sufficiently acceptable ADME properties and sufficiently acceptable toxicity properties. Drug-like molecules are defined as those compounds that have sufficiently acceptable ADME properties and sufficiently acceptable toxicity properties.

2D/3D structure: Structure of a chemical compound in two-dimensional space/ structure of a chemical compound in three-dimensional space.

Dynophore: Dynamic pharmacophore that takes into account the conformational flexibility of both the ligands and the targets derived from MD simulations.

EF (Enrichment factor): The concentration of the annotated ligands among the top-scoring docking hits compared to their concentration throughout the entire database.

Endpoint: In toxicology, endpoints are values derived from toxicity tests that are the results of specific measurements made during or at the conclusion of the test.

Enzyme: A protein catalyst that regulates the rate at which chemical reactions proceed in living organisms without itself being altered in the process. Enzymes help speed up chemical reactions in the body and regulate several functions, e. g. breathing and digestion.

Erythrocyte: A red blood cell, which (in humans) is typically a biconcave disc without a nucleus. Erythrocytes contain the pigment haemoglobin, which imparts the red color to blood, and transport oxygen and carbon dioxide to and from the tissues.

FDA: Abbreviation for Food and Drug Administration (USA).

Fragmentation of natural products: Cutting parts of relatively larger molecules to obtain the desired fragment, which might be used with the aim of keeping the pharmacophore group or testing its biological effect separately.

Fragments: small molecules for which the molecular weight (MW) < 300, the number of hydrogen bond donors is 3, the number of hydrogen bond acceptors is 3, and the clogP is 3. In addition, their number of rotatable bonds is not higher than 3 and the polar surface area (PSA) not higher than 60 $Å^2$.

Fragment-based drug design: designing new molecules based on bringing fragments together.

Fragment-like: have the characteristics of fragments. Also see "*Fragments*"

File formats: Types of files for storing chemical information that can be read and interpreted by computers.

Filters: A set of criteria used to characterize or exclude chemical compounds of a database according to specific rules.

Fingerprint: See "Molecular Fingerprints".

Flavonoid: A class of (often bitter) plant and fungus secondary metabolites, whose name is derived from the Latin word flavus meaning yellow, their color in nature. Flavonoids in plants are generally responsible for the color of leaves and flowers. They consist of a diverse group of polyphenolic compounds commonly found in the human diet. Also see "*Phenolics*".

Focused library: A set of molecules with bioactivities against a drug target or family of targets.

Foodinformatics: Application of chemical information to food chemicals.

FooDB: A free database for chemical compounds found in foods, accessible at the URL https://foodb.ca/.

Force field (or forcefield): Mathematical expression that describes the dependence of the energy of a molecule on the coordinates of the atoms in the molecule.

Functional group: A specific group of atoms or bonds within a compound that is responsible for the characteristic chemical reactions of that compound. The same functional group will behave in a similar fashion, by undergoing similar reactions, regardless of the compound of which it is a part.

Gene cluster: A group of two or more genes found within an organism's DNA that encode similar polypeptides, or proteins, which collectively share a generalized function and are often located within a few thousand base pairs of each other. Also see "BCG" and "Biosynthetic gene cluster".

Genotoxicity: The ability of a chemical compound to induce changes (damage) to genetic material.

Glucosyl: A univalent free radical or substituent structure obtained by removing the hemiacetal hydroxyl group from the cyclic form of a monosaccharide and, by extension, of a lower oligosaccharide.

Hepatotoxicity: The ability to poison liver cells.

Heterologous expression: This refers to the expression of a gene or part of a gene in a host organism, which does not naturally have this gene or gene fragment. Insertion of the gene in the heterologous host is performed by recombinant DNA technology.

High-throughput screening (HTS): Method of scientific experimentation that comprises the screening of large compound libraries for activity against biological targets via the use of automation, miniaturized assays, and large-scale data analysis.

Hit: A hit compound is a molecule that shows the desired type of activity in a screening assay. It is important to develop pharmacologically relevant screening assays for hit discovery and for the subsequent hit-to-lead selection process.

Hit identification: The first step of a drug discovery project to obtain a molecule which binds to its target.

Homology modeling: Also known as comparative modeling of protein, referring to the constructing an atomic-resolution model of the "target" protein from its amino acid sequence and an experimental three-dimensional structure of a related homologous protein (the "template").

Hydroxynitrile lyases (HNLs): An enzyme that catalyzes the cleavage of cyanohydrins into the corresponding aldehyde or ketone and hydrogen cyanide.

InChI: Acronym for IUPAC's "International Chemical Identifier", a text-based identifier for chemical substances which ensures the uniqueness of the structure represented.

InChI key: The compressed version of InChI, which is 27-characters long.

In silico: Experiments with the use of computers or experimental assay performed inside a computer.

Inhibitor: In enzymology, a compound, or even a macromolecule, that blocks the action of an enzyme by reversible attachment in such a way as to prevent binding by the substrate (competitive inhibition), or by prevention of the reaction even if the substrate can still bind (non-competitive inhibition).

Interconversion: Chemical data exchange between different file formats.

In vitro: Experimental assay conducted in the wet lab within a glass apparatus.

In vivo: Experimental assay conducted in the wet lab within a living organism, e. g. mice.

KDD (Knowledge Discovery in Databases): In data mining, refers to the process of identifying valid, novel, potentially useful and mainly understandable patterns.

k-Nearest neighbors: Used for classification of chemical compounds or prediction of their properties, such as biological activities, via machine learning. Based on similarity defined by the property of k, a number which is used for separation of various compounds into groups.

Knowledge-based predictions: Is a system for toxicity prediction that uses rules to describe the relationship between chemical structure and toxicity. The process involves assessing the predictions for a novel set of compounds and comparing them to their biological assay results as a measure of the system's performance.

LBVS (Ligand-based Virtual Screening): most popular approach for drug discovery and lead optimization in absence of 3D structure of potential drug target.

Lethal dose 50 (LD_{50}): LD_{50} is the dosage of a substance that will lead to the deaths of 50% of the dosed population.

Lead: An active chemical compound which can be optimized in order to be considered a potential drug.

Lead generation: The second step generally comes after hit identification (maybe called "hit to lead" or "H2L"), which refers to eliminating some unfavoured molecular properties to obtain developable compounds.

LELP index: Acronym for ligand efficiency dependent lipophilicity, a binding efficiency metric that combines lipophilicity, molecular size, and potency.

Library: a collection of compounds, either by physical samples or electronically for the purpose of screening. Also see "*Compound library*" and "*Library design*"

Library design: The overall procedure which takes place for the construction of a virtual chemical database, e. g. it can include the choice of tools for filtering and management tools.

Ligand: a molecule that forms, mostly via non-bonded interactions, a complex with a biomolecule

Ligand-based drug design: A rational drug design method based on a ligand presenting considerable biological activity.

Ligand efficiency (LE): Measure of the binding of a molecule in terms of free energy. LE values are often used to rank fragments and to monitor the progress of the optimization.

$$LE = -\Delta G / HAC$$

$$= -RTln(k_d)/HAC$$

a simplified version is

$$LE = 1.4(-logIC_{50})/HAC$$

where ΔG is used for Gibbs free energy, HAC stands for heavy atom count, R presents the gas constant, T shows the temperature in Kelvins, k_d represents dissociation constant, IC_{50} expresses 50% effective inhibitory concentration.

Ligand lipophilicity efficiency (LLE): Measure of a ligand-binding taking into account lipophilicity's beneficial effect. This is a metric used to monitor the lipophilicity with respect to an in vitro potency of a molecule.

$$LLE = pIC_{50}(or\ pK_i) - clogP(or\ clogD)$$

where K_i stands for inhibitory constant.

Linamarase: A beta-D-glucosidase (EC 3.2.1.21). It is an enzyme found in many plants including cassava and butter bean.

Linamarin: A cyanogenic glucoside found in the leaves and roots of plants such as cassava, lima beans, and flax. It is a glucoside of acetone cyanohydrin.

Lipinski's Rule of Five: assemble of rules useful to evaluate drug likeness. It predicts high probability of success or failure due to drug likeness for molecules complying with 2 or more of the following rules. Lipinski's rule establishes that, in general, an orally active drug has no more than one violation of the following criteria: a molecular mass less than 500 Daltons; no more than 5 hydrogen bond donors; not more than 10 hydrogen bond acceptors; an octanol-water partition coefficient logP not greater than 5. Also see *"Rule of Five"*.

Lipophilicity: The affinity of a molecule for lipophilic environments.

logD: LogP of a compound at a specified pH, mainly pH 7.4 due to the nature of the blood. However, this value may change based on the environment's acidity or basicity.

logP: A logarithm which defines the capability of a chemical compound to be dissolved in *n*-octanol (a measurement of its hydrophobicity) compared to being able to dissolve in water. Therefore, the log P value is the ratio of the concentration of the uncharged form of a compound in a nonpolar phase, traditionally water saturated octan-1-ol, to that in water. Note: Common algorithms to calculate log P are CLOGP and ALOGP. Also see *"clog P"*.

Lotaustralin: The glucoside of methyl ethyl ketone cyanohydrin and is structurally related to linamarin, the acetone cyanohydrin glucoside also found in these plants. Both lotaustralin and linamarin may be hydrolyzed by the enzyme linamarase to form glucose and a precursor to the toxic compound hydrogen cyanide.

Machine learning (ML): An algorithm used in computers which generates models for predictions after being trained with a set of carefully chosen objects. Generally speaking, ML refers to a subset of artificial intelligence which uses algorithms to build statistical models based on training data in order to make predictions or decisions without being explicitly programmed to perform the task.

Machine learning techniques: The use of various algorithms in order to classify molecules or predict properties based on their structure.

Metabolism: For the field of medicinal chemistry, this term refers to the biotransformation of drugs.

Metabolite: A product obtained due to metabolism.

Metabolomics: A branch of "omics" science dealing with the comprehensive profiling of metabolites within an animal, plant, and microbial metabolome.

Metadynamics: computational method aimed at enhancing the sampling of the configuration of space adding a time dependent repulsive bias potential function of coarse-grained variables, called collective variables.

Mimics of natural products: A variation on the pharmacophore/active NP that binds at the active site of the targeted protein. People forget that chemical structures are 3-dimensional and a molecule that in 2D does not look like the NP, in 3D has groups that bind to the active site. Examples would be the HIV protease inhibitors which use peptidomimetics to imitate the native 6-peptide substrate.

Molecular descriptors: Properties calculated from 2D/3D structure of molecules and expressed as arithmetic values in chemical databases. Molecular descriptors are terms that characterize a specific aspect of a molecule, e. g. it's molecular weight.

Molecular diversity: A measure of the spread of various properties or chemotypes within a set of compounds.

Molecular dynamics (MD): major computational tool that calculates the time dependent behavior of a molecular system.

Molecular fingerprints: Descriptions of the structure of a molecule calculated from the properties of each of its atoms and bonds that are then usually condensed into a fixed-length string by a hashing algorithm. A molecular fingerprint could be simply regarded as a binary theoretical descriptor encoding certain 2D or 3D structural features of a molecule. It is used to identify the presence or absence of a (sub)structure.

Molecular interaction: Used to refer to physicochemical contacts occurring between a biological target (generally protein) and ligand.

Molecular modeling: A method using computers and visualization techniques in order to obtain reliable 3D representations of molecules.

Molecular networking: A computational tool used to assist dereplication procedures, in which it is possible to visualize similarities in compound fragmentation (LC-MS/MS).

Molecular similarity: The degree to which two molecules resemble one another as calculated from their respective 2D or 3D properties, molecular fingerprints, fragment keys, or superimposed 3D structures that usually ranges from 1 (identical) to 0 (dissimilar). Note: Examples include Tanimoto or Tversky similarities for 2D structures and Carbo or Hodgkin for 3D structures.

MOL file: A file format which uses text-based connection tables in order to encode a chemical structure, substructure, and conformations.

Monte Carlo: A stochastic method that propagates the positions of atoms or groups of atoms in a molecule or collection of molecules through conformational space using a Boltzmann sampling of phase space.

Murcko scaffold: This refers to the most central ring system, plain ring systems, etc. The Murcko scaffold contains all plain ring systems of the given molecule plus all direct connections between them. Substituents, which don't contain ring systems, are removed from rings and ring connecting chains.

Note 1: Other terms, for example, "framework", "substructure", "cyclic system" or "fragment" are often synonymously used to refer to scaffolds.

Note 2: Also see "*Scaffold*" and "*Chemical scaffold*".

Naïve Bayesian classifier: An algorithm for supervised learning classifications. Also see "*Bayesian classifier*" and "*Bayesian*".

Natural product: Chemical substances that are produced by living organisms, used by them for specific purposes, such as protection, defense against predators, attracting mating partners, etc.

Natural Product Atlas: An open access knowledge base for microbial natural products discovery built and maintained by the Linington research group at Simon Fraser University (Canada) and curated by an international team of contributors. Available at https://www.npatlas.org/.

Natural product-likeness: Similarity with structural and physicochemical properties of a natural product.

Natural product-likeness score: A mathematical Bayesian equation that measures the NP-like score to ensure proximity to NP.

NCI: Abbreviation for National Cancer Institute, USA.

Neural network (plural neural networks): A real or virtual computer system designed to emulate the brain in its ability to "learn" to assess imprecise data. By doing so, they attempt to mimic the functioning of the network of neurons that form the brain, i. e. that function together to achieve a common purpose.

Neuron (in an Artificial Neural Network): The virtual analogous of a neuron in the human brain. It resembles the neuron's acting in the brain by receiving inputs, processing information and extracting results, and finally giving the output of the previous procedure. This is used in machine learning approaches.

Non-ribosomal peptide synthases (NRPSs): A family of multimodular enzymes, consisting of repeated modules, using regiospecific and stereospecific reactions for peptides biosynthesis.

NPs: See "Natural Products"

NRPS: Abbreviation for Non-ribosomal polyketides synthetase.

OECD: Organisation for Economic Co-operation and Development, an international organisation that creates policies to stimulate economic progress and world trade.

Open access (OA): Refers to free, unrestricted online access to research outputs such as journal articles and books. OA content is open to all, with no access fees.

Open source: The practice of providing open-source code for a product, e. g. software in general.

Orthosteric site: (the word orthosteric means "not comparable"). This describes the primary, unmodulated binding site (on a receptor) of a ligand.

OSMAC: Abbreviation of "One Strain Many Compounds".

PAINS (Pan-Assay Interference Compounds): Compounds influencing the identification of new bioactive compounds since they can originate from false activity signals. Such compounds present false positive biological activity during various tests. Compounds presenting PAINS alerts are compounds influencing the identification of new bioactive compounds since they can originate false activity signals.

Partial least squares (PLS): A https://en.wikipedia.org/wiki/Statistics method that bears some relation to https://en.wikipedia.org/wiki/Principal_component_re gression; instead of finding https://en.wikipedia.org/wiki/Hyperplane of maximum https://en.wikipedia.org/wiki/Variance between the response and independent variables, it finds a https://en.wikipedia.org/wiki/Linear_regression model by projecting the https://en.wikipedia.org/wiki/Predicted_variable and the https://en.wikipedia.org/wiki/Observable_variable to a new space.

PBPK modeling: Abbreviation of Pharmacologically based pharmacokinetics modeling, a mathematical modeling method for the prediction of ADME properties of a chemical compound in humans or animals.

PBTK/PBTD modeling: Abbreviation for physiologically based toxicokinetic and toxicodynamic models. This approach involves anatomical and physiological descriptors that modify the administration, distribution, metabolism and excretion of toxicants.

PDB (protein data bank/RCSB PDB): A database of 3D structures of macromolecules and macromolecules' complex assemblies with ligands, presenting biological interest. Available at: http://www.rcsb.org/. It is a freely available database of three-dimensional structural data of biological molecules (mostly proteins and nucleic acids) obtained typically by X-ray crystallography, NMR spectroscopy or cryo-electron microscopy.

PDB structure: A protein structure or file downloaded from the PDB, often derived by X-ray crystallography or by nuclear magnetic resonance.

Pestimep: Acronym for Pesticide Multiple Endpoint Database, collected by Madariaga & Martinez group.

Pharmacophore (ph4): The ensemble of necessary steric and electronic features which result in the optimal supramolecular interaction between a specific biological target and a molecule by triggering or blocking the relative biological response. A ph4 could be a proposal for the ensemble of steric and electronic features that define the optimal supermolecular intermolecular interaction of a ligand with a specific biological target structure with the result that it triggers or blocks its biological response.

Pharmacophore-based virtual screening: Screening virtual molecules using pharmacophore models as a query.

Pharmacophores of natural products: The base structure within a natural product that is the part of the molecule that binds to the target enzyme or protein.

Phenolics: Naturally occurring compounds containing hydroxyl a group directly linked to a phenyl ring.

Pheromones: Chemical substances produced by one species for the purpose of making other members in the same species respond for specific purposes. In the case of insects pheromones may be used to attract sexual partners (sex pheromones) or causing other members come together for collective action (aggregating pheromones), etc.

Phylogenetics: The study of evolutionary relationships among biological entities – often species, individuals or genes.

Phytochemical: A chemical substance of plant origin, i. e. naturally biosynthesized by a plant in the course of surviving, adapting to the environment or protecting itself from disease or predators.

plantiSMASH: The plant version of antiSMASH. Also see "*antiSMASH*".

Polyketides: Natural compounds which are composed of alternating carbonyl and methylene groups. The term also refers to further condensation, reduction, dehydration products of the polymethylene/keto compounds.

Polyketide synthases (PKSs): A family of enzymes consisting of some modules for the corresponding building-block of polyketides in fungi, bacteria, and plants.

Polypharmacology: Design or use of pharmaceutical agents that act on multiple targets or disease pathways.

Principal components: Non-correlated variables. Also, see "Principal component analysis".

Principal component analysis (PCA): A technique for machine learning methods used for the reduction of the variables needed to apply a classifier or an algorithm for property predictions. PCA is, therefore, a variable reduction method that operates on the correlation matrix of the variables to construct a small set of new orthogonal, i. e. non-correlated, variables (principal components) derived from linear combinations of the original variables.

Problematic functional groups: Parts of a chemical compound which contain the toxicophore or trigger off the reactions responsible for the observed adverse side effects or toxicity.

Protein crystallization: The process of formation of a regular array of individual protein molecules stabilized by crystal contacts. If the crystal is sufficiently ordered, it will diffract.

Protein target: Proteins are biochemical catalysts, i. e. chemical substances that regulate the speed of biochemical reactions. A protein that is a drug target is one for which the reaction involved is connected to a disease or biochemical function within the body. The inhibition of such a protein would affect the function of the body, hence "target" the disease.

PubChem: A free chemical database including data on chemical structures, biological activities and more, published by the National Center for Biotechnology Information (NCBI), USA. Available at: https://pubchem.ncbi.nlm.nih.gov/.

Public database: A collection of molecules freely downloadable and free to use.

PubMed: A free database including abstracts and citations from the literature, along with links to the original publications, published by the National Center for Biotechnology Information (NCBI), USA. Available at: https://www.ncbi.nlm.nih.gov/pubmed/.

Quantitative structure-activity relationships (QSAR): A method aiming at the construction of models which describe the relationship between structures and biological activities. Quantitative structure-activity relationships (QSAR) are mathematical relationships linking chemical structure and pharmacological activity in a quantitative manner for a series of compounds. Methods which can be used in QSAR include various regression and pattern recognition techniques. QSAR is often taken to be equivalent to chemometrics or multivariate statistical data analysis. It is sometimes used in a more limited sense as equivalent to Hansch analysis. QSAR is a subset of the more general term "structure-property correlations" (SPC).

Racemic: Used to describe a mixture of two substances that are identical in all respects but having opposite effects in terms of interaction with plane-polarized light. Equimolar amounts of dextrorotatory and levorotatory substances lead to a racemic mixture.

Random forest: An algorithm for the classification of compounds and for the prediction of their biological activities (machine learning technique). An ensemble learning approach that constructs a large number of decision trees, and outputs predictions that are a collection of the votes of the individual trees. A subset of the training dataset is chosen to grow individual trees, with the remaining samples used to estimate the optimal fit. Trees are grown by splitting the training set (subset) at each node according to the value of the random variable sampled independently from a subset of variables.

Rational drug design: Drug design, often referred to as rational drug design or simply rational design, is the inventive process of finding new compounds as medications based on the knowledge of a biological target. The drug is most commonly an organic small molecule that activates or inhibits the function of a biological molecule such as a protein, which in turn results in a therapeutic benefit to the patient. In the most basic sense, drug design involves the design of molecules that are complementary in shape and charge to the biomolecular target with which they interact and therefore will bind to it. Drug design frequently but not necessarily relies on computer modeling techniques. This type of modeling is sometimes referred to as computer-aided drug design.

RDKit tool: An open-source cheminformatics software (https://www.rdkit.org/), also useful for computational chemistry, and predictive modeling.

Read-across: A technique for predicting endpoint information for one substance (target substance), by using data from the same endpoint from (an)other substance (s), (source substance(s)).

Receiver Operating Characteristic (ROC) curve: A metric used as an objective way to evaluate the ability of a given test to discriminate between two populations.

Receptor: Macromolecules, usually proteins, found in cells. They recognize and make complex assemblies with specific compounds. A receptor is a protein or a protein complex in or on a cell that specifically recognizes and binds to a compound acting as a molecular messenger (neurotransmitter, hormone, lymphokine, lectin, drug, etc). In a broader sense, the term receptor is often used as a synonym for any specific (as opposed to non-specific such as binding to plasma proteins) drug binding site, also including nucleic acids such as DNA.

Regression: A method for statistical analysis. See "Regression analysis".

Regression analysis: The use of statistical methods for modeling a set of dependent variables, Y, in terms of combinations of predictors, X. It includes methods such as multiple linear regression (MLR) and partial least squares (PLS).

Rhodanese: A mitochondrial enzyme that detoxifies cyanide (CN^-) by converting it to thiocyanate (SCN^-).

ROC (Receiver Operating Characteristic) curve: A metric used as an objective way to evaluate the ability of a given test to discriminate between two populations.

Rotatable bond: The chemical bond type which can turn around itself, generally a single bond which is not in a ring system or not done with a terminal heavy atom.

Rule of Five: The rule of five states that molecules that violate two or more of the following rules are likely to have permeability problems: (1) CLOGP calculated octan-1-ol/water log P greater than 5.0; (2) molecular weight greater than 500; (3) more than five hydrogen bond donors; and (4) the sum of oxygen and nitrogen atoms is greater than 10. Note 1: Natural products, peptides and other substrates for biological transporters are exceptions. Note 2: The original authors also noted that one can calculate the octan-1-ol/water log P with the program MLOGP for which the cut-off is 4.15. Users often use log P's calculated with other programs or measured values. Also see "*Lipinski's Rule of Five*".

Rule of Three: An assembly of rules useful to evaluate drug likeness of fragments. It predicts high probability of success or failure due to drug likeness for smaller molecules complying with 2 or more of the following rules. These rules establishes that, in general, an orally active drug has no more than one violation of the following criteria: a molecular mass less than 300 Daltons; an octanol-water partition coefficient logP not greater than 3; no more than 3 hydrogen bond donors and acceptors; no more than 3 rotatable bonds.

SBVS (Structure-based Virtual Screening): A computational approach used in the early-stage drug discovery to search a chemical compound library for novel bioactive molecules against a drug target for which the 3D structure is known.

Scaffold: The representation of the core structures of the molecular framework. Molecular scaffold consists of rings connected by linkers. The widely adopted version by Bemis and Murcko is obtained from the molecule by removing all its side chains. The term Scaffold is also used in medicinal chemistry or computational chemistry to refer to core structures of natural products which may contribute to some form of desirable biological activity. This refers to the molecular core to which functional groups are attached. The method used to locate the core structure(s) depends on the chosen scaffold type. Also see "*Murcko scaffold*".

Scaffold hopping: The process for the discovery of bioactive compounds that are structurally distinct from their reference molecules.

Scaffold Hunter: A tool that helps to navigate chemical and biological spaces interactively and provide new synthetic directions based on the scaffold hierarchy.

Scaled Shannon Entropy (SSE): SE values normalized. In chemical diversity analysis, the values of SSE range between 0, where all P compounds are contained in one cyclic system, and 1.0, where each cyclic system contains an equal number of compounds. Therefore, SSE values closer to 1.0 indicate large scaffold diversity within the n most populated cyclic systems. Also, see "Shannon Entropy".

Scoring function: A mathematical function used to approximately predict the binding affinity between two molecules after they have been docked.

SDF file: Structure-Data File, a file format readable by most computational chemistry software. It uses text based connection tables in order to represent a chemical structure.

Secondary metabolism: The process resulting in the production of metabolites by routes other than the normal pathways.

Secondary metabolites: Chemical compounds produced by organisms in biochemical processes different from those life processes that are common in all living organisms. These are molecules which are not essential for the normal growth, development, or reproduction of the producer organisms but specifically modulate health-maintaining processes. They are chemical compounds produced by organisms in biochemical processes different from those life processes that are common in all living organisms.

Selectivity or sensitivity (S_e): The preference of a ligand to perform a stable complex with a specific receptor. It describes the ratio of the number of the active molecules found by a virtual screening method to the number of all active database compounds.

Self-organizing map (SOM): A type of artificial neural network that uses unsupervised machine learning to project high-dimensional data into two dimensions that are usually presented as a contour plot. Note: This is sometimes called a Kohonen map.

Sequence alignment: In bioinformatics, a sequence alignment is a way of arranging the sequences of DNA, RNA, or protein to identify regions of similarity that may be a consequence of functional, structural, or evolutionary relationships between the sequences. Aligned sequences of nucleotide or amino acid residues are typically represented as rows within a matrix. Gaps are inserted between the residues so that identical or similar characters are aligned in successive columns.

Shannon Entropy (SE): The Shannon entropy equation provides a way to estimate the average minimum number of bits needed to encode a string of symbols, based on the frequency of the symbols. In chemical diversity analysis, the Shannon entropy (SE) of a population of P compounds distributed in n systems is defined as:

$$SE = - \sum_{i=1}^{n} p_i log_2 p_i \text{ with } p_i = \frac{c_i}{P}$$

where p_i is the estimated probability of the occurrence of a specific chemotype i in a population of P compounds containing a total of n acyclic and cyclic systems, and c_i is the number of molecules that contain a particular chemotype c.

Shikimate pathway: Metabolic route found in microorganisms and plants (absent in animals) for the biosynthesis of folates and aromatic amino acids.

Similarity (Molecular Similarity): To what extent two compounds resemble based on various criteria such as 2D/3D properties or molecular fingerprints. Tanimoto similarity is commonly used.

Similarity Ensemble Approach: *In silico* method for the discovery of drug targets.

Similarity search: Screening of a database for molecules similar to a query molecule.

SMILES: Acronym for Simplified Molecular-Input Line-entry System, a specification in the form of a line notation for describing the structure of chemical species. SMILES is a string notation used to describe the nature and topology of molecular structures.

Specificity (S$_p$): The ratio between the number of inactive compounds not selected by the VS protocol, and the number of all inactive molecules included in the chemical database. Specificity ranges from 0 to 1 and denotes the percentage of truly inactive compounds. Sp = 0 defines the worst-case scenario where all inactive compounds are selected by error as actives, whereas Sp = 1 means that all inactive compounds have been correctly rejected during the screening process.

SPiDER: Self-Organizing Maps (SOM)-based prediction of drug equivalence with CATS2 and physicochemical descriptors.

Structural alert/expert rule/toxicophore: Molecular substructures associated with a particularly adverse outcome, used to inform about the toxicity of pharmaceuticals or agrochemicals.

Structure-activity landscape: See "Activity and property landscape modeling".

Structure-activity landscape index (SALI): A metric used to identify activity cliffs, defined as pairs of structurally similar compounds with large differences in bioactivity/potency.

Structure-activity relationship (SAR): model used to build up relationship between the chemical structure of a molecule and its biological activity

Structure-based drug design: A method to design a potential drug based on the 3D structure of the macromolecule target.

Structure-based virtual screening (SBVS): A computational approach used in the early-stage drug discovery to search a chemical compound library for novel bioactive molecules against a drug target for which the 3D structure is known. An example is docking various compounds against a macromolecule-target in order to investigate the interactions between them and the stability of the resulting complex.

Substructure search: Virtual screening of compound libraries in order to identify those containing the substructure used as the query.

Support vector machine (VSM): A technique used in machine learning approaches for the tasks of classification and prediction (of properties).

Sweetener: a substance that imparts a sense of sweetness in the mouthlike sugar. Sweeteners may be natural or synthetic; they may also be calorific or not. Examples of non-caloric sweeteners include stevioside(a diterpene glycoside), or some plant derived proteins like brazzein and micraculin which are taste modifying sweeteners. Synthetic sweeteners include sucralose, cyclamate, etc.

Target fishing: Discovery of the interactions between bioactive, chemical molecules and different targets.

TIGER: Target Inference Generator. Method inspired in SPiDER for target prediction.

Topological polar surface area (tPSA): An approximation to the polar surface area that is calculated from the 2D structure of a molecule.

Toxicity: The degree to which a chemical substance or a particular mixture of substances can damage an organism. This could refer to the effect on a whole organism, such as an animal, bacterium, or plant, as well as part of the organism, e. g. a cell (cytotoxicity) or an organ like the liver (hepatotoxicity). A central concept of toxicology is that the effects of a toxicant (or toxin) are dose-dependent.

Toxicophore: A functional group (structural group) which can probably cause toxic effects during metabolic activation.

Toxin: A substance (mixture or pure compound) exhibiting toxicity.

Transcript: The result copying of part of a DNA sequence into a transfer RNA (codon).

Transcription: The process by which the information provided by a specific area of a gene is copied into a transfer RNA.

Transglycosylation: A mechanism for glycosidic bond formation, particularly during polysaccharide synthesis. Nucleoside phosphate derivatives act as activated donor compounds in which the energies of their glycosidic bonds are partially conserved in the reaction products.

Virtual chemical database: A virtual collection of chemical compounds accompanied by various data (generated or derived from the literature), usually biological activities and physicochemical properties.

Virtual screening: computer-based method to predict new ligands on the basis of tridimensional biological structure, with the goal to filter enormous virtual chemical databases of small organic molecules.

X-ray structure: See "PDB structure".

ZINC (an acronym for "ZINC is not commercial"): The largest virtual collection of compounds, whose 3D structures are readily available for docking purposes. A free database of compounds which are commercially available. The current version is available at http://zinc20.docking.org/ .

References

[1]   Bajorath J. Integration of virtual and high-throughput screening. Nat Rev Drug Discov. 2002;1:882–94.
[2]   Buckle DR, Erhardt PW, Ganellin CR, Kobayashi T, Perun TJ, et al. Glossary of terms used in medicinal chemistry. Part II (IUPAC Recommendations 2013). Pure Appl Chem. 2013;85: 1725–58.
[3]   Duffus JH. Glossary for chemists of terms used in toxicology (IUPAC Recommendations 1993). Pure Appl Chem. 1993;65:2003–122.
[4]   Gu J, Gui Y, Chen L, Yuan G, Lu H-Z, Xu X. Use of natural products as chemical library for drug discovery and network pharmacology. PLoS ONE. 2013;8:e62839.
[5]   Irwin JJ, Tang KG, Young J, Dandarchuluun C, Wong BR, Khurelbaatar M, et al. ZINC20-A free ultralarge-scale chemical database for ligand discovery. J Chem Inf Model. 2020;60:6065–73.
[6]   Koulouridi E, Akone SH, Valli M, Da Silva Bolzani V, Saldívar-González FI, Pilón-Jiménez A, et al. Glossary of terms used in chemoinformatics of natural products: fundamental principles. Phys Sci Rev. 2020. DOI:10.1515/9783110579352-018.
[7]   Martin YC, Abagyan R, Ferenczy GG, Gillet VJ, Oprea TI, et al. Glossary of terms used in computational drug design, part II (IUPAC Recommendations 2015). Pure Appl Chem. 2016;88:239–64.
[8]   Moss GP, Smith PA, Tavernier D. Glossary of class names of organic compounds and reactive intermediates based on structure (IUPAC Recommendations 1995). Pure Appl Chem. 1995;67:1307–75.
21[9] Nagel B, Dellweg H, Gierasch LM. Glossary for chemists of terms used in biotechnology (IUPAC Recommendations 1992). Pure Appl Chem. 1992;64:143–68.
[10] Nordberg M, Duffus JH, Templeton DM. Glossary of terms used in toxicokinetics (IUPAC Recommendations 2003). Pure Appl Chem. 2004;76:1033–82.
[11] Wermuth CG, Ganellin CR, Lindberg P, Mitscher LA. Glossary of terms used in medicinal chemistry. Pure Appl Chem. 1998;70:1129–43.

Index

https://doi.org/10.1515/9783110668896-016

www.ingramcontent.com/pod-product-compliance
Lightning Source LLC
Chambersburg PA
CBHW080914220326
41598CB00034B/5572